Microfluidics and Lab-on-a

Microfluidics and Lab-on-a-chip

By

Andreas Manz

KIST Europe and Universität des Saarlandes, Germany
Email: manz@kist-europe.de

Pavel Neužil

Northwestern Polytechnical University, China
Email: pavel.neuzil@nwpu.edu.cn

Jonathan S. O'Connor

KIST Europe and Universität des Saarlandes, Germany
Email: j.oconnor@kist-europe.de

and

Giuseppina Simone

Northwestern Polytechnical University, China
Email: giuseppina.simone@nwpu.edu.cn

ROYAL SOCIETY
OF **CHEMISTRY**

Print ISBN: 978-1-78262-833-0
EPUB ISBN: 978-1-78801-938-5

A catalogue record for this book is available from the British Library

The Royal Society of Chemistry is a charity, registered in England and Wales, Number 207890, and a company incorporated in England by Royal Charter (Registered No. RC000524), registered office: Burlington House, Piccadilly, London W1J 0BA, UK, Telephone: +44 (0) 20 7437 8656.

Visit our website at www.rsc.org/books

Printed in the United Kingdom by CPI Group (UK) Ltd, Croydon, CR0 4YY, UK

Preface

At a time when the first hype around microfluidics or 'lab-on-a-chip' has settled, and is replaced by silent use in commercial instrumentation... At a time of dramatic chemical information needs, such as genome sequencing, drug discovery and environmental monitoring... At a time when scientific publications simply assume basic knowledge of microfluidic engineering, chip technology or 3D printing... At such a time it appears imperative to compile the essentials of microfluidics in a small, handy textbook. This will be useful to the lay reader, to the interested student and also to the non-expert professional. The highly interdisciplinary nature of microfluidics, engineering, materials, fluid mechanics and solution chemistry makes it essential for researchers, from different backgrounds, to have at least a minimal understanding of neighbouring sciences if they are to communicate effectively.

Engineering is generally targeted at human-sized or finger-sized objects, which have value for humankind and which dominated the advances in technology in the last century – engineering feats such as bridges, skyscrapers, airplanes, cars, computers, television and mobile phones, to list just a few – things we can handle, live in, drive across, *etc.* The advances in the field of biology were based more on microscopy, things we can see by eye, but in contrast chemistry and physics are mostly theory based, and build on concepts such as molecules, atoms, elementary particles, waves, fields, *etc.* Our most experienced 'teacher', Nature itself, uses a very successful size range, between microscopy and the size of our

Microfluidics and Lab-on-a-chip
By Andreas Manz, Pavel Neužil, Jonathan S. O'Connor and Giuseppina Simone

Published by the Royal Society of Chemistry, www.rsc.org

fingers. Imagine an insect or a biological cell! This book deals with microfluidics and chip technology, a part of engineering which is in that range. However, it is all about very simple fluidic circuits. Engineering is very far away from engineering something comparable to a hoverfly (Syrphidae)!

Similarly, conventional engineering deals with time scales related to human behaviour, ranging from seconds to years. Geological time scales, time scales related to plant growth or related to 'big bang' are not normally included. With mass transport and microfluidics, a time scale of milliseconds to seconds is covered. This can have an influence on obtaining information about chemical or molecular biological processes with greater efficiency, and may enhance high-throughput screening for drug discovery, or genome sequencing. However, current speeds have not yet reached those obtainable in microelectronic circuits or computers. For the given reasons, microfluidics has not yet reached its full potential, making it an interesting topic for research and education for years to come.

For many years, Professor Manz has presented a 'lab-on-a-chip' course at Imperial College of Science, Technology and Medicine (UK), the Technical University of Dortmund (Germany) and the University of Saarland (Germany). He found that most of the documentation could be found in various comprehensive textbooks, scientific literature or specifically prepared lecture materials for students. This made perfect sense, since the field started to develop in the mid-1990s and then rapidly expanded with many novel and exciting experiments in the early twenty-first century. However, it is surprising that there still is no satisfactorily compact textbook available to students at this time. In the light of these findings, it seemed appropriate to attempt to cover the now well-established field of microfluidics with a concise textbook for students of all disciplines.

This book is aimed primarily at undergraduate, graduate and postgraduate students of engineering, chemistry, biotechnology and the applied life sciences in general. It gives a simplified introduction to microfluidic engineering and its applications in chemistry or biology. The book enables the reader to gain a basic understanding of scaling laws, microfluidic theory, chip manufacture, surface modification, instrumentation and operation of a chip. It also explains the most basic standard operations such as pumping, valving, mixing, extraction, chemical reactions, separation and the handling of biological cells. The goal of the book is to provide a foundation of knowledge so that a student can understand any scientific publication that uses microfluidics and develop their understanding further. Such comprehension of the field makes for an incredibly strong starting point for laboratory work in micro total analytical systems (micro-TAS), lab-on-a-chip or general microfluidics.

The chapters are all dependent on each other in some way, and a linear sequence is not immediately clear. Therefore, for a lecture course, it may be convenient to change the sequence offered in the book. Chapters 1, 7, 9 and 13 are more theory oriented, whereas Chapters 6, 8, 10, 11, 14 and 15 are more application oriented. An engineering student might be more comfortable and gain more information from Chapters 4 and 10–14, whereas a student with a chemistry or biochemistry background will learn mostly from Chapters 1–3, 5, 6 and 9. Some chapters feature more general content, whereas others provide practical examples, such as Chapters 3 and 15. Each chapter has been compiled by a specific author and therefore the content may not necessarily represent the opinions of all of us. Some useful or neighbouring fields to lab-on-a-chip have been purposely omitted, such as general clean-room technology and microelectronics chip fabrication, analytical chemistry, the structure and chemistry of DNA and proteins, sensors, bioarrays, cell biology and assays with organoids, for example. The purpose of this book is to provide a framework of the essential concepts, which the reader can then fill out and apply the principles with their own knowledge of these neighbouring fields, or use the newly acquired information to find additional, more specific resources. Many examples are discussed within this book, but the reader is advised to use the references and read the original sources to appreciate fully the work done.

The authors hope that this book will be well received by both students and teachers, particularly at a time when microfluidics is becoming a general tool for engineering and the life sciences. The authors would like to thank Feng Jianguo, Zhang Haoqing and Zhu Hanliang for helping with some of the text and figures, Niall MacDonald for the front cover photo, Jan C. T. Eijkel, Yuliya Silina, Matthias Altmeyer, Eric Castro and Mark Tarn for being involved with the lecture course materials over the years, Leon Abelmann, Camila D. M. Campos, Baeckkyoung Sung, Rosanne Guijt, Himani Sharma, Christian Ahrberg, Wenming Wu for proofreading, as well as Chris Powell and Leon Abelmann for very helpful comments and discussions.

Andreas Manz
Pavel Neužil
Jonathan S. O'Connor
Giuseppina Simone
Saarbrücken, Naples and Xi'an

Abbreviations and Variables

[A], [B], ...	concentration of A, B, ...
<100>	Miller indices 100, crystallography
<111>	Miller indices 111, crystallography
∇	nabla operator
2D	two-dimensional
2PP	two-photon polymerization
3D	three-dimensional
3T3	mouse embryonic fibroblast cell line
A	surface area
A	absorptivity, light absorption
A	droplet elongation ratio
a	acceleration
a	radius of curvature
A	adenine
A, B, C, ...	molecular species
A1, A2, A3, ...	access reservoirs on a chip, inlets, outlets
AA	ascorbic acid
Ab	antibody
ac	alternating current
Ag	silver
AgCl	silver chloride
AKI	acute kidney injury
Al	aluminium
As^{3+}	arsenic(III) ion
ASV	anodic stripping voltammetry
Atz	atrazine
Au	gold
AUX	auxiliary electrode, counter electrode
Au_xHg_y	amalgam
B	magnetic field, magnetic field vector

Microfluidics and Lab-on-a-chip
By Andreas Manz, Pavel Neužil, Jonathan S. O'Connor and Giuseppina Simone

Published by the Royal Society of Chemistry, www.rsc.org

BHQ	black hole quencher dye
Bi	Bingham number
BICELL	bio-photonic sensing cell
BMP-NTf2	bis(trifluoromethylsulfonyl)azanide 1-butyl-1-methylpyrrolidin-1-ium; alternative name 1-butyl-1-methylpyrrolidinium bis(trifluoromethylsulfonyl)amide
Bo	Bond number
BODIPY	boron-dipyrromethene, occurs in trade names
Bp	base pairs
BSA	bovine serum albumin
C	capacitance per unit area
c	concentration
C	C-shaped turn
C	cytosine
C_0	initial concentration
c_1	concentration of a solute in phase 1
C1, C2, C3, ...	chambers on a chip
$\mathrm{C_4}$	carbohydrate molecule containing four carbon atoms
Ca	capillary number
CAD	computer-aided design
CC BY	creative commons, free to share and redistribute
CCD	charge-coupled device
CD	compact disc
C_D	frictional constant
cDNA	complementary DNA
CD-ROM	compact disc read-only memory
CE	capillary electrophoresis
CEA	carcinoembryonic antigen
CEA	contraction–expansion array
CF	carbon fibre
$-\mathrm{CF_3}$	trifluoromethyl group, at a surface
$\mathrm{CF_4}$	carbon tetrafluoride
CG	capped gold
CIF	Caltech intermediate form, data format
$\mathrm{Cl^-}$	chloride ion
CMOS	complementary metal–oxide–semiconductor
CNC	computer numerical control
CND	carbon nanodot
CNST	Center for Nanoscale Science and Technology
CNT	carbon nanotube
$\mathrm{CO_2}$	carbon dioxide
CPU	central processing unit
Cr	chromium
CR	compression ratio
CR	cutting ratio between pinching length and initial length
CRP	C-reactive protein
CSV	cathodic stripping voltammetry
CTC	circulating tumour cell
Cu	copper

CuNP	copper nanoparticle
CV	cyclic voltammetry
CVD	chemical vapour deposition
D	diffusion coefficient
d	tube diameter, capillary diameter, characteristic length
DAQ	data acquisition
dc	direct current
DCM	dichloromethane
De	Dean number
DENV	dengue fever virus
DEP	dielectrophoresis
$\mathrm{d}F/\mathrm{d}T$	fluorescence change as a function of temperature
d_{hyd}	hydraulic diameter of channel
diFET	dielectrically isolated field-effect transistor
DLW	direct laser writing
DMEM	Dulbecco's modified Eagle's medium
DMF	dimethylformamide
DNA	deoxyribonucleic acid
DNQ	diazonaphthoquinone
dsDNA	double-stranded DNA
DTPA	diethylenetriamine pentaacetate
DVD	digital versatile disc
DXF	data exchange (interchange) format
E	electric field
E	electrical potential (electrochemistry)
e	electron
E	extraction factor
E_0	standard electrical potential for the reaction
E_1	ground state (electron)
E_2	excited state (electron)
EBL	electron beam lithography
EC	electrochemistry
ECL	electrochemiluminescence
EDC	1-ethyl-3-(3-dimethylaminopropyl)carbodiimide
EDL	electrical double layer
EDTA	ethylenediaminetetraacetic acid
E_{ECL}	applied potential, electrochemiluminescence
EIS	electrochemical impedance spectroscopy
ELISA	enzyme-linked immunosorbent assay
EOF	electroosmotic flow
Ery	erythrocytes
EWOD	electrowetting-on-dielectric
F	Faraday constant
F	fluorescence intensity
F	force
FACS	fluorescence-activated cell sorting
FAM	6-carboxyfluorescein
FAS-17	1*H*,1*H*,2*H*,2*H*-perfluorodecyltriethoxysilane
F_{cen}	centrifugal force
F_{DEP}	dielectrophoretic force

Fe^{2+}	ferrous ion, iron(II) ion
Fe^{3+}	ferric ion, iron(III) ion
FEA	finite element analysis
f_I	force of inertia
FIB	focused ion beam
FITC	fluorescein isothiocyanate
F_m	magnetic force
Fo	Fourier number
FOTS	(tridecafluoro-1,1,2,2-tetrahydrooctyl)trichlorosilane
F_S	shear-gradient lift force
FT-IR	Fourier transform infrared (spectroscopy)
f_V	shear force
F_W	wall-induced lift force
G	thermal conductance
g	acceleration due to gravity
G	guanine
GA	glutaraldehyde
GAPDH	glyceraldehyde-3-phosphate dehydrogenase
Gd	gadolinium
$GdCl_3$	gadolinium(III) chloride
GDSII	graphic data system II, data format
GFAP	glial fibrillary acidic protein
GFP	green fluorescent protein
GIS	Genome Institute of Singapore
GNP	gold nanoparticle
H	height, layer thickness
h	height, layer thickness
H	system heat capacity
h	Planck's constant
H	H-shaped channel configuration
H_2O	water
H_2O_2	hydrogen peroxide
H_2SO_4	sulfuric acid
H5N1	avian influenza virus subtype, 'bird flu' strain
H7N9	avian influenza virus subtype, 'bird flu' strain
HA	haemagglutinin
HEMA	2-hydroxyethyl methacrylate
HepG2	a human liver cancer cell line
Hg	mercury
Hg_2Cl_2	mercurous chloride, mercury(I) chloride
HMDS	hexamethylenedisilazane
HNO_3	nitric acid
HOMO	highest occupied molecular orbital
HPLC	high-performance liquid chromatography
HRP	horseradish peroxidase
HUVEC	human umbilical vein endothelial cells
HV	applied high voltage
I	electric current
I_0	intensity of incident light
$i_{0.5\nu}$	current measured at $\nu/2$

$i_{2\nu}$	current measured at 2ν
IBE	ion beam etching
I_C	charging current
IC	integrated circuit, electronic chip
iDEP	insulator-based dielectrophoresis
IDT	interdigitated transducer
IFS	integral field spectrometer
IgG	immunoglobulin G
I_K	kinetic current
IL	interference lithography
IMTEK	Institut für Mikrosystemtechnik (Department of Microsystems Engineering), University of Freiburg, Germany
IoT	Internet of things
I_R	reversible current
IRT	immunoreactive trypsinogen
I_t	intensity of transmitted light
ITO	indium tin oxide
ITP	isotachophoresis
i_ν	current measured at nominal scan rate ν
i_{WE}	current at the working electrode
J	molecular mass flux
j, g, ...	emission lines of an atom
K	consistency index
k	Boltzmann constant
k	spinner constant
K^+	potassium ion
$K_3Fe(CN)_6$	potassium ferricyanide, potassium hexacyanoferrate(iii)
$K_4Fe(CN)_6$	potassium ferrocyanide, potassium hexacyanoferrate(ii)
KCl	potassium chloride
K_D	partition ratio, partition coefficient
Kn	Knudsen number
KOH	potassium hydroxide
L	length
L	length of plug, droplet or fluid section
L	optical pathlength
L	L-shaped turn
L1, L2	flow streamlines from left
LAMP	loop-mediated isothermal amplification
LB	Langmuir–Blodgett, LB film
LCD	liquid crystal display
LD	laser diode
LED	light-emitting diode
LIF	laser-induced fluorescence
LiF	lithium fluoride
LOC	lab-on-a-chip
LOD	limit of detection
LSPR	localized surface plasmon resonance
LUMO	lowest unoccupied molecular orbital
M	molar mass

m	mass
M	buffer, inlet
M^0	metal (no charge)
MCA	melting curve analysis
MCU	microcontroller unit
MEGA	microfluidic emulsion generator array
MeIM	methylimidazole
MEMS	microelectromechanical system
Mg^{2+}	magnesium ion
MIMIC	micromoulding in capillaries
miRNA	microRNA, micro-ribonucleic acid
M^{n+}	metal cation with charge n
Mn^{2+}	manganese(II) ion
MNP	magnetic nanoparticle
MS	mass spectrometry
MZI	Mach–Zehnder interferometer
n	number of electrons transferred
n	number of steps
N	north magnetic pole
N_2	nitrogen gas
NA	neuraminidase
Na^+	sodium ion
NaCl	sodium chloride
NaOH	sodium hydroxide
nDEP	negative dielectrophoresis
ne^-	number of transferred electrons
NF_3	nitrogen trifluoride
$-NH_2$	amine group, at a surface
NHS	*N*-hydroxysuccinimide
Ni	nickel
NiCr	nichrome, nickel–chrome alloy
NIST	National Institute of Standards and Technology
NP	nanoparticle
NT	Nanolithography Toolbox
NW	nanowire
O	oxygen
O_2	oxygen gas
-OH	hydroxyl group, at a surface
Oh	Ohnesorge number
OLED	organic light-emitting diode
OPD	organic photodetector
P	Joule heat, power
p	photoresist solids content
p	pressure
PB	parallelogram barrier
Pb^{2+}	lead(II) ion
p_{backward}	backward pressure
p_{forward}	forward pressure
PBS	phosphate-buffered saline
PC	personal computer

PCB	printed circuit board
P_{cen}	pressure by centrifugal force
PCI	phenol–chloroform–isoamyl alcohol
PCR	polymerase chain reaction
Pd	Palladium
PD	Photodiode
pDEP	positive dielectrophoresis
PDMS	polydimethylsiloxane
Pe	Péclet number
PET	poly(ethylene terephthalate)
PET	positron emission tomography
PFD	perfluorodecalin
pH	acidity scale, logarithmic hydrogen ion concentration
PID	proportional integrative derivative
PMMA	poly(methyl methacrylate)
PMT	photomultiplier tube
PNIPAM	poly(*N*-isopropylacrylamide)
POC	point-of-care
PR	photoresist
PR	preconcentration
PS	polystyrene
PSA	pressure-sensitive adhesive, self-adhesive
Pt	platinum
PTFE	polytetrafluoroethylene
PVC	poly(vinyl chloride)
PWM	pulse-width modulation
PZT	lead zirconate titanate
Q	electrical charge
Q	volumetric flow rate
Q1	a transistor
qPCR	quantitative PCR, real-time PCR
R	electrical resistance
R	radius
R	universal gas constant
r	coordinate r
r	hydrodynamic radius (molecule)
R1, R2	flow streamlines from right
R1, R2, R3, ...	access reservoirs on a chip, inlets, outlets
RBC	red blood cell
R_{ct}	charge-transfer resistance
Re	Reynolds number
REF	reference electrode
REM	replica moulding
Res	flow resistance
RF	radiofrequency
RIE	reactive ion etching
RIU	refractive index unit
RNA	ribonucleic acid
$R–NH_2$	amine, R = rest of the molecule
R–NP	resonant nanopillar

ROI	region of influence
rpm	rotations per minute
RS-232	a standard for serial communication
R–SH	thiol, R = rest of the molecule
$R–SiCl_3$	silanizing agent, R = rest of the molecule
RT	reverse transcription
RTD	resistive temperature detector
RT-PCR	reverse transcriptase PCR
$Ru(bpy)_3^{2+}$	tris(2,2′-bipyridine)ruthenium(ii) ion
Ru^{2+}	ruthenium(ii) ion
Ru^{3+}	ruthenium(iii) ion
S	sample, inlet
S	south magnetic pole
SAM	self-assembled monolayer
SARS	severe acute respiratory syndrome
SAW	surface acoustic wave
Sb	antimony
SF_6	sulfur hexafluoride
Si	silicon, silicon substrate
Si_3N_4	silicon nitride
$–Si–O^-$	deprotonated hydroxyl group at surface
SiO_2	silicon dioxide, quartz
–SiOH	hydroxyl group at surface
SIP	surface-initiated polymerization
SLM	spatial light modulator
SPE	screen-printed electrode
SPL	scanning-probe lithography
SPR	surface plasmon resonance
SSAW	standing surface acoustic wave
ssDNA	single-stranded DNA
s_t	striation thickness
ST	stripping
s_{t0}	initial striation thickness
STD	standard
SU-8	an epoxy resin, photoresist
SV	stripping voltammetry
SW	sample waste, outlet
SWCNT	single-walled carbon nanotube
T	temperature
t	time
T	thymine
T	T-shaped channel configuration
T3	temporary transfer tattoo
Taq	*Thermus aquaticus*, a thermophilic bacterium
TAS	total analysis system, chemical monitor
TCR	temperature coefficient of resistance
TER	tuneable elastic reversible
TFAA	trifluoroacetic anhydride
THP-1	a human monocytic cell line
Ti	titanium

T_M	melting temperature
TMBA	trimethyl-*N*-butylammonium
TMOS	tetramethyl orthosilicate
U	speed, linear velocity
u	mean velocity
U	U-shaped turn
UA	uric acid
US	Ultrasound
USB	universal serial bus
UV	ultraviolet
u_x	velocity in *x*-direction
u_y	velocity in *y*-direction
u_z	velocity in *z*-direction
V	volume, particle volume
V	voltage
v	scan rate, cyclic voltammetry
V	V-shaped channel configuration
V_0	initial volume
V_1	volume of phase 1
V_{ECL}	voltage output of PMT, electrochemiluminescence
VRC	virtual reaction chamber
V_{REF}	voltage at the reference electrode
w	rotational speed
W	waste, outlet
W	width
w	width
We	Weber number
WE	working electrode
W_{TD}	width by Taylor dispersion
x	average distance, distance travelled
x	*x*-coordinate
XeF_2	xenon difluoride
y	*y*-coordinate
Y	Y-shaped channel configuration
YPD	yeast extract–peptone–dextrose
Z	impedance
z	*z*-coordinate
z	number of electrons, number of charges
ZIF	zeolitic imidazolate framework
Z_{im}	imaginary compound of impedance
ZPAL	zone-plate-array lithography
Z_{re}	real component of impedance
ZrO_2	zirconium dioxide
Γ	shear rate
Γ	surface tension
γ_0	surface tension with no potential applied
γ_{AG}	surface tension between aqueous and gas phases
γ_{lv}	surface tension between liquid and vapour phases
γ_{SA}	surface tension between solid and aqueous phases
γ_{SG}	surface tension between solid and gas phases

γ_{sl}	surface tension between solid and liquid phases
γ_{sv}	surface tension between solid and vapour phases
Δ	difference
Δp	pressure difference
ΔP_c	capillary pressure in a pore
Δt	time
ΔV	volume change
Δx	distance
ε	molar absorption coefficient, molar absorptivity (optics)
ε	permittivity, dielectric constant
ε_0	permittivity of vacuum
ε_m	dielectrophoretic constant of the medium
ε_p	dielectrophoretic constant of the particle or cell
ε_r	relative permittivity
ζ	zeta potential
η	kinematic viscosity
η	pump efficiency
θ	contact angle
θ	θ-coordinate
θ_0	contact angle with no potential applied
θ_a	contact angle, advancing droplet
θ_d	dynamic contact angle
θ_r	contact angle, receding droplet
θ_s	static contact angle
λ	free path
λ	thickness of dielectric
λ	wavelength
λ_D	electrical double-layer thickness
λ_{Si}	thermal conductivity of silicon
μ	dynamic viscosity
μ_0	magnetic permeability of vacuum
µCOR	microfluidic-controlled optical router
µCP	microcontact printing
μ_{eo}	electroosmotic mobility
µTAS	micro total analysis system
µTM	microtransfer moulding
ξ	distance to the wall
π	3.14159
ρ	specific mass
σ	electrical conductivity
τ	characteristic time
τ	thermal time constant
T_i	characteristic time
T_s	force per unit surface area
T_v	yield stress
v	light frequency
v'	emitted light frequency

φ	electrical potential
φ_w	wall potential
χ_m	magnetic susceptibility of the medium
χ_P	magnetic susceptibility of the particle
ω	frequency
ω	rotation speed
δ	critical gap parameter, droplet splitting

Contents

1 Theory of Microfluidics 1

2 Device Fabrication 23

Microfluidics and Lab-on-a-chip
By Andreas Manz, Pavel Neužil, Jonathan S. O'Connor and Giuseppina Simone

Published by the Royal Society of Chemistry, www.rsc.org

1 Theory of Microfluidics

1.1 Introduction

Early scientists may have known more about microfluidics and nanotechnology than we are aware of. An example by Isaac Newton shows the observation of an optical interference pattern ('Newton's rings') on a glass device (Figure 1.1a) and a very nice experimental demonstration of capillary action (Figure 1.1b).[1]

Much later, in 1967, G. K. Batchelor coined the term 'microhydrodynamics' to describe the flow regime.[2] Today, the term microhydrodynamics has been replaced by the term 'microfluidics', which embodies an independent discipline of fluid dynamics referring to all phenomena observable in a scale range between 1 μm and a few hundred μm. The most important new topic introduced by the small scale of microfluidic devices is the significant role of surface forces (surface tension, electrical effects, van der Waals interactions and surface roughness), in addition to complicated three-dimensional geometries. Chip fabrication and clean-room use were derived from microelectronic chip manufacturing. The desire to explore very small spaces for engineering is often deduced from a famous talk 'There's plenty of room at the bottom' by Richard Feynman,[3] and the experimental take-off was triggered by needs in environmental monitoring, in drug discovery, in clinical diagnostics and by the human genome project. The concept of bringing a microfluidic circuit very close to the point of measurement was dubbed 'micro total analysis system' or μTAS, see Figure 1.2.[4]

Microfluidics and Lab-on-a-chip
By Andreas Manz, Pavel Neužil, Jonathan S. O'Connor and Giuseppina Simone

Published by the Royal Society of Chemistry, www.rsc.org

a

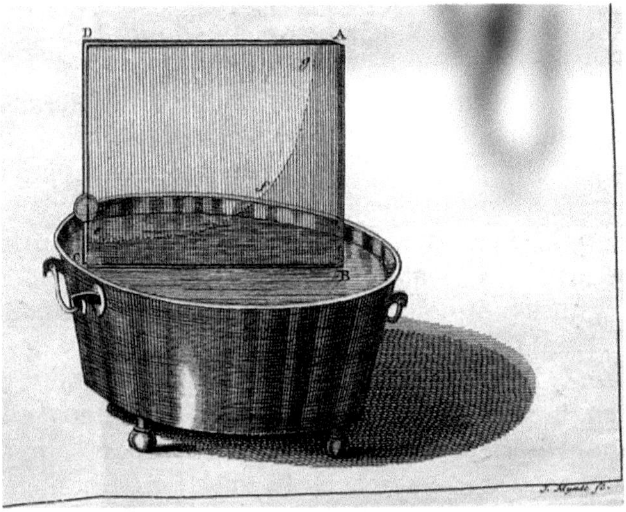

b

Figure 1.1 Textbook illustrations from 1747 showing (a) the observation of optical interference between two glass plates and (b) capillarity as a function of the gap distance between two glass plates.

At the sub-millimetre scale, the dynamics of fluids is characterized by the prevalence of viscous over inertial forces. To understand the reasons for this, we can consider a swimmer and a bacterium, both moving in the water. The swimmer, owing to their large dimensions, shows a significant inertial force while moving in water whereas the bacterium, having smaller dimensions and extremely low momentum,

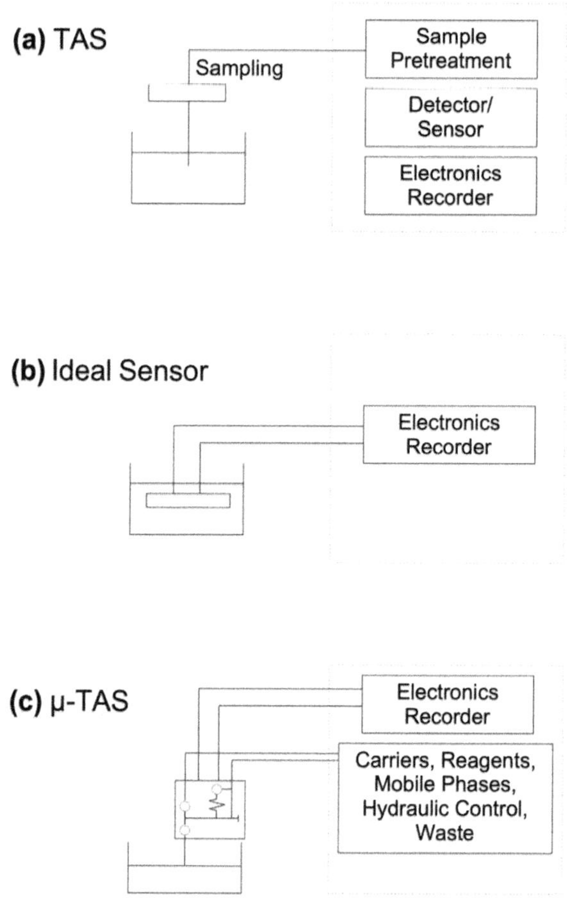

Figure 1.2 Schematic of different approaches to monitoring a chemical compound in solution. (a) A total analysis system (TAS) with a sampling device; (b) an ideal sensor; (c) the concept for a micro total analysis system (μ-TAS), where the sample is handled very close to the point of sampling. Reproduced from ref. 4 with permission from Elsevier, Copyright 1990.

stops within microseconds, when the flagellum motors stop. The very small momentum is such that for the bacterium the water appears as viscous as honey to us.

1.2 Definition of Fluids and Fluid Properties

Unlike molecules in a gas, which have a non-zero free path, the molecules in a liquid are so close to each other that momentum is always exchanged between them. Unlike solid matter, a fluid will always change shape according to the shape of the container. As per definition, liquid is a matter that cannot sustain shear stress in the absence of motion. The fluids can be gases (air), liquids (water, oil, syrup) and more complex systems consisting of several phases (blood, suspensions, emulsions). Such fluids can deform continually under shear stress and consequently, they can flow with no rigid restrictions. In fluids, discrete quantities such as mass and force give way to continuous fields such as specific mass, ρ, dynamic or kinematic viscosity, μ or η, and pressure, p. The specific mass is defined as the mass, m, per unit volume, V. Values of ρ for selected fluids are given in Table 1.1.

The pressure in the liquid is dependent only on the depth and it is not affected by the shape of the vessel containing the liquid. In a system with characteristic dimensions from a few micrometres to a few hundred micrometres, pressure differences can be neglected. It is worth noting that when referring to open microchannels, which have inlets and outlets, any pressure difference induced externally at these openings is transmitted to every point in the liquid, thereby inducing the liquid to flow. The relationship between flow and pressure is mainly defined by the viscosity of the fluids and the geometry of the channel.

A typical example is presented to introduce and define the viscosity. The scheme of the experiment is displayed in Figure 1.3. Two infinite plates are separated by a fluid layer of thickness H. One plate is moved in a straight line relative to the other at a constant speed, U, and the required viscous drag force F amplitude is measured. After an initial transient, F approaches a constant value.

The movement of the upper plane first sets the immediately adjacent layer of liquid molecules into motion; this layer transmits the action to the subsequent layers underneath it because of the intermolecular forces between the liquid molecules. In a steady state, the velocities of these layers range from U (the layer closest to the moving plate) to 0 (the layer closest to the stationary plate).

Table 1.1 Specific mass values ρ of some common fluids (in g cm^{-3}) as a function of temperature.[5]

Fluid	0 °C	20 °C
Water	0.999	0.998
Air	0.0013	0.0012
Ethanol	0.81	0.79
Glycerol	1.26	1.26

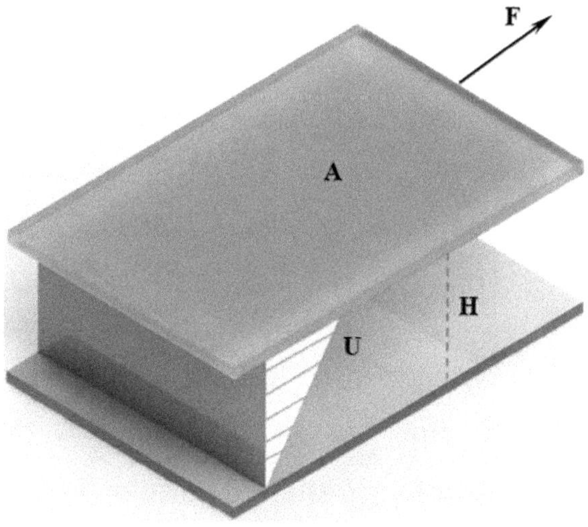

Figure 1.3 Schematic model explaining the viscosity of fluids in a shearing experiment.

$$\tau_s = \frac{F}{A} = \mu \frac{U}{H} \tag{1.1}$$

The force per unit surface area, the ratio of F to A, defines the shear stress τ_S. The shear stress shows that forces on fluids arise from fluid stress forces per unit area exerted on the element surfaces, in addition to externally applied body forces exerted on the bulk of the fluid. Common sense suggests that this force may be proportional to the area and velocity, while being inversely proportional to the distance between the plates as expressed in eqn (1.1), where the constant of proportionality is a property of the fluid, defined to be the dynamic viscosity.

In conclusion, the dynamic viscosity μ is given by the ratio between shear stress and shear rate and is measured in pascal seconds, Pa s. For a Newtonian fluid, the relationship between the shear stress and shear rate is always linear. However, Figure 1.4 shows the plot of shear stress *versus* shear rate for different fluids: Newtonian, shear thinning, shear thickening and viscoelastic.

The last three items in the list are defined as non-Newtonian fluids as they do not show a linear dependence of the shear stress on the shear rate. In particular, polymeric systems have a 'shear thinning' (*i.e.* decreasing) viscosity–shear rate behaviour, whereas slurries and suspensions often tend to be 'shear thickening'. Finally, in polymeric liquids a fluid- and solid-like behaviour can be observed, depending on the time scale of the experiment. These are denoted viscoelastic fluids, which are characterized by a viscosity and a relaxation time. Table 1.2 gives dynamic viscosity values for the liquids for which specific mass (relative density) values were reported in Table 1.1.

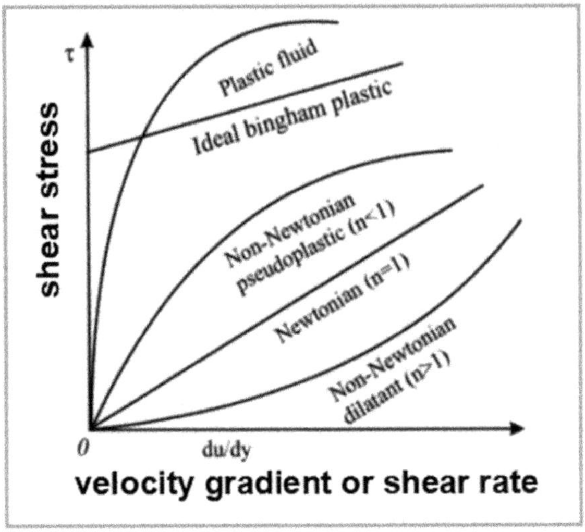

Figure 1.4 Shear stress as a function of shear rate for Newtonian and non-Newtonian fluids.

Table 1.2 Dynamic viscosities of some common fluids in mPa s.[2]

Fluid	0 °C	20 °C
Water	1.79	1.00
Air	0.017	0.018
Ethanol	1.79	1.18
Glycerol	13 000	1400

1.3 Reynolds Number and Its Physical Meaning

The characterization of intrinsic fluid properties is a fundamental premise to understand flow behaviour. Dimensionless numbers can be useful to describe fluid flow.

A Newtonian fluid has a constant density and viscosity flows inside a circular tube with a diameter d and length l as shown schematically in Figure 1.5. The fluid has a volumetric flow rate Q and the flow is generated by the difference in pressure, Δp, existing between the inlet and the outlet of the microfluidic channel. The mean velocity u, which is normally used instead of the flow rate, is a function of the flow rate of the fluid and also of the cross-section of the channel, $u = 4Q\pi^{-1}d^{-2}$. An accurate dimensionless analysis, based on the degrees of freedom of the problem, results in a single dimensionless independent number that can describe the fluid properties.[6,7] The dimensionless group is the following:

$$\mathrm{Re} = \frac{u\rho d}{\mu} \tag{1.2}$$

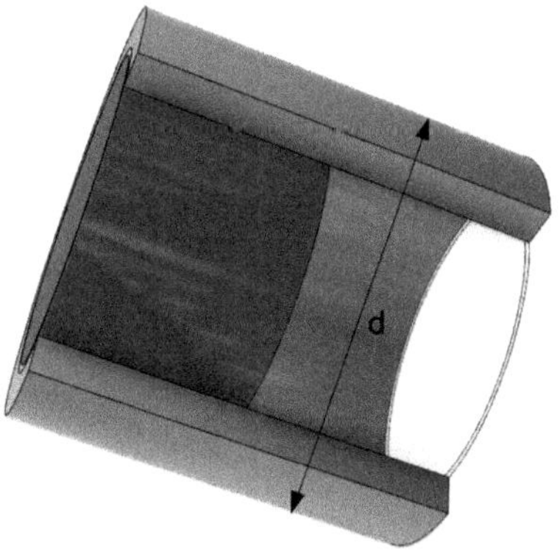

Figure 1.5 Fluid moving in a circular tube by the difference in pressure between the inlet and the outlet.

This description of fluid flow is defined as the Reynolds number. A low Reynolds number (Re < 2100) is characteristic of low flow rates and small geometries resulting in laminar flow. After a transient region (2100 < Re < 4000), a turbulent flow regime takes place.

Owing to the dimensions of the channels and the flow velocity, microfluidics falls in the laminar flow regime. In fact, assuming a microfluidic channel with a circular cross-section with a diameter of 100 μm and with water being pumped through the microfluidic channel at an average velocity of 1 cm s^{-1} (the density and viscosity can be found in Tables 1.1 and 1.2, respectively), the value of Re is ~1, hence there is laminar flow inside the channel. If the velocity of the fluid is increased to 1 m s^{-1} and the channel has a larger cross-section, the behaviour of the water flow inside may change and, analogously, the situation can be predicted if the water is replaced with air or glycerol.

To understand further the physical meaning of Re, we can consider a fluid of mass ρ moving with velocity u in a channel with a certain curvature (Figure 1.6).

If the fluid were to be decelerated to zero by hitting a wall at a distance Δx away, the fluid would exert a force on the wall and the wall would exert an equal and opposite force on the fluid. As the fluid has inertia, it needs a force to change the uniform motion. This force is equal to the rate of change of momentum, where the momentum changes from ρu to zero in the time Δt required to move a distance Δx; the change of rate of momentum is given by $\rho u/\Delta t$, where $\Delta t = \Delta x/u$. In conclusion, the force of inertia can be written as

$$f_I \sim \rho u^2 (\Delta x)^2 \tag{1.3}$$

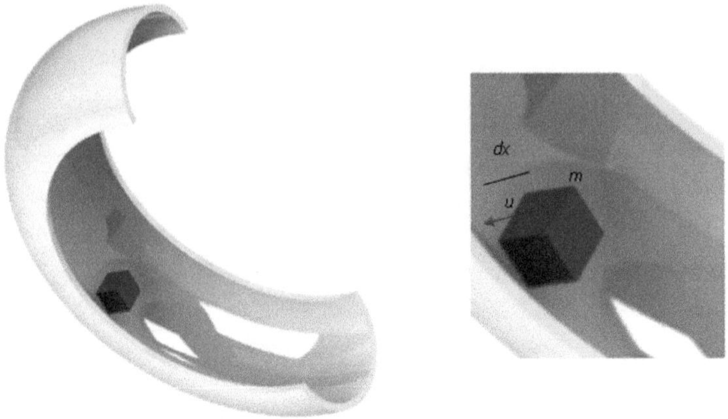

Figure 1.6 Schematic of the motion of a fluid element near a wall.

If the same element of fluid starts to move parallel to the wall, it experiences a shear stress given by $\mu u/\Delta x$ and the shear force is given by

$$f_v \sim \rho u \Delta x \tag{1.4}$$

The magnitudes of inertial and viscous force densities have to be compared. The ratio of the forces results in the dimensionless Reynolds number:

$$\frac{f_I}{f_V} = \frac{\rho u \Delta x}{\mu} \equiv \mathrm{Re} \tag{1.5}$$

which gives back the Re value and provides the comparison of the inertial and viscous forces.

1.4 Scaling Down Fluid Dynamics. A Navier–Stokes Equation Simplified for Microscale Fluid Flow

To determine the fluid velocity, it might be tempting to use Bernoulli's equation, which is derived from the conservation of energy statement by assuming first that the fluid has no viscosity and second that the energy dissipation due to shear stresses in the fluid is negligible. Both are inappropriate assumptions for low Reynolds number flows.

Instead of using energy conservation to determine a low Re flow response to applied pressure, it is more accurate to use conservation of momentum, which leads to the Stokes or Navier–Stokes equations.

First, it is useful to refer to the continuity equation, which describes the time rate of change of the fluid density at a fixed point in space, which can be written by using vector notation as follows:

$$\frac{d\rho}{dt} = -\nabla \cdot (\rho u) \tag{1.6}$$

A very important special form of the continuity equation is that for a fluid of constant density (incompressible fluid), where the term on the left is zero and the continuity equation assumes the particularly simple form

$$\nabla \cdot (u) = 0 \tag{1.7}$$

No fluid is perfectly incompressible, but frequently in engineering and biological applications the assumption of constant density results in considerable simplification with insignificant errors.

For the same volume of liquid, the second law of mechanics $F = ma$ under the continuum hypothesis assumes the form of the momentum equation also defined by the Navier–Stokes equation:

$$\rho\left(\frac{\partial u}{\partial t} + u \cdot \nabla u\right) = -\nabla p + \mu \nabla^2 u + \rho g \tag{1.8}$$

where inertial acceleration terms appear on the left and the all the forces on the right.

Introducing the absolute pressure $P = p + \rho g$, the equation can be also written in compact mode:

$$\rho\left(\frac{\partial u}{\partial t} + u \cdot \nabla u\right) = -\nabla P + \mu \nabla^2 u \tag{1.9}$$

Analysing the Navier–Stokes eqn (1.9), the linear unsteady term sets the inertial time scale T_i required to establish steady flows. This time scale can be estimated by balancing the unsteady inertial force density with the viscous force density giving the characteristic time:

$$\Gamma_i = \frac{\rho d x^2}{\mu} \tag{1.10}$$

The inertial time scale can be interpreted as the time required for vorticity to diffuse a distance dx.

Because of time scales in microfluidics and the characteristic properties of working fluids (*i.e.* water density, viscosity), inertia rarely plays a significant role in microfluidic systems, so that without the inertial non-linearity, straightforward microfluidic systems have regular flow and therefore the left-hand term in eqn (1.9) can be neglected.

The same result can be obtained by considering the dimensionless variables and assuming that the only characteristic time of the system is the one required for the element of fluid to move through a characteristic distance, eqn (1.9) can be rewritten as follows:

$$\frac{\partial u}{\partial t}+u\cdot\nabla u=-\nabla P+\frac{1}{\text{Re}}\nabla^2 u \tag{1.11}$$

which states that when the amplitude of Re is very small, as in the case of viscous flow and microflow, the non-linear terms disappear, resulting in linear and predictable Stokes flow (or creeping flow) and the Stokes equation:

$$0=-\nabla P+\mu\nabla^2 u \tag{1.12}$$

The boundary conditions state the no-slip condition, which means $u = 0$ at the wall; in reality, the velocity of the fluid at the wall is almost negligible.

It is worth noting that there are some circumstances where the inertial forces become apparent at the microscale. One of the causes can be derived from the geometry of the channel. The no-slip boundary condition and heterogeneous flow profile cause heterogeneous centrifugal forcing and give the secondary 'Dean flow'. A very high Re value can also be a cause of non-linear inertial forces that destabilize the flow.

1.5 Plane Poiseuille Flow and Parabolic Velocity Profile

After introducing the Navier–Stokes equation and the simplification in microfluidics, we can reconsider the flow of an incompressible Newtonian fluid caused by a difference in pressure between the two extremes of the channel with a rectangular cross-section (Figure 1.7a).

The flow is in the x-direction, while H/W and H/L are both small numbers and, as mentioned, the flow is caused by a pressure difference between $x = 0$ and $x = L$.

As consequence of the fact that H/L is small, the first hypothesis is 'the flow is fully developed' due to the entrance region being significantly shorter than the length of the channel. This means that $u(x)$ is constant, $u(z)$ is constant and, consequently, the velocity vector will depend only on y $[u = u(y)]$.

As H/W is small, it can be derived that the second hypothesis is 'the flow field is infinite in expanse in the z-direction'. As a consequence, there is no distinction between one z location and the flow field should not change in the z-direction, $u_x(z) = \text{constant}$, $u_y(z) = \text{constant}$ and $u_z(z) = \text{constant}$. In particular, here the analysis refers to the flow in tubing with a rectangular cross-section (see Figure 1.7a).

In the absence of motion normal to the direction of mean flow and with the flow lines parallel to the walls, the velocity has two null components ($u_y = u_z = 0$) and the third component varies along the orthogonal direction, $u_x = u_x(y)$.

With those simplifications and taking into account that u_x is a function only of y, the Navier–Stokes eqn (1.12) along the three axes can be written as

$$0 = -\frac{\partial P}{\partial x} + \mu \frac{\mathrm{d}^2 u_x}{\mathrm{d} y^2}; \quad 0 = -\frac{\partial P}{\partial y}; \quad 0 = \frac{\partial P}{\partial z} \tag{1.13}$$

From this, it derives that the pressure is a function only of x and the system of eqn (1.13) can be rewritten as

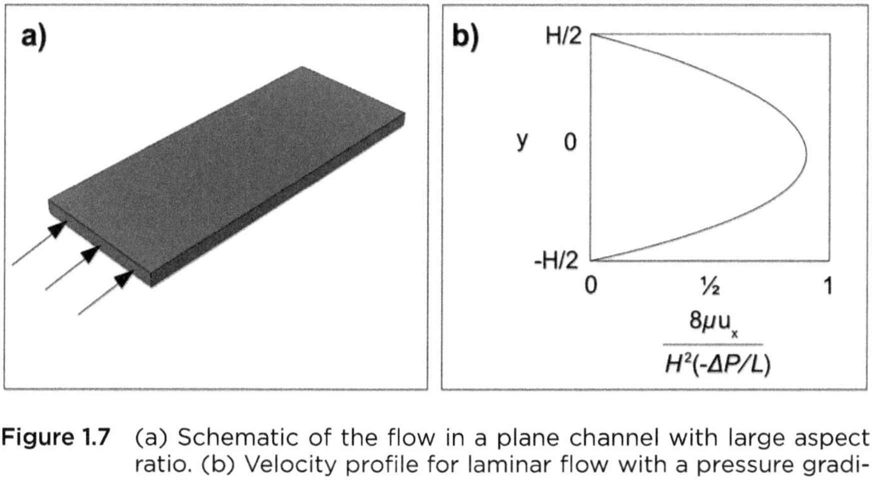

Figure 1.7 (a) Schematic of the flow in a plane channel with large aspect ratio. (b) Velocity profile for laminar flow with a pressure gradient between infinite stationary planes.

$$0 = -\frac{\mathrm{d}P}{\mathrm{d}x} + \mu\frac{\mathrm{d}^2 u_x}{\mathrm{d}y^2} \tag{1.14}$$

Assuming that both terms in eqn (1.14) are independent of x and y, we can write $\mathrm{d}P/\mathrm{d}x = \Delta P/L$ and finally the integration of the equation with condition of no slip ($u_x = 0$ at $y = \pm H/2$) provides the velocity profile:

$$u_x = \frac{H^2}{8\mu}\left(-\frac{\Delta P}{L}\right)\left[1-\left(\frac{2y}{H}\right)^2\right] \tag{1.15}$$

which is a parabola, so the profile of velocities is parabolic, with a maximum at the centreline, as shown in Figure 1.7b. The average velocity defined by the expression

$$\langle u_x \rangle = \frac{1}{H}\int_{-H/2}^{H/2} u_x\,\mathrm{d}y = \frac{H^2}{12\mu}\left(-\frac{\Delta P}{L}\right) \tag{1.16}$$

which is fundamental for estimating the flow rate $Q = WH\langle u_x \rangle$.

Owing to the geometry of the channel, the walls exert significant friction on the fluid, which in turn requires a source of high energy in order to flow. In analogy with Ohm's law in electricity, where electrical resistance is defined as $R = V/I$, fluidic resistance is defined by the ratio between the applied pressure, Δp, and the volumetric flow rate, Q: $\mathrm{Res} = \Delta P/Q$.

For the channel analysed here, the resistance can be defined as

$$\mathrm{Res} \approx \frac{3\mu L}{4WH^3} \tag{1.17}$$

Table 1.3 summarizes the results for channels of rectangular and circular cross-section.

Table 1.3 Equations of fluid flow.

Equation	Rectangular cross-section (x, y, z) L, H, W	Circular cross-section (r, z, θ) L, R
Poiseuille	$0 = -\frac{\mathrm{d}P}{\mathrm{d}x} + \mu\frac{\mathrm{d}^2 u_x}{\mathrm{d}y^2}$	$0 = -\frac{\mathrm{d}P}{\mathrm{d}z} + \mu\frac{1}{r}\frac{\mathrm{d}}{\mathrm{d}r}\left(r\frac{\mathrm{d}v_z}{\mathrm{d}r}\right)$
Velocity	$u_x = \frac{H^2}{8\mu}\left(-\frac{\Delta P}{L}\right)\left[1-\left(\frac{2y}{H}\right)^2\right]$	$u_z = \frac{R^2}{4\mu}\left(-\frac{\Delta P}{L}\right)\left[1-\left(\frac{r}{R}\right)^2\right]$
Average velocity	$\langle u_x \rangle = \frac{H^2}{12\mu}\left(-\frac{\Delta P}{L}\right)$	$\langle u_z \rangle = \frac{R^2}{8\mu}\left(-\frac{\Delta P}{L}\right)$
Resistance	$\mathrm{Res} \approx \frac{3\mu L}{4WH^3}$	$\mathrm{Res} \approx \frac{8\mu L}{\pi R^4}$

1.6 Mass Transport

Within microfluidic systems, two different types of transport can occur: active and statistical transport. Active transport is transport controlled by exerting work on the fluid; the work results in a volume flow of the fluid, where the flow can usually be characterized by its direction and a flow profile.

Statistical transport is entropy driven, which means that transport occurs only if a fluid is more disordered after transport than before. A typical situation is at an interface between a liquid with a high concentration of one type of molecule and a liquid with zero concentration of the same molecule. This orderly situation leads spontaneously to statistical transport of molecules from the side with high concentration to the side with zero concentration. Eventually, the concentration is equal in both liquids, which are therefore evenly mixed, and the situation is less ordered than before.

1.7 Diffusion

In fluids under conditions with no applied external forces, each molecule moves in a certain direction for a certain time until it is hit by another molecule and changes direction. In a fluid composed of identical molecules, no net flow is observable. Once there are two types of molecules, diffusion occurs and it is observable when there is a concentration gradient of one kind of molecule within a fluid.

Areas of high concentration contain more molecules than areas of low concentration, so that many molecules move randomly in one direction, but only a few molecules move randomly backwards.

The statistical movement of a single molecule in a fluid can be described as a random walk. The movement is characterized by the 1-dimensional Einstein–Smoluchowski equation:[8]

$$x = \sqrt{2Dt} \tag{1.18}$$

where x is the average distance moved after an elapsed time between molecule collisions and D is the diffusion constant (measured in $m^2\ s^{-1}$). The average distance moved varies with the square root of time, in contrast to directed movement, where the distance is directly proportional to time.

In the laminar flow regime, at the steady state, the mass flow rate (the variation of velocity per unit time) of the fluid per unit area (or mass flux) is proportional to the mass fraction difference divided by the width of the channel, which for a differential element of fluid inside the microfluidics is the one-dimensional form of Fick's first law of diffusion:

$$J = -D\frac{dc}{dx} \tag{1.19}$$

where J is the mass flux in the positive x-direction and D is the characteristic diffusion coefficient of a given molecule in solution (two liquids, solute/liquid).

Fick's first law states the profile of concentration when two different species flow together inside a microfluidic channel with a rectangular cross-section and underscores that diffusion is a statistical transfer mechanism. See also Figure 1.8.

The continuity eqn (1.6) for molecules dissolved in a solvent will need to include diffusion. It is given by

$$\frac{\partial c}{\partial t} = -\frac{\partial J}{\partial x} \tag{1.20}$$

The continuity equation and Fick's first law can be combined to give Fick's second law of diffusion, which can be expressed in one dimension as

$$\frac{\partial c}{\partial t} = D\frac{\partial^2 c}{\partial x^2} \tag{1.21}$$

The solution of eqn (1.21) is given by

$$c = c_0 \cdot erfc\left[\frac{x}{2\sqrt{Dt}}\right] \tag{1.22}$$

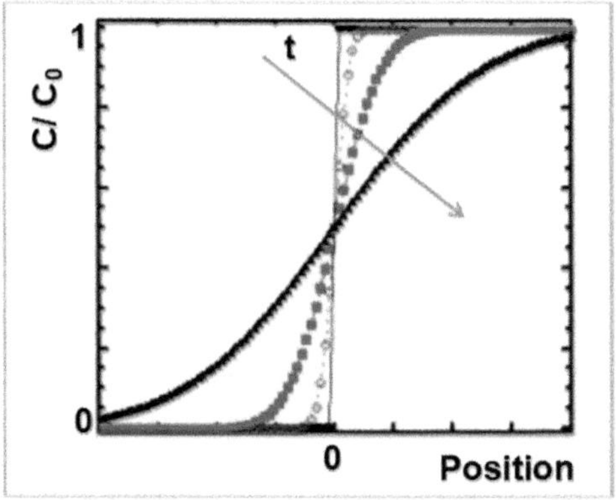

Figure 1.8 Example of development of diffusion front according to the time (increasing in the direction of the arrow).

This solution breaks down when the concentration at the centre plane starts to decrease from its initial value, C_0, or when the time is longer than $\tau = d/D$, a simplc cstimatc to define the time needed for particles or molecules to diffuse across the entire channel. The profiles of concentrations are illustrated in Figure 1.8. In the beginning, the profile is still flat at the centre with a concentration C_0. At times t longer than τ, the concentration at the centre plane decreases below the value C_0.

If an area of high concentration directly borders an area of zero concentration, then the sharp border fades slowly over time. The relationship between the diffusion behaviour in the form of the diffusion constant and a microscopic property of the molecules in the form of the hydrodynamic radius, r, can be expressed by the Stokes–Einstein relation, which can be modified according the following relation:

$$D = \frac{kT}{C_{\mathrm{D}}} \tag{1.23}$$

where C_{D} is a frictional constant known from the Stokes equation, k is the Boltzmann constant and T is the thermodynamic temperature. Eqn (1.23) shows the temperature dependence of the diffusion constant and allows us to estimate the diffusion constant of molecules when the hydrodynamic radius is known.

1.8 Péclet Number

The mixing of two or more components in laminar fluid flows is forced to occur by diffusion alone, resulting in long mixing times. For some applications, it is extremely important to control the velocity of the mixing and a slow process is essential. However, for other applications, the mechanism of diffusion is too slow and the mixing of the two components has to be promoted using different approaches.

To understand how the solute is homogenized inside the channel, it can be observed that during this time, the actual profile moves a distance Y, hence the number of channel widths required for complete mixing would be of the order

$$\frac{Y}{d} \approx \frac{ud}{D} = \mathrm{Pe} \tag{1.24}$$

Simply put; the Péclet number, Pe, compares the convective and the diffusion mechanisms. For small values of the Péclet number, $\mathrm{Pe} < 10$, which can be measured at low flow velocity, the diffusion remains the predominant mixing mechanism.

It is worth noting that the ratio between the Péclet number and the Reynolds number is actually the ratio between momentum transport and diffusive mass transport. For diffusion coefficients ranging from 10^{-5} to 10^{-7} cm^2 s^{-1}, the Péclet numbers are in the range $100 < \text{Pe} < 10\,000$, hence convective mass transport dominates over diffusive transport in almost all microfluidic applications.

The average diffusion time over the characteristic mixing length, also called the striation thickness, is represented by the Fourier number, Fo:[9]

$$\text{Fo} = \frac{Dt_{\text{diff}}}{L_{\text{mix}}^{2}} \tag{1.25}$$

The Fourier number is usually in the range $\text{Fo} = 0.1 - 1$. For a simple T-mixer with two streams in a microfluidic channel of mixer length L and width W, the residence time should be the same as the average diffusion time, allowing the necessary channel length for complete mixing to be calculated.

In general, fast mixing can be achieved with smaller mixing paths (channel diameter) and larger contact surface area. If the channel geometry is very small, the fluid molecules collide more often with the channel wall and not with other molecules. In this case, the diffusion process is called Knudsen diffusion. The ratio between the distance of molecules and the channel size is characterized by the dimensionless Knudsen number (Kn):

$$\text{Kn} = \frac{\lambda}{d_{\text{hyd}}} \tag{1.26}$$

where λ is the mean free path and d_{hyd} is hydraulic diameter of the channel. The free path λ is a function of the Boltzmann constant ($k = 1.38066 \times 10^{-23}$ J K^{-1}), the absolute temperature, the pressure and the molecular diameter of the diffusing species. The Knudsen number for liquids is small, because the mean free path of liquids is on the order of a few Ångströms. In gases, the mean free path is on the order of 100 nm to several micrometres. For example, at room conditions, the mean free path of hydrogen is 0.2 μm. Knudsen diffusion may occur in microchannels with diameters on the order of a few micrometres.

1.9 Dispersion

According to previous analysis, a low value of Re and laminar flow retard the mechanism of mixing. A low velocity of mixing can be desirable in some cases, for example when solute samples or biological cells are transported in a fluid stream, one after another, and should not interfere with each other. A special case of this is used to separate molecules, such as electrophoresis and chromatography, which will be discussed later.

Taylor dispersion describes the role of convection in dispersing heterogeneous flows. An important hypothesis of Taylor theory is that the dispersion acts in the direction of the flow (axial) and not in the perpendicular direction.[10] Taylor dispersion is not observed in microfluidic mixing devices, where steady-state conditions are maintained.

To explain Taylor dispersion, it is useful to refer to the example in Figure 1.9a. At the inlet of the microfluidic channel, where $x = 0$, the fluid flows in a laminar regime, a pulse of mass m is introduced over a very short period near time $t = 0$ and the progress of the pulse through the tube is analysed on the time axis.

A short distance downstream from the inlet, a given sample volume, for example, an injected plug of dissolved molecules, will be distorted in shape. The parabolic profile of velocity observed in Poiseuille flow stretches the volume element into a parabolic form, but two different conditions need to be considered, dispersion with diffusion and dispersion without diffusion.

Considering the mean flow velocity, u, it can be considered that the tracer stripe is stretched to $\sim ut$, each stripe within the plug is convectively stretched and a wide band broadening of the initial pulse can be observed (Figure 1.9b).

The mechanism of molecular diffusion across the channel promotes mixing with a time scale

$$\tau_D = \frac{W^2}{D} \tag{1.27}$$

at which molecular diffusion across the channel smoothes the parabolic stripe into a uniform plug of width

$$W_{TD} = \frac{uW^2}{D} \tag{1.28}$$

of the channel. Each stripe within the plug is convectively stretched and then diffusively smeared to a width W_{TD} after a time t_D, as demonstrated in Figure 1.9c. Note that diffusion is effective in both the axial and radial directions. Taylor theory describes a solution valid for long times after injection. When the pulse dimension W_{TD} attains this range, sufficient time has elapsed that the initial shape of the pulse no longer matters. Furthermore, after flowing for a certain length, the

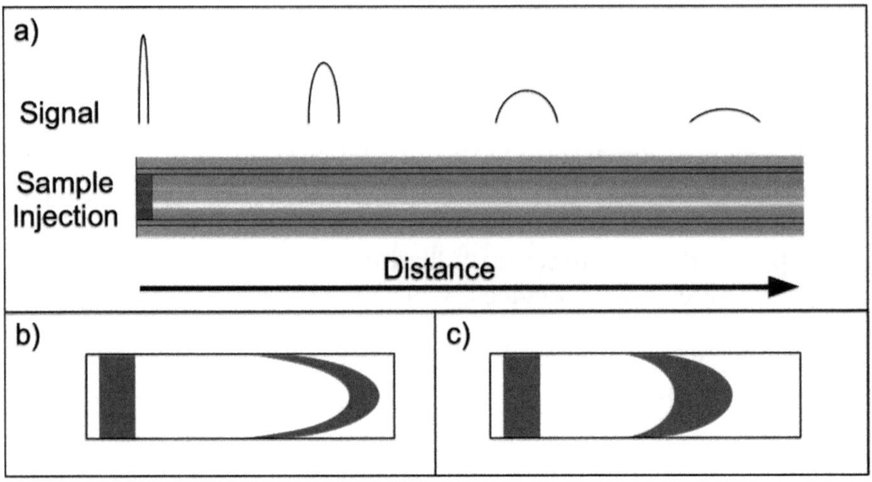

Figure 1.9 (a) Fluid flows in a laminar regime when a pulse of mass is introduced. (b) Taylor dispersion without diffusion. (c) Taylor dispersion with diffusion.

initially thin pulse evolves into a Gaussian distribution. The tracer distribution thus grows diffusively as $t^{1/2}$ with an effective long-time axial diffusivity

$$D_z \approx \frac{u^2 W^2}{D} \approx \mathrm{Pe}^2 D \tag{1.29}$$

occurring in addition to the molecular diffusivity. Taylor dispersion is valid only when the following points are valid. First, the axial Taylor dispersivity D_z is valid only on long time scales $t > W^2/D$ or after a downstream distance $L \gg \mathrm{Pe}W$. In fact, Taylor began by neglecting the axial molecular diffusion term and subsequently showed that this is permissible if the Péclet number is on the order of 70 or greater and if the length of the region occupied by the pulse is >170 times the channel radius. Second, Taylor dispersion may appear to enhance dispersion for small solutes, since $D_z \approx D^{-1}$.

Without Taylor dispersion, however, convection would stretch tracer spots indefinitely with time. This is the case for very slowly diffusing matter, such as particles, biological cells or large proteins. Therefore, electroosmotic flow, electrophoresis, segmented two-phase flow and droplets are more favourable in those cases.

1.10 Capillary Number

For liquids, the surface energy and the surface tension are equivalent and both have dimensions of force per unit length (*e.g.* N m^{-1}, dyne cm^{-1}) or, equivalently, energy per unit area (*e.g.* J m^{-2}).

A dimensionless parameter, the capillary number, Ca, is defined as the competition between the interfacial stresses and the viscous stresses:

$$\text{Ca} = \frac{\mu u}{\gamma} \tag{1.30}$$

where μ is the dynamic viscosity of the moving fluid, u is its velocity and γ is its surface tension (N m^{-1}). Interfacial force dominates over inertial force at low capillary numbers. The balance between the surface tension (aka capillary forces) and viscous forces rules the contact angle formed between a flowing liquid front and a solid surface. The expression for the dynamic contact angle is:

$$\vartheta_\text{d} - \vartheta_\text{s} = \frac{1}{3}\frac{A\text{Ca}}{\vartheta_\text{s}^{\,2}} \tag{1.31}$$

where A is a dimensionless proportionality factor, and ϑ_d and ϑ_s are the dynamic and static contact angles, respectively. Eqn (1.31) shows that $\vartheta_\text{d} - \vartheta_\text{s}$ is of the same sign as Ca, confirming that the advancing contact angle is larger than the static contact angle and the receding contact angle is smaller than the static contact angle.

1.11 Scaling Laws for Molecules in Solution

Based on the above theory, the process of miniaturization can be simplified. This is not a fundamental law, but rather a strategy for experiments at a small scale. As an experiment of thought, let us assume we take an existing experiment, such as the flow in a channel, from a large scale (known) to a small scale.[11] Useful scaling laws can be derived easily if the experiment is miniaturized by the same factor in all three dimensions of space. A scaling factor d applies to all lengths involved, such as the capillary diameter, the capillary length or the optical path-length across the channel, for example. If the time scale is treated like a surface, as shown in 'scaling game #2' in Figure 1.10, miniaturization by a factor of 10 in x, y and z implies that time must be shortened

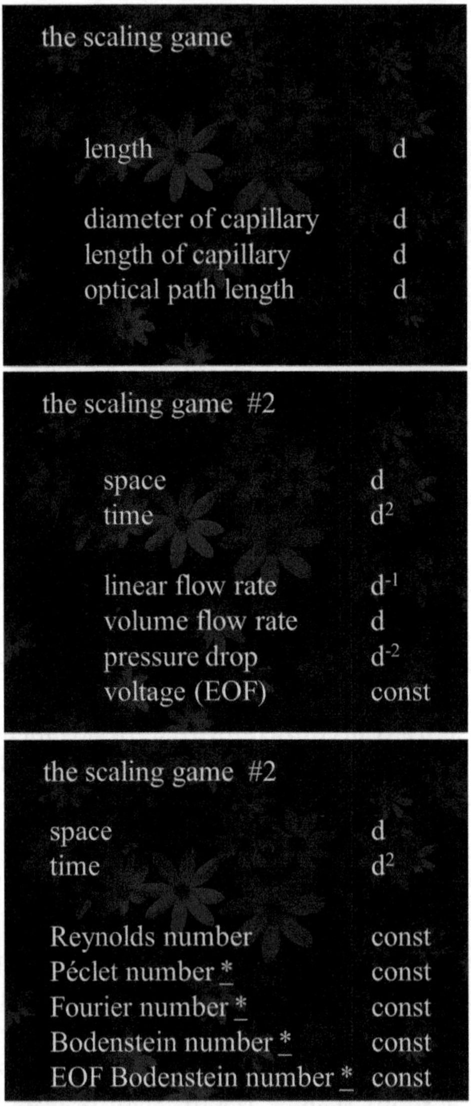

the scaling game
length d
diameter of capillary d
length of capillary d
optical path length d
the scaling game #2
space d
time d2
linear flow rate d-1
volume flow rate d
pressure drop d-2
voltage (EOF) const
the scaling game #2
space d
time d2
Reynolds number const
Péclet number * const
Fourier number * const
Bodenstein number * const
EOF Bodenstein number * const

Figure 1.10 Experiment of thought 'scaling game' for shrinking an existing experiment to a smaller size. The trivial case whereby time is kept a constant (scaling game #1) is not shown, because it is not helpful for our considerations. (Top) Length means all sorts of lengths in three spatial dimensions; (middle) in the case that time is treated like a surface (scaling game #2), the consequences for the flow rate and pressure drop are shown; (bottom) several dimensionless parameters are independent of scale if time is treated like a surface.

100-fold. This would lead to a 10-fold higher linear flow rate, a 10-fold lower volume flow rate and a 100-fold higher pressure drop for a given channel. Interestingly, several dimensionless parameters are independent of scale under such conditions. This kind of miniaturization maintains the similarity of large and small systems, which can be very useful for predicting the behaviour of a microfluidic device.

Furthermore, a 'back of an envelope' calculation can provide an idea of throughput for a chemical reaction (or a bioassay) carried out in a given volume. Figure 1.11 illustrates such an example for a micro well plate (1 μL) and a simple diffusion-controlled chemical binding assay. High throughput can be obtained by parallel processing of such volumes in an array, represented by the number of volumes on a square centimetre. In addition, the throughput can be obtained by serially performing the same assay in the same location over and over again, represented by the diffusion time across the given volume. The total throughput is the product of the serial and parallel processing. It can be seen that picolitre volumes would allow millions of chemical reactions per second per square centimetre, a value that has not yet been demonstrated experimentally.

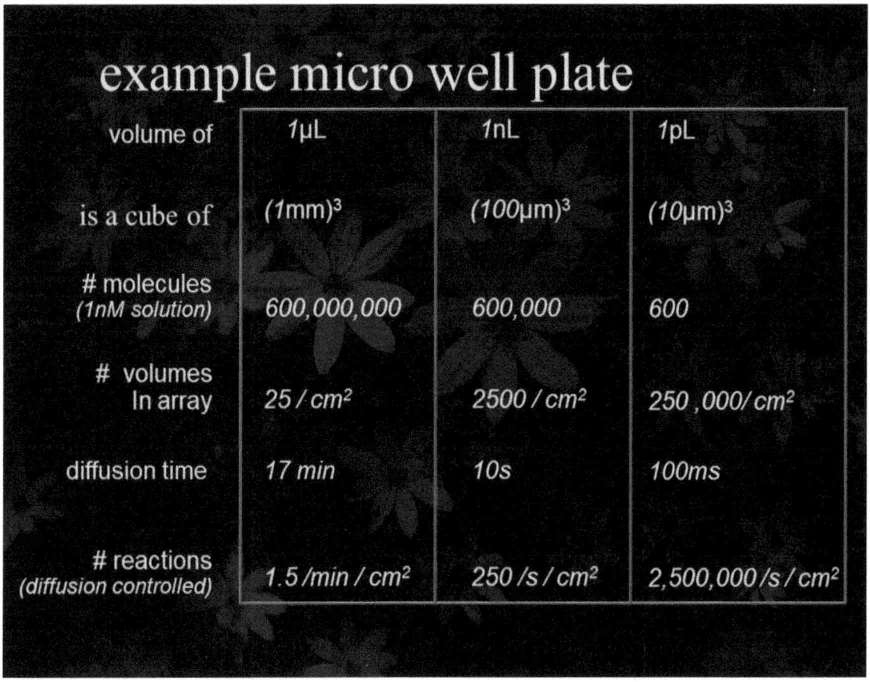
example micro well plate

volume of	1μL	1nL	1pL
is a cube of	$(1mm)^3$	$(100μm)^3$	$(10μm)^3$
# molecules (1nM solution)	600,000,000	600,000	600
# volumes In array	25 / cm^2	2500 / cm^2	250 ,000/ cm^2
diffusion time	17 min	10s	100ms
# reactions (diffusion controlled)	1.5 /min / cm^2	250 /s / cm^2	2,500,000 /s / cm^2

Figure 1.11 Calculated values for throughput as a function of volume, using a micro well plate for a diffusion-controlled chemical binding reaction. The diffusion coefficient is 10^{-9} m^2 s^{-1} for a small molecule in water.

References

1. W. J. s'Gravesande (Latin original 1721), translation into English by J. T. Desaguliers, *Mathematical Elements of Natural Philosophy*, confirm'd, by Experiments: or, an introduction to Sir Isaac Newton's philosophy, 6th edn, London MDCCXLVII. vol. II, p. 267, plate 120; and vol. I, p. 23f, plate 3.
2. G. K. Batchelor, *An Introduction to Fluid Dynamics*, Cambridge University press, 1967.
3. R. P. Feynman, talk given at the annual American Physical Society meeting at Caltech, December 29, 1959, Engineering and Science (Caltech), 5 February 1960, vol. 23, pp. 22–36.
4. A. Manz, N. Graber and H. M. Widmer, *Sens. Actuators, B*, 1990, **1**, 244–248.
5. M. Denn, *Process Fluid Mechanics*, Prentice Hall, 1979.
6. R. B. Bird, W. E. Stewart and E. N. Lightfoot, *Transport Phenomena*, John Wiley, 2nd edn, 2006.
7. O. Gerschke, H. Klank and P. Telleman, *Microsystem Engineering of Lab on a Chip Devices*, Wiley, 2008.
8. H. Bruus, *Theoretical Microfluidics*, Oxford. 2010.
9. E. L. Cussler, *Diffusion Mass Transfer in Fluid Systems*, Cambridge University Press, New York, 1996.
10. D. Janasek, J. Franzke and A. Manz, *Nature*, 2006, **442**(7101), 374.
11. Lectures given by A. Manz on many occasions; https://speakerdeck.com/amzamz.

2 Device Fabrication

2.1 Photolithography

Photolithography originates from compounding three ancient Greek words, meaning light–stone–writing, and refers to patterns printed on top of a stone/metal plate with a smooth surface. The 'photo-' means that these techniques focus on transferring patterns to the surface of the light sensitive coating, which is on top of a substrate wafer, using optical methods.

2.1.1 Lithography

Original photolithography techniques, dating back to the 1800s, used Bitumen of Judea and strong acid etching of metal, glass or stone plates, which would then be used in printing. Since then, far more sophisticated microlithography techniques have been developed. However, the essential principle of transferring a pattern onto an ultraviolet (UV) light-sensitive material called photoresist (PR), followed by removal of softened or uncured material *via* a cleaning solution, called developer, to fabricate a pattern upon the substrate is constant throughout photolithography techniques.

2.1.2 Photoresist (PR)

PR typically comes in a liquid form to be spread evenly onto a substrate, then exposed with a desired pattern and developed for subsequent processing. There are two types of PR, negative and positive.

Microfluidics and Lab-on-a-chip
By Andreas Manz, Pavel Neužil, Jonathan S. O'Connor and Giuseppina Simone

Published by the Royal Society of Chemistry, www.rsc.org

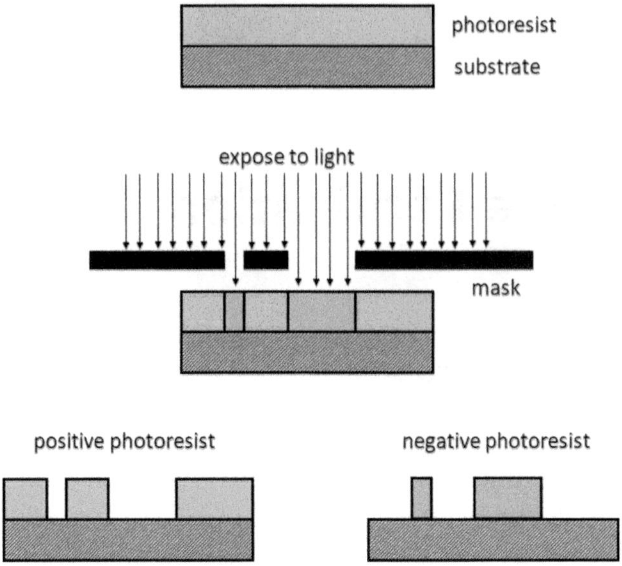

Figure 2.1 Comparison of positive and negative photoresist.

Negative resist is hardened by UV exposure so it cannot be removed by the developer, whereas positive PR is damaged by UV exposure, making the exposed areas easier to remove by the developer (see Figure 2.1). The type of resist used can depend on the materials of the wafer, the desired structure and ease of removal. In general, positive PRs are easier to handle, hence their greater use, but again the selection of a PR is dependent on how it will be used. A common positive PR is diazonaphthoquinone and novolac resin (DNQ-novolac) and a common negative PR is SU-8, an epoxy resin. Because of the coating and exposure method, photolithography provides a binary pattern transfer as there is essentially nothing in between. There are no colours, greyscale or depth to the image, just masked or unmasked.

Practical photolithography consists in the following steps; surface cleaning, priming, photoresist coating, pre-bake, exposure, post-bake and development. The following sections will provide an in-depth description of the complete photolithography process.

2.1.3 Surface Cleaning

In order to optimize the PR binding, the substrate must be extremely clean. Any dirt, residue or dust can result in incomplete binding and surface irregularities. Similarly to the way in which surgeons work in a scrupulously clean operating room and must use clean instruments to avoid infecting their patients, so we must maintain the photolithography fabrication facilities clean to avoid ruining our attempts. First, the substrate, such as Si, glass or SiO_2, is cleaned, typically with piranha solution (H_2SO_4–H_2O_2) at 120 °C for 10–20 min, to remove all organic residue. Substrates containing materials that are incompatible with the piranha cleaning process, such as metals, must be also cleaned but with an alternative solution, *e.g.* 99% HNO_3, which passivates the surface of Al and does not etch it. Post-wash, the substrate is heated to 120–150 °C to remove traces of water.

2.1.4 Priming

The clean wafer is covered with a PR adhesion promoter, such as hexamethyldisilazane (HMDS), to form a covalent bond-based cross-linker, Si–O–Si, between the substrate surface and the PR. HMDS is deposited using a process called chemical vapour deposition (CVD), forming a self-assembled monolayer (SAM). The substrate is placed inside a vacuum oven, set to a temperature between 90 and 150 °C, then the oven is evacuated, flushed with N_2 and evacuated again to remove any O_2. These steps are repeated three times, then a stream of HMDS vapour is introduced for 50–60 s. The oven is finally evacuated and flushed a few times with N_2. By the end of this step, we have a SAM of the HMDS on the substrate surface.

2.1.5 Photoresist Coating

The wafer is placed on a spinner chuck held by vacuum from the back of the wafer. PR is dispensed onto the surface while the wafer is spinning at 3000–6000 rpm for 30 s. The thickness, d, of the PR layer is set predominantly by the resist viscosity and also by the spinner rotational speed:

$$d = \frac{kp^2}{\sqrt{w}} \tag{2.1}$$

where k is the spinner constant, typically 80–100, p is the percentage PR solids content and w is the spinner rotational speed in rpm/1000. PR thicknesses between 1 and 12 μm are normal and are rarely exceeded. There is also another technique of PR deposition called spray coating. This is more suitable for very uneven surfaces, but lacks the accuracy and reproducibility of a flat spin-coated wafer.

2.1.6 Pre-bake

The substrate is then pre-baked to remove solvents from the PR. Pre-baking is essential to avoid mask contamination by the PR. It is typically conducted at 110 °C for 1 min on a hotplate. It can be also performed in an oven for 10–20 min.

2.1.7 Exposure

The wafer, now coated in PR, is placed on an aligner and aligned with a patterned mask and exposed to UV radiation. There are three exposure systems: contact, proximity and projection. Contact exposure places the mask directly on the PR, proximity exposure leaves a small gap between the two and projection exposure uses a partial mask, called a reticle or field, which is projected onto the PR many times to create the structure. Contact and proximity patterning are commonly used for microfluidics, as the resolution is sufficient and structures with no regular need for sub-micrometre dimensions.

2.1.8 Post-bake and Development

The wafer, now coated in exposed PR, is post-baked at 110 °C for 1 min on a hotplate and then developed with a KOH- or tetramethylammonium hydroxide-based developer for 1 min. This step removes the damaged (positive PR) or uncured material (negative PR). The wafer is then washed with deionized water and finally dried in a stream of N_2. The wafer is now ready for use or further processing.

2.2 Maskless Lithography

Maskless lithography is a technique that transfers designed patterns directly to the PR without any masks. Compared with masked photolithography, maskless lithography has the advantages of higher resolution and flexibility. Maskless lithography is typically divided into the following types: electron beam lithography (EBL),[1] focused ion-beam (FIB) lithography,[2] interference lithography (IL),[3] scanning probe lithography (SPL)[4] and zone plate array lithography (ZPAL).[5]

2.2.1 Electron Beam Lithography (EBL)

EBL is a method of patterning designed structures using a focused electron beam (e-beam) (Figure 2.2). The structure is commercially designed using computer-aided design (CAD) software. It is then

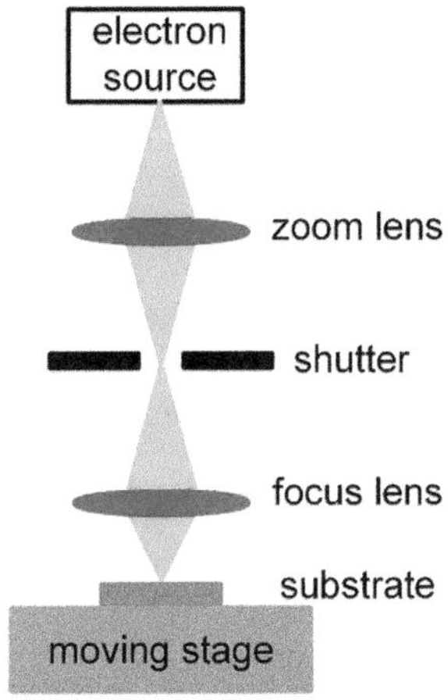

Figure 2.2 Schematic of the EBL principle.

converted to a format available for the e-beam writer and loaded into the writer. A focused e-beam is scanned over a layer of electron-sensitive resist on top of a substrate. The bombardment of electrons changes the chemical or physical properties of the resist, and the bombarded areas can then be washed away with chemical reagents depending on the properties of the material. The pattern resolution based on EBL is <10 nm, depending on the energy of the electrons. However, pattern placement during EBL is affected by temperature gradients and stray electromagnetic fields, causing the beam to deviate from its set position.

2.2.2 Focused Ion Beam (FIB) Lithography

FIB lithography is a variation of EBL, and is very similar except that it utilizes a focused ion beam instead of an electron beam. FIB lithography provides higher resolution than EBL as ions are much heavier than electrons. However, the pattern placement accuracy of FIB lithography is poorer than that of EBL.

2.2.3 Laser Interference Lithography (IL)

IL is performed based on laser light interference (Figure 2.3). A beam of coherent UV laser light is expanded when it passes through a spatial filter. A substrate coated with a polymer resist and a mirror are placed perpendicularly to each other. The expanded beam shines simultaneously on the mirror and the polymer resist. The light reflected from the mirror also shines on the resist. As a result, the incident light and

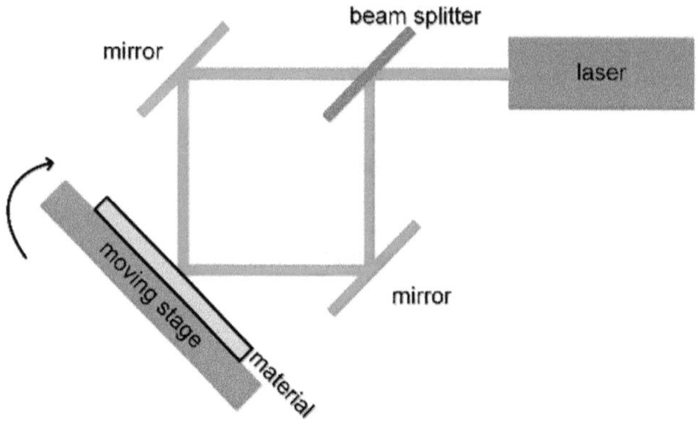

Figure 2.3 Schematic of the IL principle.

reflected light interfere on the resist and generate a standing wave, leading to bright and dark spots on top of the resist surface. IL is only applied to generate simple repeated structures such as lines and dots in small areas. It is also utilized as a tool to make masks for other technologies such as microcontact printing.

2.2.4 Scanning Probe Lithography (SPL)

SPL is a direct-write maskless lithography method suitable for prototyping applications using calixarene-based molecular glass resist thin films (Figure 2.4). The resist layer below the scanning nano-probe can be removed by a current flow of low-energy electrons within a high, non-uniform electric field. SPL avoids the diffraction limit and achieves a pattern resolution below 10 nm. Three-dimensional nanometre-scale structures are formed by the mechanical movement of a microscopic or nanoscopic tip over the surface of the resist. The nano-probe movement is comparably slow, ~20 mm s^{-1}, but a number of inscription probes can be performed in parallel to improve the working efficiency.[6]

2.2.5 Zone Plate Array Lithography (ZPAL)

ZPAL is performed based on the principle of diffraction (Figure 2.5). A laser beam passes through a spatial filter and a collimating lens, generating a uniform incident beam on a spatial light modulator (SLM). The SLM divides the beam into independent beamlets. The SLM is controlled by a computer, allowing individual beamlet control. These

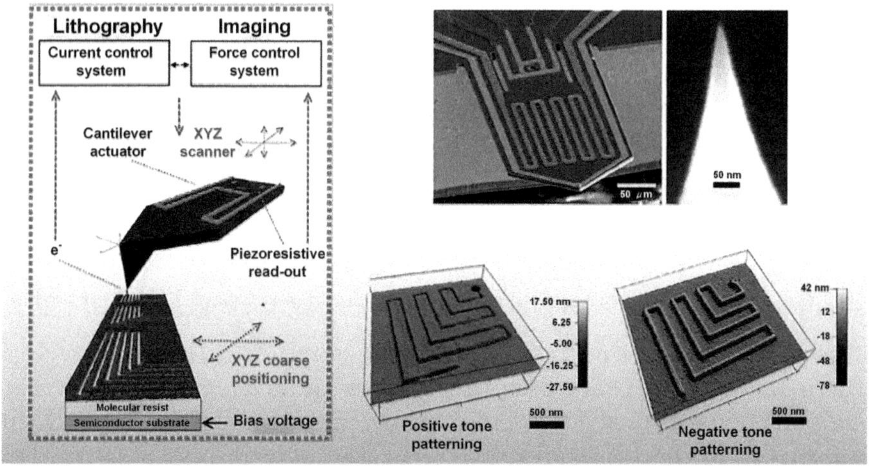

Figure 2.4 Schematic of the SPL principle. Reproduced from ref. 4 with permission from Society of Photo-Optical Instrumentation Engineers, Copyright 2014.

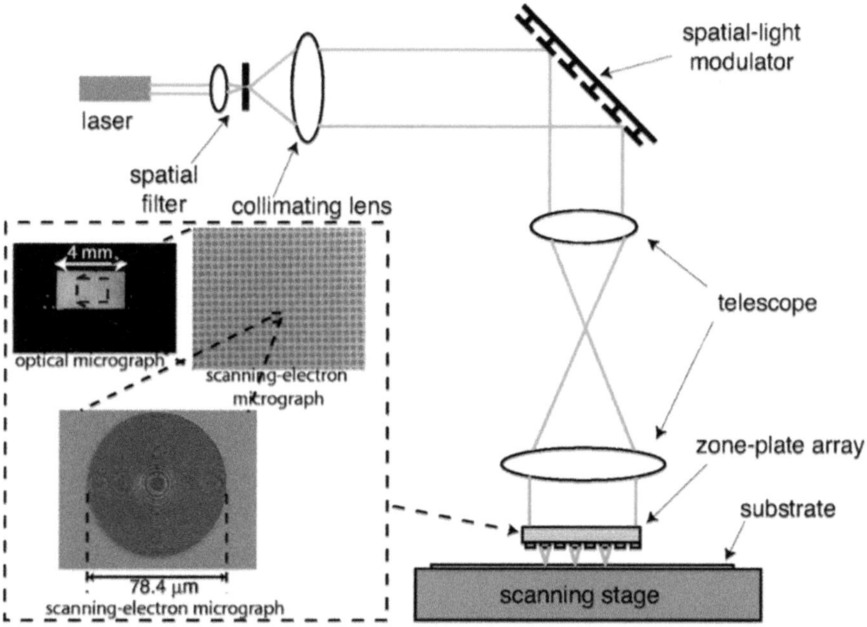

Figure 2.5 Schematic of the ZPAL principle. Reproduced from ref. 7 with permission from Optical Society of America, Copyright 2006.

beamlets pass through a telescope with a Fourier filter to prevent cross-talk between beamlets. These then shine onto the zone plate array, which subsequently focuses the shine onto the PR. The desired pattern is printed in a dot-matrix fashion as the SLM could control the light intensity of each lens. ZPAL has the advantage of high throughput using the array of zone plates, but the process is limited by the precision of the zone plate array.[7]

2.2.6 Light-emitting Diode (LED) Array-based Lithography

New tools for performing lithography emerged with the development of modern LEDs operating in the deep blue and ultraviolet range. They are based on a vast array consisting of LEDs, and the pattern is formed by activating selected corresponding LEDs. The light is focused on the substrate, which is then altered due to the exposure. There are distinct advantages as there is no mechanical scanning of a light beam. Also, LEDs have a long lifetime, typically 50 000 h, in comparison with the 2000 h lifetime of a typical UV laser light source, making the LED light source much more economical than the classical UV laser.

2.3 Etching

Etching is a technique for transferring the pattern from a PR layer to the substrate layer. The process is carried out by removing regions left unprotected by the PR layer of the substrate. Etching technology is typically divided into dry etching and wet etching. Moreover, the etching can also be divided into isotropic etching and anisotropic etching, in which the etching rate is either the same in all directions or different in different directions, respectively. Figure 2.6 demonstrates this difference in etching rates. Isotropic etching results in under-etching of the PR, whereas anisotropic etching does not.

2.3.1 Dry Etching

Dry etching, which is divided into non-plasma-based and plasma-based etching, removes material by bombardment with particles.

Non-plasma-based dry etching uses gases containing fluorine, such as fluorides or interhalogens, to react chemically with Si, forming volatile or gaseous reaction products. The etching rate is nearly the same

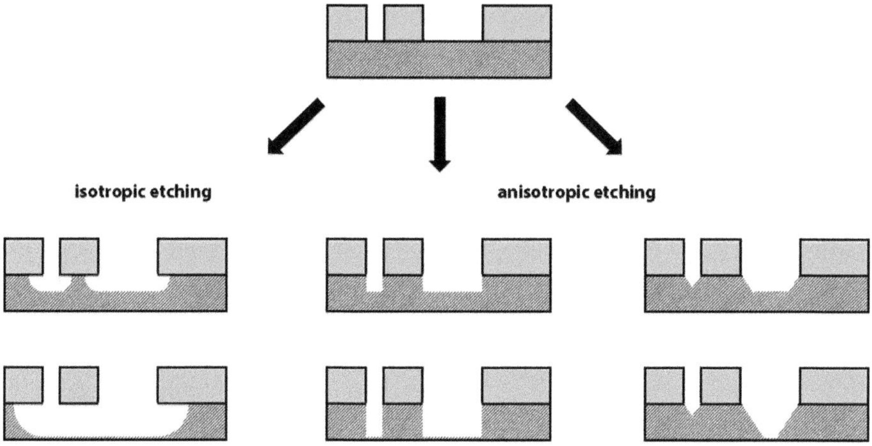

Figure 2.6 The principle of isotropic (left) and anisotropic (centre and right) etching.

in each direction of Si, also known as isotropic etching. The typical gas for performing Si etching is XeF_2, reacting with Si as follows:

$$2XeF_2 + Si \rightarrow 2Xe + SiF_4 \quad (2.2)$$

Plasma-based dry etching is a process in which high-energy radicals are generated using a radiofrequency (RF) electromagnetic field. These radicals are then accelerated towards the surface of the substrate and erode its material. It is an anisotropic or semi-anisotropic etching as the material is only removed along the direction of the particle beam. There are three types of plasma-based dry etching: physical etching, chemical etching and reactive ion etching (RIE).[8]

Ion beam etching (IBE) is a physical dry etching method (Figure 2.7). A beam of argon ions with high energy in the approximate range 1–3 keV is radiated onto the substrate surface in a vacuum chamber. IBE is a highly anisotropic etching method as the angle between the beam and wafer can be adjusted based on the requirements. However, insufficient vacuuming results in the re-deposition of residual material on top of the wafer.

Plasma etching is a chemical dry etching method (Figure 2.8). Chemically reactive species such as free radicals are produced by plasma from an inert molecular gas by an RF field and then absorbed on the surface of the wafer. As a result, these radicals react with the surface molecules of the wafer, forming gaseous by-products. A vacuum pump then extracts the by-products.

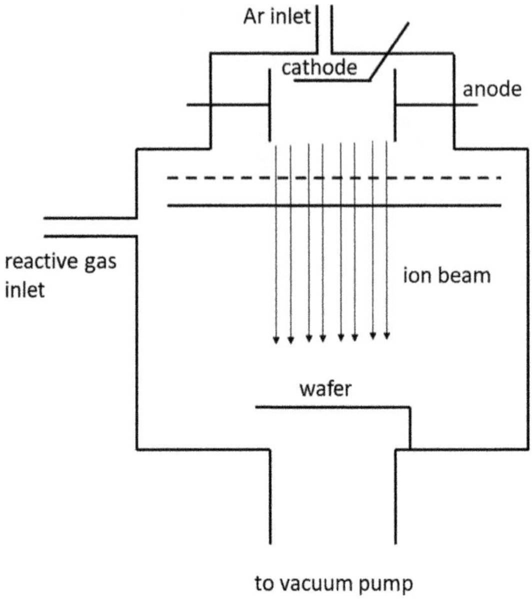

Figure 2.7 Schematic of the principle of ion beam etching.

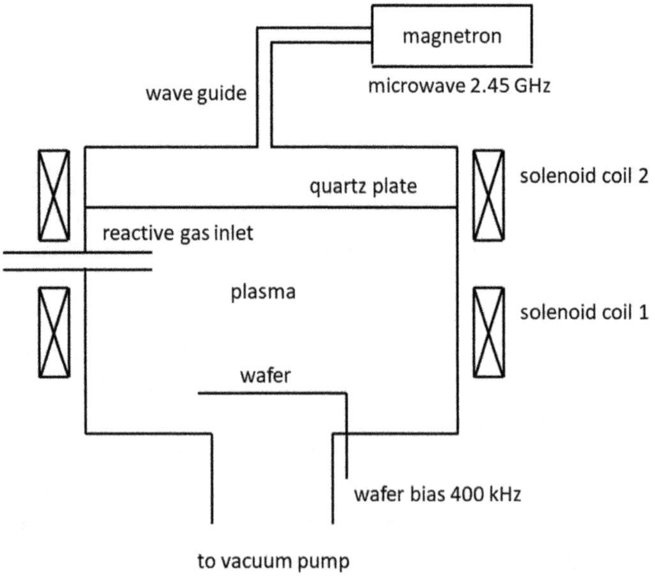

Figure 2.8 Schematic of the principle of plasma etching.

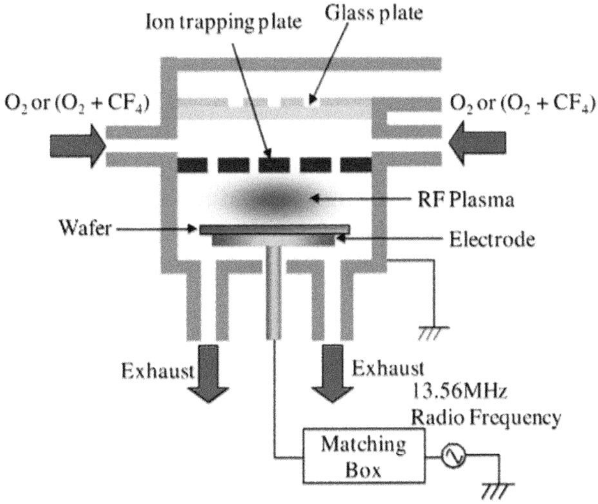

Figure 2.9 Schematic of the principle of RIE. Reproduced from ref. 8 with permission from Elsevier, Copyright 2012.

RIE is an etching process that combines both physical and chemical etching, resulting in high etch rates (Figure 2.9). Gases in the chamber are activated by an RF field to generate a plasma consisting of radicals of the active compound. The plasma accelerates radicals, by a negative bias, towards the wafer, resulting in bombardment of the exposed wafer surface. The additional energy subsequently enhances the effectiveness and speed of the chemical reaction. Typical RIE gases for Si are CF_4, NF_3 and SF_6. RIE has the advantages of both being strongly anisotropic and having rapid etching rate, making it one of the most widely used dry etching methods. The one difficulty with RIE is selecting the correct masking material, as different materials require different operational layer thicknesses, limiting the resolution.

2.3.2 Wet Etching

Wet etching processes use liquids to etch the unprotected surface; most are isotropic etching in nature. The only exception is the wet etching of Si using KOH or a similar solution, where the etching in the <111> plane is significantly slower than that in the <100> plane. Wet etching is a low-cost process, but it is challenging to control as the etching rate is dependent on the complexity of the structures and the area contact with the chemical solutions.

2.4 Soft Lithography

Soft lithography represents a conceptually different approach to the rapid prototyping of various types of both microscale and nanoscale structures and devices on variable materials and complex microstructures. The term 'soft' refers to the fabricated material, such as organic and polymeric, which is softer than silicon and glass fabricated by lithography. A large number of patterning techniques have been developed for the microfabrication of microfluidic devices, such as replica moulding (REM),[9] microcontact printing (μCP),[10,11] micromoulding in capillaries (MIMIC),[12] hot embossing[13] and microtransfer moulding (μTM).[14]

2.4.1 Replica Moulding (REM)

REM is an inexpensive method for fabricating microfluidic devices completed by taking sequential moulds. The intended microstructure is fabricated into the original mould, made of silicon or SU-8. A second mould, which is made of an elastomer material such as polydimethylsiloxane (PDMS), is produced. Sometimes this PDMS structure is used directly for microfluidic experiments. However, sometimes a liquid monomer of another polymer is then poured over the second mould and cured, resulting in a replica of the original microstructure (Figure 2.10). This method has been demonstrated to be particularly suitable for various biomedical materials.

2.4.2 Microcontact Printing (μCP)

μCP is used to transfer a uniform molecular or inorganic ink from a patterned elastomeric stamp to the substrate (Figure 2.11). The standard μCP process involves fabrication of a patterned PDMS stamp, applying target ink to the stamp and finally using the stamp to apply the ink to the surface of the substrate. This technique offers a method to pattern a surface with molecular-level detail using self-assembled monolayers of ink with nanoparticles or biomolecules, such as metal particles, proteins and nucleic acids.

2.4.3 Micromoulding in a Capillary (MIMIC)

MIMIC is used to pattern micrometre- and submicrometre-scale structure substrate surfaces. It is based on the spontaneous filling of capillaries formed between two surfaces in conformal contact, one of which

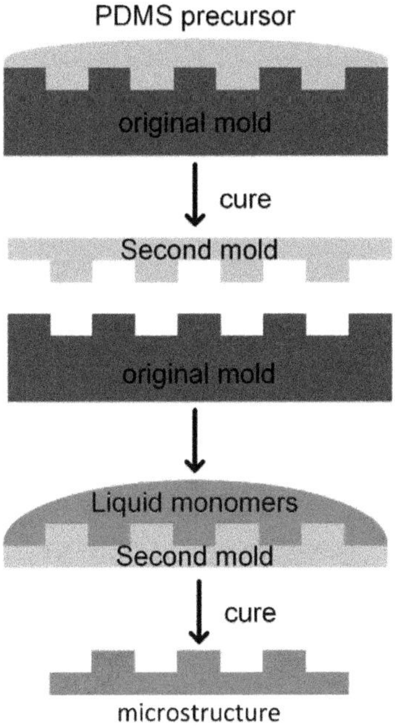

Figure 2.10 Schematic illustration of the replica moulding steps from an Si master to a polymer replica using a PDMS mould.

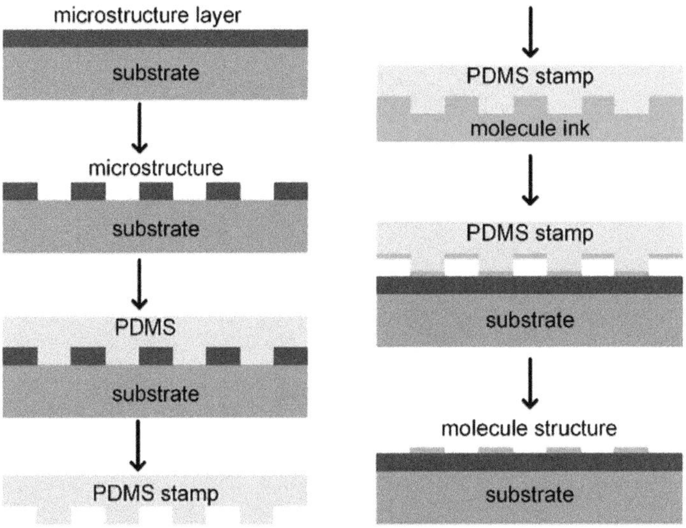

Figure 2.11 The patterning procedure using μCP.

has a recessed relief structure (Figure 2.12). The filled fluid could be a liquid prepolymer, a solution or suspension of the materials to be formed or precursors of these materials. The standard MIMIC process typically consists in (1) a fabricated mould with recessed relief structure, either polymer or silicon; (2) placing the mould on a substrate

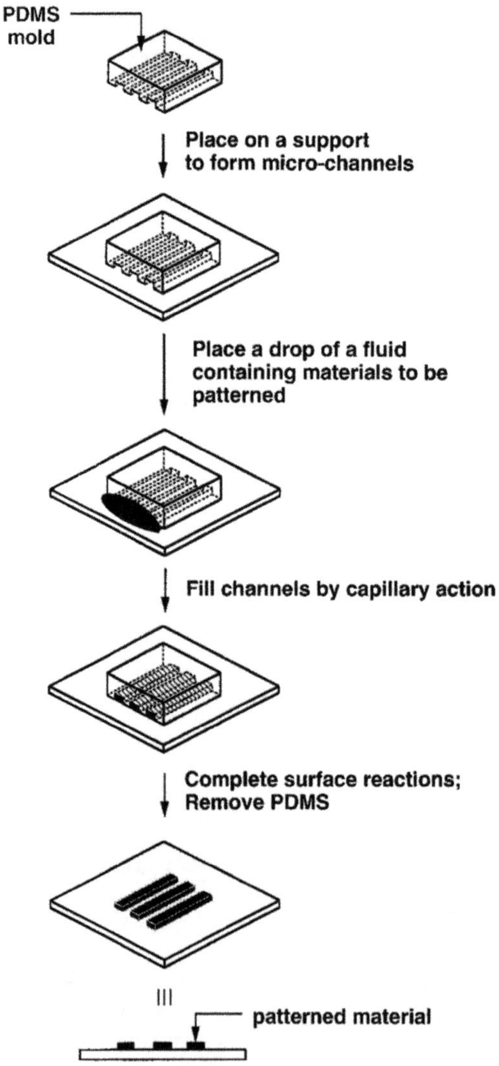

Figure 2.12 Schematic of the MIMIC procedure. Reproduced from ref. 12 with permission from American Chemical Society, Copyright 1996.

to form capillary channels; (3) a fluid, such as a liquid prepolymer, a solution or suspension of the materials to be patterned, filling these channels *via* capillary action; and (4) removing the mould after the material in the fluid has cross-linked, cured, adhered to or deposited onto the surface of the substrate. MIMIC is used to fabricate microstructures of organic polymers, inorganic and organic salts, ceramics, metals and crystalline microparticles.

2.4.4 Hot Embossing Lithography

The hot embossing technique fabricates high-precision and high-quality plastic microstructures (Figure 2.13). Unlike optical lithography, this technique is not limited by the diffraction of light, and thin resist films can be patterned down to 10 nm. A thin thermoplastic resist film is spin-coated onto the substrate. The resist has a thickness slightly greater than the required structure height. The thermoplastic film is then heated and pressed into the master, shaping the desired structure into the resist surface. The excess thin polymer layer is then removed with oxygen plasma, reducing the resist to the desired structure.

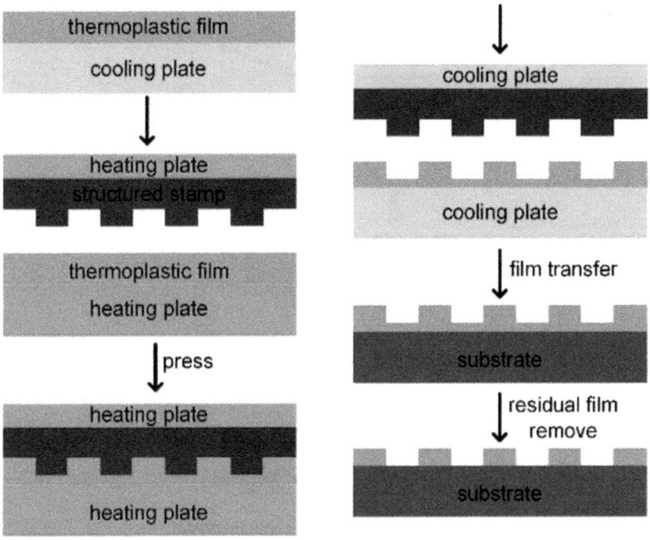

Figure 2.13 Hot embossing lithography: imprint replication in polymer followed by window opening.

2.4.5 Microtransfer Moulding (μTM)

In the μTM process, microstructures are formed by filling microchannels on the surface of an elastomeric mould with a liquid precursor and then bringing the mould into contact with a planar or contoured substrate (Figure 2.14). The liquid precursor either thermally or photochemically solidifies *in situ*. The elastomeric mould is then peeled away, leaving the resulting microstructures on the surface of the substrate. μTM is used for forming complex, 3D microstructures of organic polymers and ceramics.

2.4.6 Xurography

Xurography was first proposed in 2005 as an alternative technique for rapidly prototyping microstructures for microfluidic applications. This technique uses a cutting plotter to cut microstructures into variable microfabricated films, such as thin polymer films, thermal laminating films and pressure-sensitive adhesive-coated films. Unnecessary parts of the film are then removed after pattern cutting. These adhesive-backed films, often vinyl, come in rolls laminated on a release liner. The patterned film is versatile enough to bond with

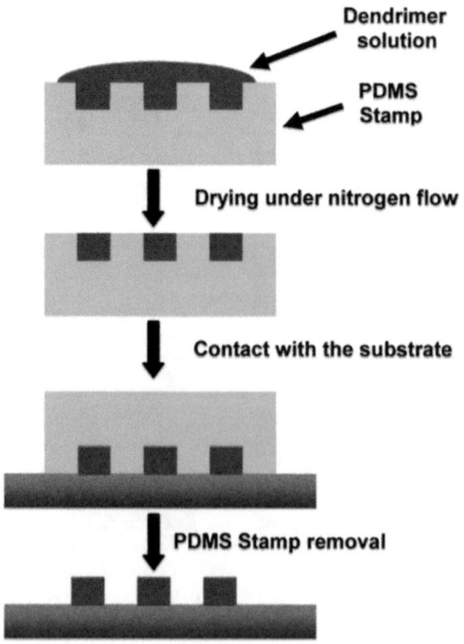

Figure 2.14 Schematic of the μTM procedure.

substrates such as silicon, glass and polymers. Xurography has the advantages of short processing times, low cost and no cleanroom requirement, but it is less precise and has a lower resolution, thus limiting its utilization.[15]

2.4.7 3D Printing and Other 3D Processes

3D printing, also named additive manufacturing, covers a variety of processes in which material is joined or solidified under computer control to create a 3D object, with materials typically added layer by layer. 3D printing technology utilizes CAD software, hence there is no need for a mask. The demand for faster and lower-cost microfluidic chip fabrication has seen 3D printing technology being gradually applied to biomaterials and microstructures. Suitable printing techniques are being developed for microfluidics and biotechnology, such as fused deposition modelling, direct light processing, stereolithography, selective laser sintering, laminated object manufacturing and inkjet-based printing.[16]

The two-photon polymerization (2PP) technique is a rapid, single-step, straightforward, direct laser writing method to fabricate complicated 3D microstructures based on non-linear processing of photosensitive materials (Figure 2.15). The principle of 2PP technology is based on two-photon absorption of a liquid transparent

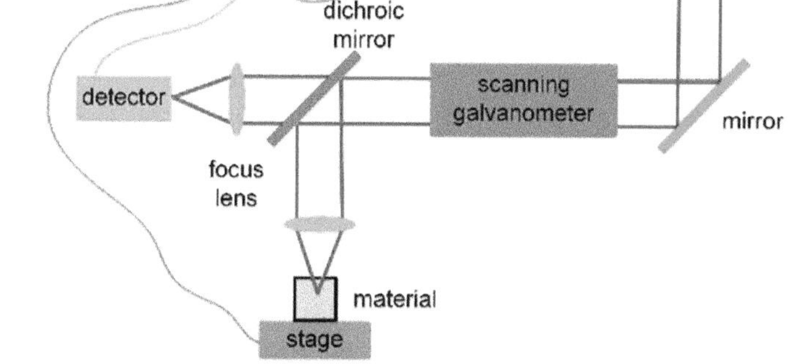

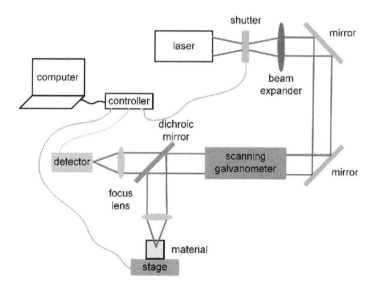

Figure 2.15 Setup of two-photon polymerization technique.

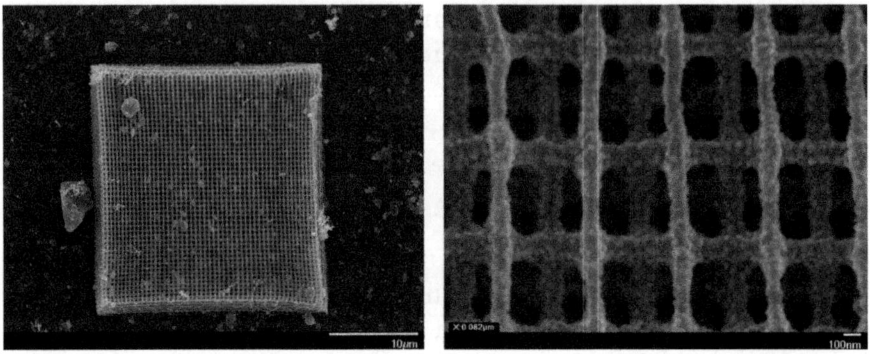

Figure 2.16 SEM images of a photonic crystal structure fabricated by 2PP. Reproduced from ref. 17, https://doi.org/10.1155/2012/927931, under the terms of a CC BY 3.0 license, https://creativecommons.org/licenses/by/3.0/.

polymeric material polymerizing with the energy of the laser beam at the focused area. Microstructure fabrication is performed by controlling the movement of the laser-focused area and the stage, and removing the unpolymerized material with an organic solvent. A simple modification of the input CAD model alters the structure fabricated, allowing for rapid prototyping. The resolution of this method can be fine-tuned by applying appropriate focusing optics and laser irradiation parameters. Structures possess feature sizes ranging from <100 nm to hundreds of μm.

The experimental setup for 2PP processing consists of a second-harmonic generation unit, acoustic–optical modulator, polarization rotating element, polarization-sensitive beamsplitter, power meter and beam expander. The photonic crystal shown in Figure 2.16 was built up from individual rods with a diameter of 100 nm and a spacing between the rods of 500 nm.[17]

2.4.8 Direct Laser Writing

Direct laser writing (DLW) for 3D microfabrication is based on using a laser to create a localized chemical change. The thermal energy from a laser beam is absorbed by the wafer/bulk material and forms a molten, vaporized or chemically changed state, which can then be removed. In DLW systems, the laser beam is focused on the target spot by a lens and the beam is adjusted by computer-controlled scanning mirrors. Once the area accessible by the scanning mirrors has been processed, the sample is moved by its stage and another area is patterned until the entire substrate patterning step is completed. DLW has the advantage of eliminating masks but the patterning is rather slow. If more than a single substrate is to be patterned, conventional lithography is a faster process.

2.4.9 Computer Numerical Control (CNC) Machining

CNC machining is a manufacturing method to micromachine a microfluidic structure on the substrate, mostly using a conventional drilling tool. The microstructures are directly fabricated into the substrate by physically cutting away at the material following a CAD without the need for a mask, just as in DLW. CNC machining is suitable for making prototypes with millimetre to micrometre features in both plastics and metals and has been applied in multiple fields. CNC methods can produce microfluidic chips rapidly and cheaply, but this fabrication process cannot match the resolution of classical lithography methods.

2.5 Bonding

The technique of bonding two wafers or two individual chips together was developed to close simple channels or to build more complex microsystems, sometimes even using several layers bonded together. Often two different materials are used, one to make a microfluidic system and another to cap it to form closed channels and chambers. Originally, it was developed to build pressure sensors by bonding a silicon substrate with a pressure sensor membrane to a glass substrate.

Bonding forms a permanent or temporary connection between two or more wafers or devices. It requires force, heat, chemistry or electrical current depending on the substrate material and process utilized. If both sides are structured, the wafers need to be aligned with each other. There are some exceptions, such as making all structures on one substrate including fluid inlets/outlets and no structures on the other substrate. There is then no need to align the substrates with each other.

There are known problems, such as the difference in thermal expansion coefficients of the two substrates, the formation of an even bond across the entire substrate, the formation of strong and stable bonding and, especially, the preparation of clean, particle-free surfaces.

The most commonly used bonding is called anodic bonding. It can be used to bond Pyrex glass wafers to silicon and it forms permanent bonding. The substrates are placed in contact, then heated to temperatures up to 550 °C while applying a voltage of up to 2000 V. The Na^+ ions in Pyrex become mobile at the applied temperature and due to electric field they travel to the Pyrex–Si interface. Electrons are attracted from the Si and with Na^+ they form a double layer at the interface, forming a permanent bond between the Pyrex and Si. This process is limited to glass–Si combinations.

Thermocompression bonding is another popular technique. It is used for wafer bonding of Si and other substrate materials with intermediate layers requiring mechanical pressure. Typically, it is utilized with frit glass eutectic materials or adhesives where the materials to be bonded are not restricted. Eutectic bonding is also a popular technique. A pure metal is deposited on one substrate and placed in contact with another at elevated temperature. A typical example is Au–Si eutectic bonding at 363 °C, which is the Au–Si eutectic temperature.

Si can also be bonded to Si by a technique called fusion bonding. It requires high temperatures in a furnace. Silicon wafers are oxidized, then their surface is hydrated using H_2SO_4–H_2O_2 solution, placed in contact to utilize van der Waals forces to form pre-bonding and annealed at 1100 °C to form permanent bonds. The bond strength is very good, but the method is restricted to oxidized silicon wafers and is very sensitive to surface contamination.

A final method is adhesive bonding. It does not depend on the materials, surface roughness or planarity of the bonding surfaces. Unfortunately, the bonding strength is not as high as that of other processes. Often wax, photoresist or Parylene is used for this technique. Adhesive bonding is mostly used for PDMS–PDMS and PDMS–glass. A more detailed review of bonding and interfacing the chip to the rest of the lab was presented by Temiz *et al.*[18]

References

1. P. Mali, A. Sarkar and R. Lal, *Lab Chip*, 2006, **6**(2), 310.
2. J. Melngailis, *Nucl. Instrum. Methods Phys. Res., Sect. B*, 1993, **80**, 1271.
3. Q. Xie, M. H. Hong, H. L. Tan, G. X. Chen, L. P. Shi and T. C. Chong, *J. Alloys Compd.*, 2008, **449**(1–2), 261.
4. M. Kaestner, K. Nieradka, T. Ivanov, S. Lenk, Y. Krivoshapkina, A. Ahmad, T. Angelov, E. Guliyev, A. Reum and M. Budden, *Proc. SPIE 9049*, Alternative Lithographic Technologies VI, 2014, 90490C.
5. D. J. Carter, G. Gil, R. Menon, M. K. Mondol, H. I. Smith and E. H. Anderson, *J. Vac. Sci. Technol., B: Microelectron. Nanometer Struct.--Process., Meas., Phenom.*, 1999, **17**(6), 3449.

6. M. Kaestner, K. Nieradka, T. Ivanov, S. Lenk, Y. Krivoshapkina, A. Ahmad, T. Angelov, E. Guliyev, A. Reum, M. Budden, T. Hrasok, M. Hofer, C. Neuber and I. W. Rangelow, *Proc. SPIE 9049*, Alternative Lithographic Technologies VI, 2014, 90490C.
7. R. Menon, D. Gill and H. I. Smith, *J. Opt. Soc. Am. A*, 2006, **23**, 567.
8. C. Wang and T. Suga, *Microelectron. Reliab.*, 2012, **52**(2), 347.
9. Y. Zhang, C.-W. Lo, J. A. Taylor and S. Yang, *Langmuir*, 2006, **22**(20), 8595.
10. S. A. Lange, V. Benes, D. P. Kern, J. K. H. Hörber and A. Bernard, *Anal. Chem.*, 2004, **76**(6), 1641.
11. V. Santhanam and R. P. Andres, *Nano Lett.*, 2004, **4**(1), 41.
12. E. Kim, Y. Xia and G. M. Whitesides, *J. Am. Chem. Soc.*, 1996, **118**(24), 5722.
13. L. J. Heyderman, H. Schift, C. David, J. Gobrecht and T. Schweizer, *Microelectron. Eng.*, 2000, **54**(3–4), 229.
14. C. Thibault, C. Severac, E. Trévisiol and C. Vieu, *Microelectron. Eng.*, 2006, **83**(4–9), 1513.
15. D. A. Bartholomeusz, R. W. Boutté and J. D. Andrade, *J. Microelectromech. Syst.*, 2005, **14**(6), 1364.
16. E. Gal-Or, Y. Gershoni, G. Scotti, S. M. E. Nilsson, J. Saarinen, V. Jokinen, C. J. Strachan, G. Boije af Gennäs, J. Yli-Kauhaluoma and T. Kotiaho, *Anal. Methods*, 2019, **11**, 1802.
17. G. Bickauskaite, M. Manousidaki, K. Terzaki, E. Kambouraki, I. Sakellari, N. Vasilantonakis, D. Gray, C. M. Soukoulis, C. Fotakis, M. Vamvakaki, M. Kafesaki, M. Farsari, A. Pikulin and N. Bityurin, *Adv. OptoElectron.*, 2012, 927931.
18. Y. Temiz, R. D. Lovchik, G. V. Kaigala and E. Delamarche, *Microelectron. Eng.*, 2015, **132**, 156.

3 Layout of Microfluidic Chips

3.1 Design Graphics

Introduced in 1979, the first modern microfluidic device[1] brought an awareness of a new application of planar technology. The actual expansion of microfluidic technologies started approximately 10 years later with the introduction of the concept of miniaturized microfluidic chips[2–4] The first applications were conducted with traditional analytical chemistry, capillary electrophoresis (CE),[5] flow-through polymerase chain reaction (PCR)[6] and many others. The device layout was first drawn as a computer-aided design (CAD), *i.e.* AutoCAD, Solidworks and AutodeskInventor, or using different software developed for the design of integrated circuits (ICs) such as Oasys-RTL™, Nitro-SoC™, Calibre® InRoute™, L-Edit, DW2000, CleWin, LASU and K-Layout; Figure 3.1 illustrates examples of the two methods.[7]

Microfluidic layouts, and also ICs, typically consist of repetitive blocks called cells. Each cell is defined only once and then there is a definition of the particular cell location forming higher-level cells with practically unlimited levels of hierarchy, and dedicated software reflects this. The industrial standard data format for these designs is graphic data system II (GDSII). Exporting the layout data *via* Caltech Intermediate Form (CIF) or data exchange (interchange) format (DXF) can lead to severely problematic dimension incompatibility or open polygons.

Microfluidics and Lab-on-a-chip
By Andreas Manz, Pavel Neužil, Jonathan S. O'Connor and Giuseppina Simone

Published by the Royal Society of Chemistry, www.rsc.org

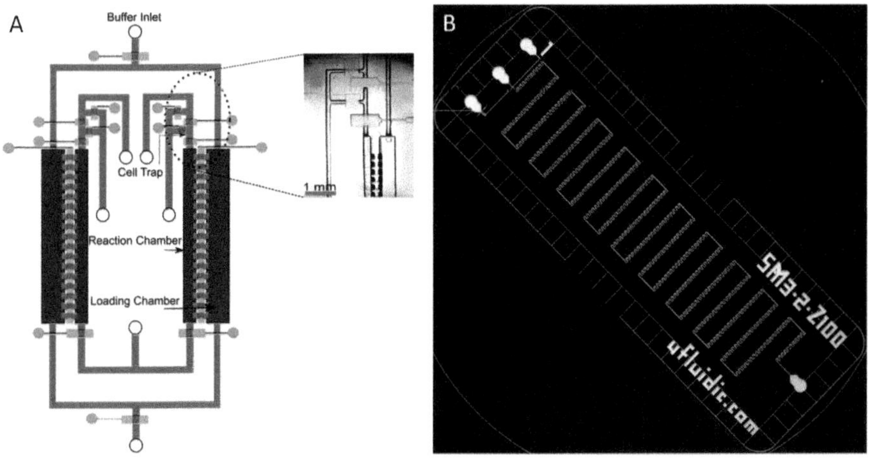

Figure 3.1 (A) Layout generated from CAD. (B) Layout generated from L-Edit. (A) Image courtesy of uFluidic Inc.; (B) Reproduced from ref. 7 with permission from the Royal Society of Chemistry.

IC design software often features a user-friendly drawing interface and uses a hierarchical layout structure. These structures are created at different set layers for the convenience of the subsequent chip fabrication. Layers can consist of interlinked blocks called cells, making editing a simple task. However, these are optimized for ICs, which predominately use rectangles, polygons and wires and use a limited selection of non-orthogonal features, such as circles, arcs and doughnut shapes. Further, microfluidic devices require a diverse set of shapes with different lengths, widths and number of repetitions, hence there is a feature called parameter. You, the designer, then must choose a suitable cell and the type-associated parameters to generate the layout. This method makes the subsequent layout editing work fast and efficient. Nevertheless, the particular cells have to be made per demand.

Microfluidic devices typically need much more complex shapes, such as funnels, spirals, splitters and channels that have sections of different widths with particularly smooth transitions between objects and sections to eliminate singularities and dead volumes. Some of these, such as spirals, can be created mathematically by generating the polygon vertex points using a mathematical formula in a spreadsheet application. The polygon can then be converted into a CIF format and uploaded into a cell of the IC layout editor software. This procedure is tedious and subsequent layout editing requires several repetitions of the same procedure. In addition, the mathematical description of the desired shape is often rather complicated, eliminating this layout-generating option.

The progress of microfluidic technology requires the design of specific elements such as funnels, spirals, splitters or customized objects of arbitrary complexity with a smooth transition between these elements, sometimes with hierarchical organization. A novel layout-generating tool called the Nanolithography Toolbox (NT),[8] based on parameterized cells, was developed by Ilic and co-workers at the Center for Nanoscale Science and Technology (CNST) of the National Institute of Standards and Technology (NIST) at Gaithersburg, MD, USA, to make layout generation an easy task. This NT software also contains a family of cells specifically created for microfluidic devices.

This new, platform-independent concept of generating a layout is scripted based on parameterized structures (Figure 3.2A). Users create a layout by choosing existing objects/shapes from a library. Each basic building block (cell) is represented by a single line of text with corresponding parameters. The text file, containing the entire design, can be hierarchical like any design created by professional chip layout editors, making the total file short and easily editable. Users can also create customized objects of arbitrary complexity (Figure 3.2B).

The NT software is a JAVA-based platform consisting of more than 400 parameterized fundamental blocks (Figure 3.3). Users are required to choose the most suitable parameterized building blocks, identify their parameters and locations, form cells and place them into the hierarchy of their design in the same way as when using conventional layout editing software.

The basic primitive blocks that are of interest for microfluidics contain spirals for the following applications: mixing and cell separations; injectors for forming segmented flow; Bezier curves for

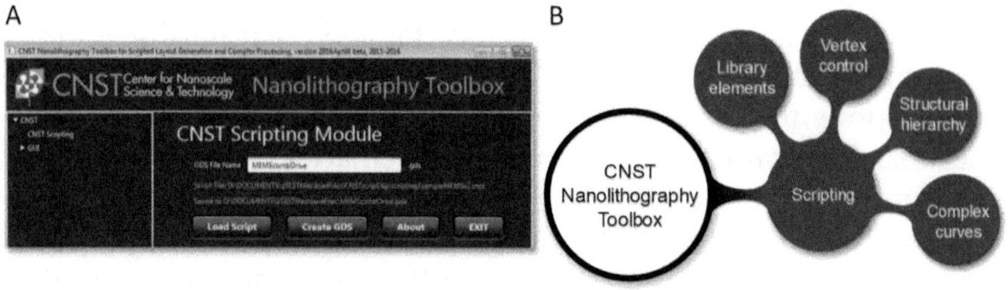

Figure 3.2 (A) User interface of the NT. (B) Principle of the NT.

Figure 3.3 Parts of basic blocks of the NT.

smooth connections between individual blocks; funnels for smooth transition elements from narrow to wide channels and *vice versa*; Y-bends and fractals for fluid distribution into chambers; rectangular, hexagonal and polar arrays used for centrifugal force-based devices; rounded path for complex microfluidic channels; and function plots-based shapes for mixers (Figure 3.3). Once the blocks' locations and other parameters have been identified and the layout hierarchy is complete, the created text file is loaded into the NT software and converted into GDSII format. The GDSII file can be checked in a GDSII viewer and, if there are any inconsistencies with the original idea of the designer, then the parameters in the script file can be corrected and a new GDSII file is generated. The most important

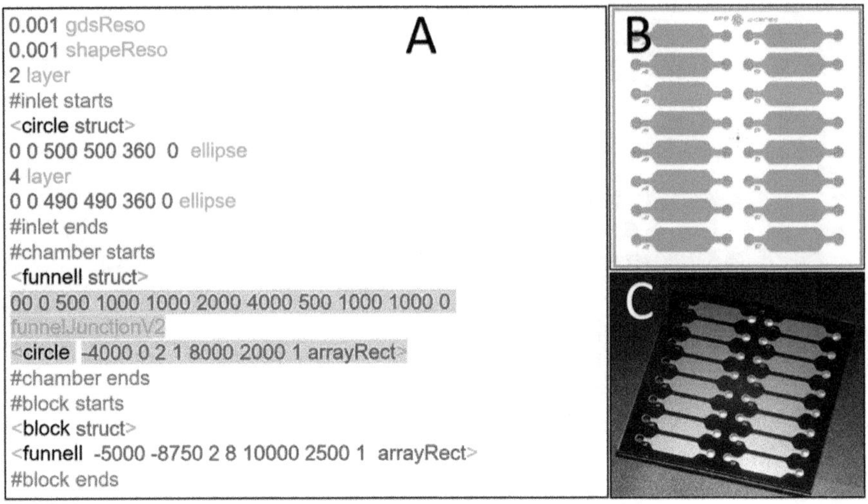

Figure 3.4 Simple array of microfluidic chambers on a chip with dimensions 22 × 22 mm. (A) The chamber is part of the NT library funnel-JunctionV2. Here, we also created the inlets/outlets, called 'circle' and the chamber with inlet/outlet was stepped into an 8 × 2 array. (B) layout of the chip. (C) Photograph of the fabricated device.

features are simplicity, platform independence and the fact that no programming skills are required.

Each design created using the NT starts with the resolution definition of the layout and the actual layout description then starts. In the following sections, we demonstrate classical microfluidic devices designed and fabricated by the authors, which were designed in the NT, as examples of the functionality and efficiency of the NT. The following were fabricated in glass/silicon using standard microfabrication techniques or in polydimethylsiloxane (PDMS) using soft lithography.

3.2 Simplifying the Design Process

Simple microfluidic structures or arrays can be accomplished by a few lines of scripts instead of complex drawing and location definition. A simple chamber for biological applications was designed and fabricated (Figure 3.4) consisting of a basic block, funnelJunctionV2, which is part of the NT library. Users can select the funnel shape and define the widths and lengths of the inlet channel and size of the chamber. There is an option for the inlet channel to have a different

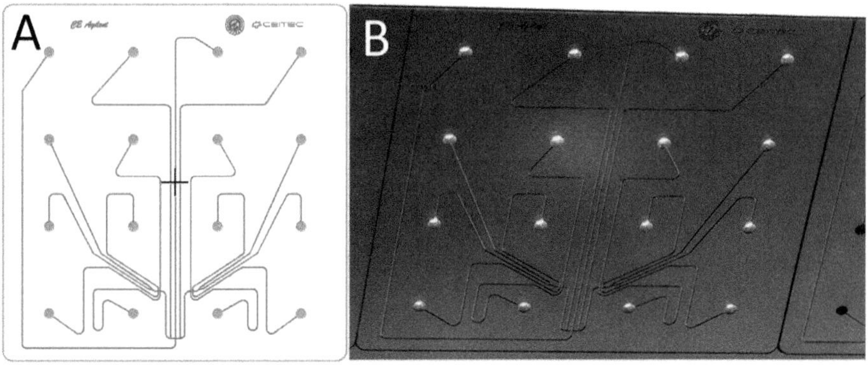

Figure 3.5 (A) Layout of a chip for capillary electrophoresis (originally by Caliper but now used by Agilent) with a size of 18 × 18 mm. The channel is made by the function roundPath. The entire design consists of six lines with corner coordinates and channels and width size description. (B) Fabricated device from Si.

width than the outlet. We formed a cell containing the chamber and two inlet/outlet circles and created an 8 × 2 array. We then added the chamber identification, alignment marks and the logo in the centre to complete the layout.

3.3 Improving the Design's Efficiency

The drawing of the complex structures of the microfluidic chip is time consuming and it is easy to make mistakes. Subsequent debugging is also a headache for users. These inconveniences negatively affect the design's efficiency and the NT is an option to improve them. Example one (Figure 3.5) is probably the first commercialized microfluidic device, which is a chip for capillary electrophoresis using a cross-junction injector capable of sequentially performing up to 12 samples. This device was originally introduced by Caliper and later transferred to Agilent.[9] The channels were designed by a primitive block called roundPath, which is defined by corner coordinates, a radius at each corner, its precision (number of vertexes per round segment) and channel width. The entire layout of this rather complex chip is then simplified into defining corner coordinates for six channels.

Example two (Figure 3.6) is a device utilizing the inertial (Dean) forces for cell separation and sorting.[10] The key primitive block is an Archimedes spiral, spiralArchST, containing straight connections

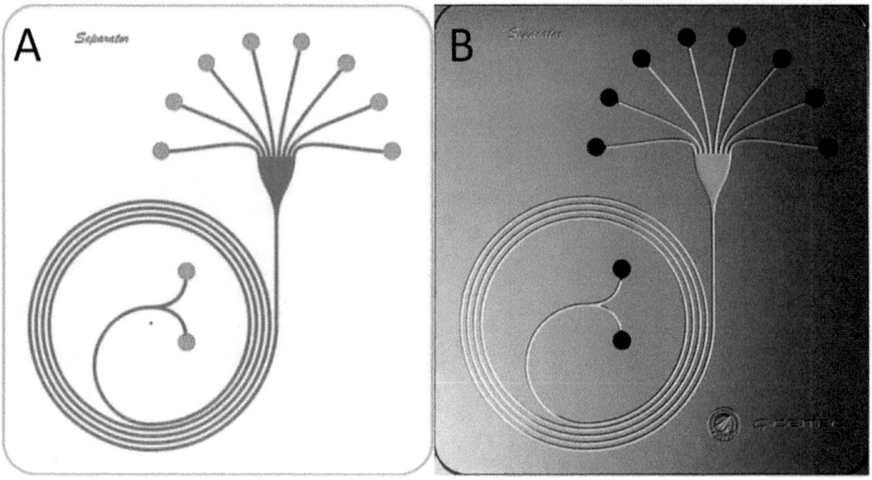

Figure 3.6 (A) Inertial force-based cell sorter using spiral and the Bezier curves function bezierCurve. (B) 13 × 13 mm fabricated device.

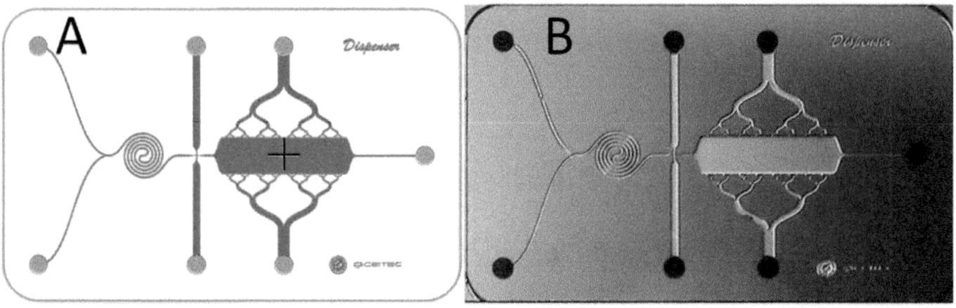

Figure 3.7 (A) A combination of two fluid mixers with the symmetrical droplet generator funnelJunctionV1R and a chamber with side inlets with the size of 13 × 8 mm. These inlets were made by the four-level fractal function curvedTree to distribute the fluid evenly into the chamber. The fractal block forming the distributed inlet consists of a single line with five parameters of which two are the block location, making this layout generation extremely simple. (B) Fabricated chip from Si.

with a parameterized length. The spiral is linked with two inlets connected to the spiral *via* channels made by Bezier curves with eight outlets. These are connected *via* channels with smooth shapes defined by the Bezier function bezierCurve into a single block made of a funnel, the sBendFunnel, to match the final channel width with one of the spirals.

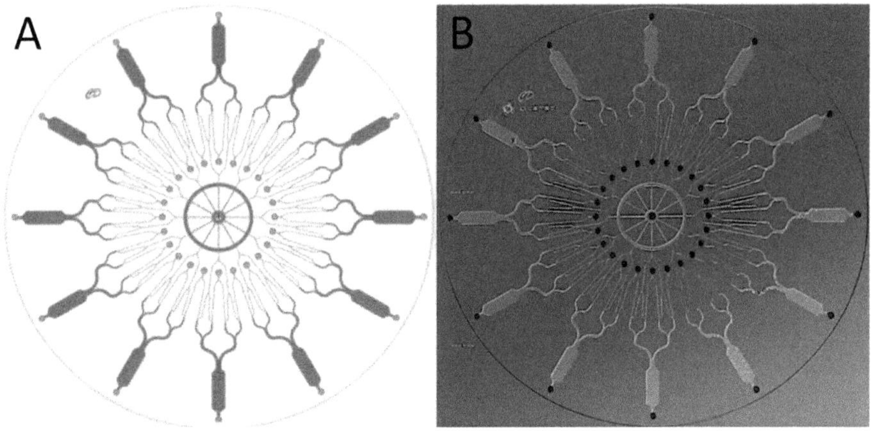

Figure 3.8 (A) CD-ROM-based system to generate segmented flow for further processing, such as digital PCR and single cell analysis with a diameter of 32.5 mm. The layout is made using the roundPath and arrayPolar features to step the basic block 12 times with 30° rotations. (B) Fabricated chip using double DRIE process and anodic bonding.

Example three is a combined microfluidic device consisting of a double spiral-based mixer based on a double Archimedes spiral, spiralDelayLineArchV2, a symmetrical droplet generator, funnelJunctionV1R, and a fractal-based uniform fluid distributor into a chamber (Figure 3.7). Uniform fluid distribution into a microfluidic chamber can be carried out with the help of conventional layout editing software, but it is a tedious and time-consuming process.[10] Here, we show this block formed by the four-level fractal function curvedTree to distribute the fluid evenly into the chamber. The block has only five parameters, two of which are the X, Y locations of the block, one is the number of fractal levels and two identify the location of the curve tree size, which makes the generation of this complex layout extremely simple.

Several research groups utilized centrifugal force with the compact disc read-only memory (CD-ROM) format to move fluids in microfluidic chips and process biological samples,[11] such as conducting real-time PCR[12] or generating droplets with microfluidic emulsion generator array (MEGA) devices.[13] Layouts created by the NT provide smooth transitions between channels and sections to eliminate dead volumes. In example four, we designed the MEGA device (Figure 3.8). Here, we utilized a basic roundPath primitive block and stepped the cell using the arrayPolar feature.

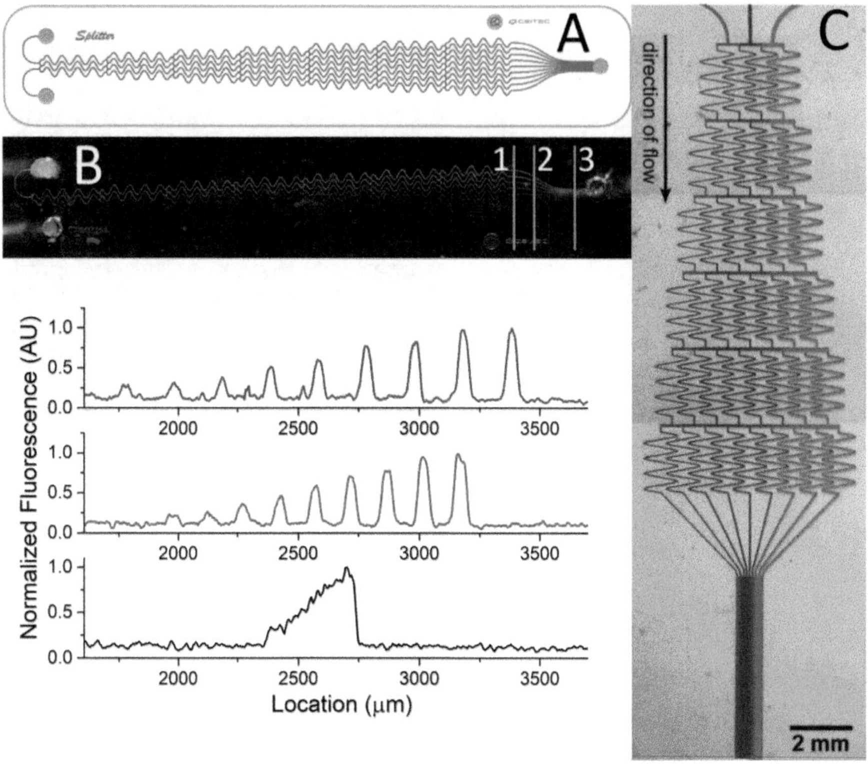

Figure 3.9 (A) Splitter/gradient generator made by a set of channels generated by the 150*Math.sin(*x*/80) sine shape function and the second set made by the rounded path with a size of 20.5 × 4 mm. (B) Fluorescence image of a chip fabricated by the double moulding technique from PDMS filled with two solutions with two different concentrations of fluorescein. Profile of fluorescence intensity as a function of the locations of points 1, 2 and 3 in blue, red and black, respectively. (C) Original splitter with a zig-zag-shaped design. (C) Reproduced from ref. 15 with permission from American Chemical Society, Copyright 2001.

3.4 Optimization

The NT is also an effective tool for designers as it provides smooth transitions in the library to avoid dead volumes in microfluidic channels. An example is a gradient generator/mixer that we originally designed using a simple zig-zag-shaped design.[14] We modified the original layout to make the structure more compact by introducing channels in the shape of a sine function (Figure 3.9). The NT has a primitive block called Function Plot, which allows the formation of channels using different mathematical functions. Here, the basic channel block was a sine-shaped function, 150*Math.sin(x/80), in which the parameters define the amplitude and frequency, *i.e.* size in X and Y directions. Interconnections were made using the rounded path to eliminate dead volume and shape singularities.

The functionality of the generator/mixers was tested showing the formation of a concentration gradient. We pumped two water-based solutions containing fluorescein at ~2×10^{-4} and ~2×10^{-5} kg m^{-3}, respectively, with both pumping speeds set to ~2 μL min^{-1}. The smooth connection between channels removed the dead volumes shown in the original work.

References

1. S. C. Terry, J. H. Jerman and J. B. Angell, *IEEE Trans. Electron Devices*, 1979, **26**, 1880.
2. A. Manz, N. Graber and H. M. Widmer, *Sens. Actuators, B*, 1990, **1**, 244.
3. A. Manz, J. C. Fettinger, E. Verpoorte, H. Ludi, H. M. Widmer and D. J. Harrison, *Trends Anal. Chem.*, 1991, **10**, 144.
4. A. Manz, D. J. Harrison, E. M. J. Verpoorte, J. C. Fettinger, H. Ludi and H. M. Widmer, *Chimia*, 1991, **45**, 103.
5. D. J. Harrison, K. Fluri, K. Seiler, Z. H. Fan, C. S. Effenhauser and A. Manz, *Science*, 1993, **261**, 895.
6. M. U. Kopp, A. J. de Mello and A. Manz, *Science*, 1998, **280**, 1046.
7. M. Sarma, J. Lee, S. Ma, S. Li and C. Lu, *Lab Chip*, 2019, **19**, 1247.
8. K. C. Balram, D. A. Westly, M. Davanco, K. Grutter, Q. Li, T. Michels, C. H. Ray, L. Yu, R. Kasica, C. B. Wallin, I. Gilbert, B. A. Bryce, G. Simelgor, J. Topolancik, N. Lobontiu, Y. Liu, P. Neuzil, V. Svatos, K. A. Dill, N. A. Bertrand, M. Metzler, G. Lopez, D. A. Czaplewski, L. Ocola, K. Srinivasan, S. Stavis, V. Aksyuk, J. A. Liddle, S. Krylov and B. R. Ilic, *J. Res. Natl. Inst. Stand. Technol.*, 2016, **121**, 464.
9. A. R. Kopf-Sill, in *Micro Total Analysis Systems 2000: Proceedings of the μTAS 2000 Symposium, Held in Enschede, The Netherlands, 14–18 May 2000*, ed. A. van den Berg, W. Olthuis and P. Bergveld, Springer Netherlands, Dordrecht, 2000, p. 233, DOI: 10.1007/978-94-017-2264-3_54.
10. S. S. Kuntaegowdanahalli, A. A. S. Bhagat, G. Kumar and I. Papautsky, *Lab Chip*, 2009, **9**, 2973.
11. J. Ducree, S. Haeberle, S. Lutz, S. Pausch, F. von Stetten and R. Zengerle, *J. Micromech. Microeng.*, 2007, **17**, S103.
12. M. Keller, S. Wadle, N. Paust, L. Dreesen, C. Nuese, O. Strohmeier, R. Zengerle and F. von Stetten, *RSC Adv.*, 2015, **5**, 89603.
13. Y. Zeng, R. Novak, J. Shuga, M. T. Smith and R. A. Mathies, *Anal. Chem.*, 2010, **82**, 3183.
14. S. K. W. Dertinger, D. T. Chiu, N. L. Jeon and G. M. Whitesides, *Anal. Chem.*, 2001, **73**, 1240.

4 Engineering Surfaces

4.1 Introduction

For miniature devices, the surface area-to-volume ratio drastically increases. Any physical or chemical phenomenon, *e.g.* wettability, undesired chemical coatings (fouling) or desired chemical binding (sensing, extraction), occurring at the surface will be much more dominant. Hence the surface chemistry of the substrate will behave differently compared with the bulk. Therefore, it is important to discuss the manipulation of the surface properties of a microfluidic device.

The molecules at the surface of a material have different free energies, reactivities and structures from those in the bulk and as such need specific designs and analysis techniques. In the case of antifouling surfaces, the ideal surface modification aims to eliminate non-specific protein adsorption while being cost-effective and robust. Another example, chemical sensing layers, is influenced by the quality of the bio-interface. Similarly to antifouling layers, specific sensing layers seek to eliminate non-specific binding to improve the signal-to-noise ratio.

The selection of the right protocol is essential when modifying the surface. The first step is to shortlist the molecule(s) to be targeted, then to identify which molecule the surface should present in order to interact accordingly, followed by the deposition of that molecule. The deposition can be passive or active. Passive deposition takes place by hydrophobic or ionic interactions (physical adsorption), whereas active deposition requires covalent immobilization (chemical reaction). Covalent immobilization often allows better bioactivity and better stability, because the molecules preserve their native conformation

Microfluidics and Lab-on-a-chip
By Andreas Manz, Pavel Neužil, Jonathan S. O'Connor and Giuseppina Simone

Published by the Royal Society of Chemistry, www.rsc.org

and are not denatured. The selection of the protocol also needs to consider that the substrate can be made of different materials, such as silicon, glass, polymers and hybrid materials, or that a patterned surface needs to be obtained.

A preliminary step for the modification of the surfaces includes the formation of a monolayer consisting of silanes or thiols, in the case of glass or gold substrates. These molecules could then form the interface between the substrate and a biomolecule, which could then alter the wettability.

This chapter introduces some important methods for forming functional layers and active bio-surfaces and some of the most common methods for creating micropatterned surfaces.

4.2 Radicalization of Surfaces by a Gas Plasma

Several feed gases can be used inside the plasma generator to create the plasma selected according to the intended application, *e.g.* oxidizing or reducing plasma atmospheres. Gaseous molecules are ionized inside the reactor by a plasma source, *e.g.* glow discharges, radiofrequencies and gas arcs, that generate an electromagnetic field, forming both neutral and charged active species – ions, electrons, radicals, excited species and photons – in a low-pressure chamber. Then, the excited molecules hit the surface with an energy that is comparable to the chemical bond energies of the organic compounds (covalent bonding about 10 kJ mol^{-1}, ionic bonding about 1 kJ mol^{-1} and van der Waals bonding about 0.1 kJ mol^{-1}), breaking the bonds and oxidizing or reducing the 'exposed' terminal groups.

The most common application of the plasma reactor is to render hydrophobic surfaces, such as polydimethylsiloxane (PDMS), into hydrophilic surfaces. An oxidizing plasma is created by using O_2 as a feed gas, resulting in the generation of charged oxygen species (O^+, O^-, O^{2+} and O^{2-}), neutral species (O_3 and O), ionized ozone, metastable excited oxygen and free electrons.[1] These plasma components impact the surface, breaking the surface chemical bonds and oxidizing the surface. Oxygen plasma treatment is used for PDMS, SiOH, SU-8 photoresist and poly(methyl methacrylate) (PMMA). About 50% of the surface groups are converted into OH groups, increasing the hydrophilicity. Oxygen produced by plasma treatment does not exceed a depth of a few hundred nanometres. The effect induced by the gas plasma is only temporary and decays with time, hence the need for rapid treatment or binding after plasma treatment.

As discussed in Section 2.5 in Chapter 2, bonding between activated surfaces or deposition of functional molecular films is possible. For the treatment of fluorinated polymers or fluorocarbon polymers such as polytetrafluoroethylene (PTFE), a reducing plasma is used. This is created with a mixture of hydrogen, argon and ammonia. Tetramethyl orthosilicate (TMOS) is used as a feed gas for the deposition of silicon dioxide coatings. These techniques are used to obtain surfaces with permanent wettability.

Plasma treatment permits molecule depositions; an example is PDMS and the deposition of 2-hydroxyethyl methacrylate (HEMA),[2] or alternatively the polymerization of acrylic acid to a polymer surface.[3]

CO_2 plasma treatment is possible, with such treatment of polypropylene resulting in an increase in the surface roughness and an increase in the alcohol, ketone and acid functional groups present.[4]

4.3 Formation of Functional Films

4.3.1 Self-assembled Monolayers

Self-assembled monolayers (SAMs) are highly ordered molecular assemblies formed by the adsorption of molecules on a solid surface. Molecules that form SAMs include alkanethiols and alkylsilanes, as shown in Figure 4.1a, normally used with gold and silicon substrates,

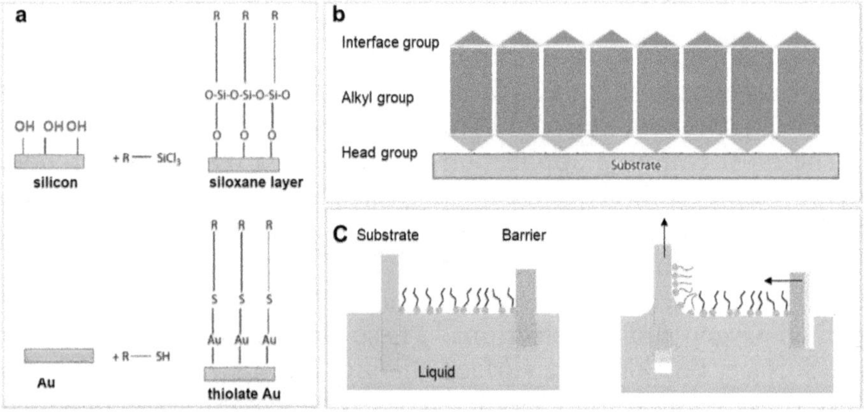

Figure 4.1 (a) Top, SAM of silane on silicon; bottom, SAM of thiol on gold. (b) Thiol or silane molecules having the two extremities of the chains made by two functional groups: an interface group pointing away from the substrate and head groups oriented towards the substrate. (c) Method of monolayer deposition *via* Langmuir-Blodgett deposition.

and have the role of interfacing the substrate to other molecules (*e.g.* proteins, DNA, antibodies, glycans). Alkanethiols and alkylsilanes are characterized as surfactants, possessing two functional groups: a non-polar interface chain pointing away from the substrate and polar head groups bonded to the substrate surface (Figure 4.1b). Surfactants can also be defined as amphiphilic molecules containing both hydrophilic and hydrophobic groups. Hydrophilic groups are polar, whereas hydrophobic groups typically have non-polar hydrocarbon structures.

Alkylthiolate monolayers were first produced by Nuzzo and Allara at Bell Laboratories in 1983 and have been a powerful tool for studying the fundamentals of surface physics while also providing a method to make synthetic materials more compatible with biologically relevant systems.[5]

After the activation (*e.g.* by plasma treatment of the Au surface) and dilution of the alkanethiol in a suitable solvent, the self-assembly of the monolayer takes place almost immediately, but ordering and packing of chains occur over several hours. A tilt angle of approximately 30° from the normal surface maximizes the van der Waals interactions between the carbon chains.

Monolayers of organosilicon derivatives can be successfully prepared on hydroxylated surfaces, such as silicon dioxide, aluminium oxide, quartz and glass. Coupling of an organofunctional alkoxysilane to the hydroxyl (OH) groups of a substrate is also common, often referred to as silanization. These procedures can be performed at ambient temperatures and result in biocompatible surfaces with tuneable wettability.[6–8] Polymers can also be grafted onto hydroxylated surfaces by silane molecules and are used for changing the surface properties, especially the wettability,[9] or for imparting antifouling properties,[10] in addition to non-specific adsorption.[11] Molecules with both terminal thiol and silane groups exist, such as the (3-mercaptopropyl)trimethoxysilane, which was used as an adhesive between vapour-deposited gold and a glass substrate.[12]

4.3.2 Langmuir–Blodgett Monolayers

Langmuir–Blodgett (LB) films are monolayers formed by the two-dimensional compression of surface-active molecules (surfactants) and subsequent transfer from the surface of water onto a solid substrate. The LB film is a quasi-two-dimensional solid with the molecules closely packed and aligned with the hydrophilic head groups on the surface of water. The monolayer is transferred to a solid substrate by slowly pulling the substrate from a bath containing the molecules, as shown in Figure 4.1c. During the pulling, the film present on the liquid/air interface is transferred onto the solid substrate by van der Waals interactions.

4.4 Deposition of Biomolecules for Modifying Surfaces

Several studies have been performed to improve the biocompatibility of microfluidic devices. Peptides have been utilized for surface treatments.[13,14] To reach the binding sites, the peptide must extend out from the polymer surface and requires the use of spacer molecules to create the required distance between the solid surface and the active peptide. The presence of the spacer can improve the control over the peptide surface density and surface distribution.

The immobilization of proteins onto solid supports is crucial for the generation of high-density protein microarrays for functional proteomics. Even though the difficulties associated with the generation of protein microarrays have prevented their widespread adoption, there is still active interest in the field.[15]

The simplest methods for protein immobilization consist in the reaction between the amino groups of the protein and the aldehydic functionality of the solid support. However, there are different methods for immobilizing a protein molecule onto a surface. The following are three of the most common (Figure 4.2a):

(i) activation of surface amines with glutaraldehyde to form an aldehyde-functionalized surface;
(ii) formation of an anhydride monolayer from a carboxylic acid;
(iii) formation of an active ester from a carboxylic acid.

These three methods are not specific, so any peptide or protein could bind to the site, hence the protein solution must be very pure, to avoid the immobilization of competitive impurities. If a high selectivity is needed, often biotin–avidin complexation can be used.[16] Biotin can be grafted onto an alkylated surface, which acts as a spacer. Avidin complexes with the biotin and, *via* a biotin linker, with an enzyme. The enzyme is specific to a component in the sample.[17]

Another crucial aspect related to the binding of the proteins is that once immobilized on the substrate, the proteins need to maintain their native biological functions, so there are more restrictions

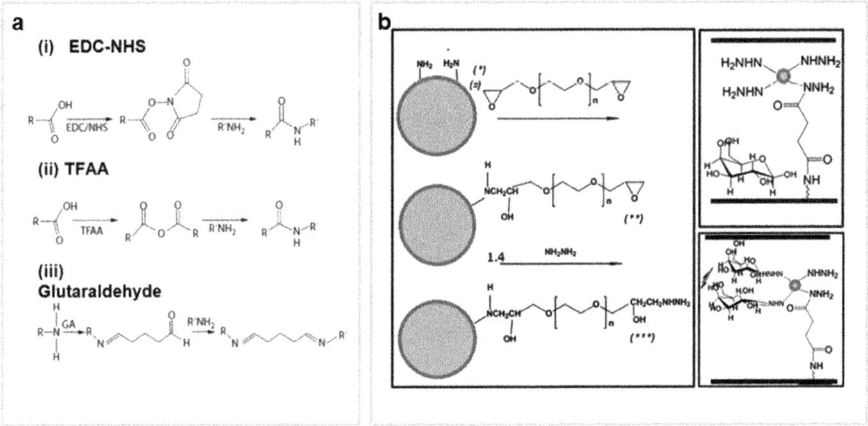

Figure 4.2 (a) Protocol examples for depositing proteins on substrates after SAM formation. (b) Spacer between the substrate and the active molecule. Reproduced from ref. 18 with permission from the Royal Society of Chemistry.

on the immobilization methods. The spacer molecules or particles built between the surface and the protein permit the control of the final orientation of the protein, its distance from the surface and its affinity to the surface (Figure 4.2b).[18] While covering the surface of the nanoparticles with the proteins using spacers, the biomolecules could still have some rotational freedom around the axis connecting them to the nanoparticle.

Biomimetic, biocompatible and biodegradable layers are fundamental in tissue engineering applications. The main challenge is that the specific interaction between the functional layer and cells should not inhibit the cellular functions, while not compromising the cell anchorage and signal transduction. Functional layers are required for stimulating a positive cell response that will help tissue regeneration.

Other parameters influence the interaction between the substrate and the cells, such as the geometry, the topology of the functional layer and the roughness and also wettability of the surface. To improve the hydrophilicity of a biomaterial, the surface can be treated with water-soluble polymers, which can swell, retain the aqueous solution or media and enhance cell proliferation. There are many interesting natural materials that can be used for this purpose. For example, polycaprolactone has good biocompatibility properties, but it requires modification to improve the cytocompatibility. Other molecules such as collagen, gelatin and chitosan have also been immobilized onto polymeric surfaces.[19–21] The naturally occurring polymers improve the hydrophilic properties of the surface and provide the sites necessary for further immobilizations. Most of the natural gels mentioned above can also be used because of their rapid transport and exchange of the materials and have been used as microfluidic microgels for sustaining the metabolism of embedded cells. Chitosan is an example of a pH-responsive polymer that can be assembled at readily addressable sites in microfluidic channels from their aqueous environment. At low pH, chitosan is soluble in aqueous solution whereas at neutral to high pH, the amine group becomes deprotonated and uncharged, making chitosan insoluble.

4.5 Methods for Biomimetic Surfaces

4.5.1 Micropatterning

The selective deposition of the active molecules on a substrate is denoted micropatterning. A standard technique of micropatterning is to use a photoresist mask to block SAM formation, as shown in Figure 4.3a. The photoresist pattern is made by photolithography followed by developing the pattern, as described in Chapter 2. Changes in wettability are measured by the change in the contact angle before and after washing. Measurements are also repeated before and after the formation of the monolayer. Micropatterning by photolithography can be used to deposit proteins (Figure 4.3b).[22]

The protocol based on the photolithographic technique includes the following steps:

- Deposition of aminosilane SAM.
- Pattern bathed in glutaraldehyde, a cross-linker that binds to amines on both ends; one of the ends of the cross-linker could presumably be free to bind amine groups on proteins. As an alternative, the glutaraldehyde simply cross-links amino groups on the aminosilane SAM.

4.5.2 Microstamping

The microstamping method is shown in Figure 4.4a. Microstamping requires dry samples, which can be problematic for some biomolecules. For instance, some dried proteins form crystals, thus irreversibly losing the protein tertiary structure and bioactivity. Other parameters influence the procedure; for example, the choice of substrate,

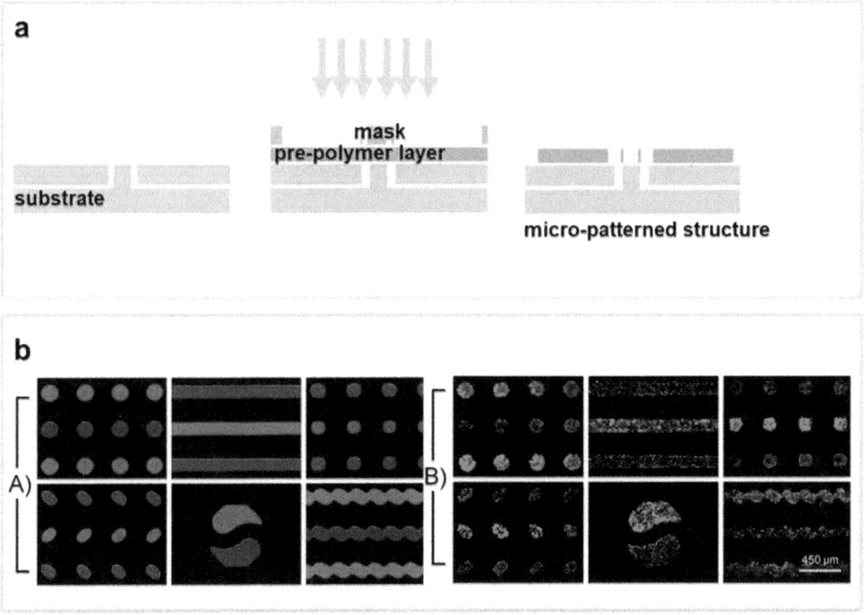

Figure 4.3 (a) Micropatterning technique for creating an island of the polymer layer. (b) Co-micropatterning of different proteins/cells. (A) Immobilization of both FITC-labelled BSA (bovine serum albumin) and Alexa Fluor 594-labelled CEA (chicken egg albumin). (B) Localization of both HUVEC-C cells (red) and HepG2 cells (green) with various shapes in the same chamber. Reproduced from ref. 22 with permission from the Royal Society of Chemistry.

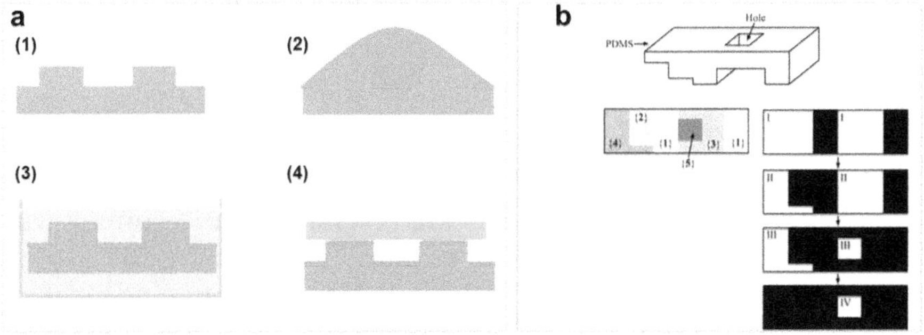

Figure 4.4 (a) Microstamping: (1) silicon stamp; (2) ink deposition; (3) excess removal; (4) transfer of the ink to the substrate. (b) An application of the microstamp for the patterning with a five-level polydimethylsiloxane (PDMS) membrane. Reproduced from ref. 24 with permission from National Academy of Sciences, USA, Copyright 2002.

ambient humidity, surface conditions, the hydrophobicity of the stamp, the hydrophobicity of the stamped surfaces and the charges on the sample surface are strong determinants of the attraction of the proteins by the surface. The deposition of the proteins depends on the difference in wettability between the stamp and the surface. When the stamp consists of a very hydrophobic surface (*e.g.* derivatized with $-CF_3$), the proteins are easily transferred to moderately hydrophilic surfaces, but if the stamp is derivatized with a hydrophilic functionality (*e.g.* $-NH_2$), then the protein is only transferred to an even more hydrophilic surface. All of this is a consequence of the fact that the proteins are amphiphilic, so either the polar or non-polar adhesion forces must be overcome.

Normally the stamps are produced using elastomers, such as PDMS, which deforms under stamping pressure, so the features are difficult to align on top of each other when several layers are being patterned. The most obvious solution is to make the PDMS stamp more rigid, either by increasing the cross-linking or by mounting a thin slab on a thick glass support. Another issue with stamping is that the surface of the PDMS elastomer cannot act as a reservoir for the ink, so the stamp needs to be loaded with ink after every use. To solve this issue, a hydrogel stamp was proposed. The hydrogel acted as reservoir for the ink, thus allowing repetitive stamping without re-inking at resolutions down to 2 μm.[23] The most interesting result is that it is also possible to print gradients of different proteins, using the diffusion mechanism for creating gradients on the stamp, which was completely impossible with the silicone (Figure 4.4b).[24] To summarize, microstamping is a simple and common method to transfer a pattern onto a surface but the direct microstamping of proteins is not trivial.

4.5.3 Microfluidic Patterning

Silicone microfluidic channels can be used for patterning proteins (Figure 4.5a). The microchannels are reversibly put in contact with a substrate and then filled with a solution containing the protein of interest. The solution is then dried and the protein adsorbs on the substrate, after which the microfluidic master can be peeled off. Despite the simplicity of the method, there are several difficulties associated with patterning proteins using microfluidic devices. The resolution of microfluidic patterning is limited mostly by the ability to fill the channels with the protein solution of interest (surface tension, capillary forces) and to flush them out after adsorption is complete (pressure drop). Capillary filling works well for micrometre-sized channels that have been made hydrophilic, but on the other hand, if the microchannel layer is not completely bonded to the substrate, the solution will spread between layers and coat an unwanted part of the surface. Once the channels are filled, the solutions cannot be exchanged easily. If resolution and reagent costs are not a problem, wider/deeper channels can be used to allow continuous perfusion for simple exchange. An example of a result is given in Figure 4.5b.[25]

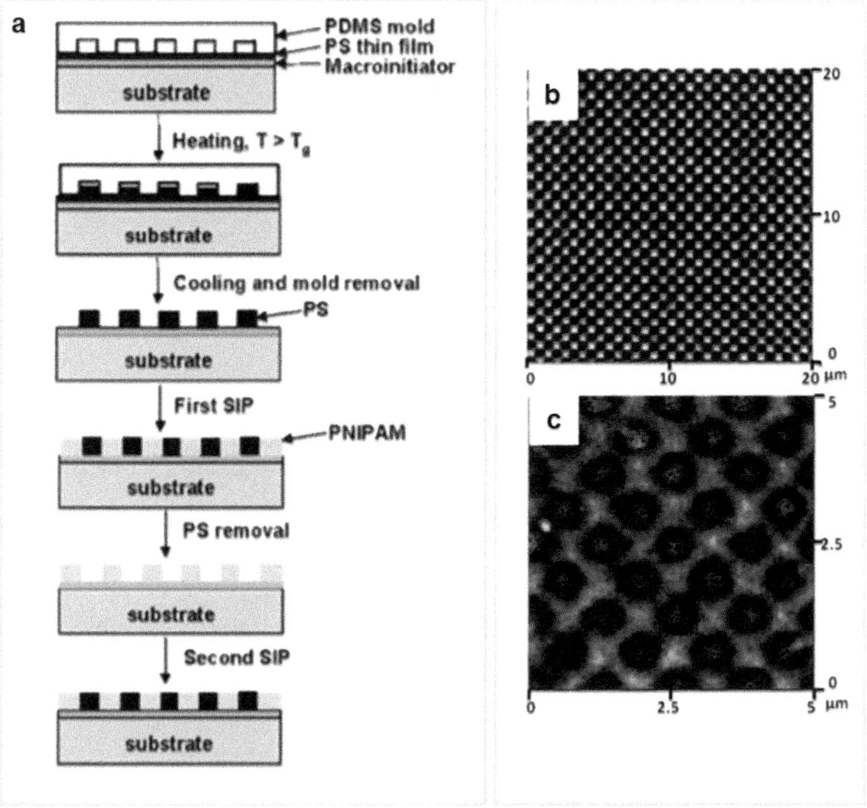

Figure 4.5 (a) Schematic diagram of patterned surfaces created by capillary force lithography (CFL) and surface-induced polymerisation (SIP). (b) AFM topography image of a polystyrene (PS) pattern (20 × 20 μm, AFM roughness 7 nm). (c) AFM image of PNIPAM brush pattern created by CFL and SIP (20 × 20 μm, AFM roughness 6 nm). Reproduced from ref. 25 with permission from American Chemical Society, Copyright 2006.

References

1. A. Pizzi and K. Mittal, *Handbook of Adhesive Technology, Revised and Expanded*, CRC Press, Boca Raton, FL, 3rd edn, 2017.
2. D. Bodas and C. Khan-Malek, *Microelectron. Eng.*, 2006, **83**(4–9), 1277.
3. R. Bitar, P. Cools, N. De Geyter and R. Morent, *Appl. Surf. Sci.*, 2018, **448**, 168.
4. M. Aouinti, P. Bertrand and F. Poncin-Epaillard, *Plasmas Polym.*, 2003, **8**(4), 225.
5. R. G. Nuzzo and D. L. Allara, *J. Am. Chem. Soc.*, 1983, **105**(13), 4481.
6. J. A. Howarter and J. P. Youngblood, *Langmuir*, 2006, **22**(26), 11142.
7. J. B. Brzoska, I. B. Azouz and F. Rondelez, *Langmuir*, 1994, **10**(11), 4367.
8. M. E. McGovern, K. M. R. Kallury and M. Thompson, *Langmuir*, 1994, **10**(10), 3607.
9. K. M. R. Kallury, M. Cheung, V. Ghaemmaghami, U. J. Krull and M. Thompson, *Colloids Surf.*, 1992, **63**(1–2), 1.
10. H. Zhang and M. Chiao, *J. Med. Bioeng.*, 2015, **35**, 143.
11. J.-K. Kim, D.-S. Shin, W.-J. Chung, K.-H. Jang, K.-N. Lee, Y.-K. Kim and Y.-S. Lee, *Colloids Surf., B*, 2013, **33**(2), 67.
12. C. A. Goss, D. H. Charych and M. Majda, *Anal. Chem.*, 1991, **63**(1), 85.
13. B. D. Plouffe, D. N. Njoka, J. Harris, J. Liao, N. K. Horick, M. Radisic and S. K. Murthy, *Langmuir*, 2007, **23**(9), 5050.
14. J. A. Hubbell, S. P. Massia, N. P. Desai and P. D. Drumheller, *Bio/Technology*, 1991, **9**, 568.
15. J. M. Rosenberg and P. J. Utz, *Front. Immunol.*, 2015, **6**, 138.
16. L. M. Shamansky, C. B. Davis, J. K. Stuart and W. G. Kuhr, *Talanta*, 2001, **55**(5), 909.
17. S. Fornera, P. Kuhn, D. Lombardi, A. D. Schlüter, P. S. Dittrich and P. Walde, *ChemPlusChem*, 2012, **77**, 98.
18. G. Simone, P. Neuzil, G. Perozziello, M. Francardi, N. Malara, E. Di Fabrizio and A. Manz, *Lab Chip*, 2012, **12**(8), 1500.
19. J. W. Park, S. C. Na, T. Q. Nguyen, S.-M. Paik, M. Kang, D. Hong, I. S. Choi, J.-H. Lee and N. L. Jeon, *Biotechnol. Bioeng.*, 2015, **112**(3), 494.
20. W. Lan, S. Li, J. Xu and G. Luo, *Biomed. Microdevices*, 2010, **12**, 1087.
21. P.-C. Chen, R. L. C. Chen, T.-J. Cheng and G. Wittstock, *Electroanalysis*, 2009, **21**(7), 804.
22. J. C. Wang, W. Liu, Q. Tu, C. Ma, L. Zhao, Y. Wang, J. Ouyang, L. Pang and J. Wang, *Analyst*, 2015, **140**(3), 827.
23. N. W. Bartlett, M. T. Tolley, J. T. B. Overvelde, J. C. Weaver, B. Mosadegh, K. Bertoldi, G. M. Whitesides and R. J. Wood, *Science*, 2015, **349**, 161.
24. J. Tien, C. M. Nelson and C. S. Chen, *Proc. Natl. Acad. Sci. U. S. A.*, 2002, **99**(4), 1758.
25. Y. Liu, V. Klep and I. Luzinov, *J. Am. Chem. Soc.*, 2006, **128**, 8106.

5 Forces in Microfluidics

5.1 Introduction

Pressure or voltage (electrophoresis) is normally used for generating flows in microfluidic channels. This is discussed and applied in several of the coming chapters (Chapters 6–11), and especially in Chapter 7, which is dedicated to pumping. Here, the focus is mainly on additional forces that may be applied to particles or cells in suspension.

Microfluidics is an essential technique for the manipulation of particles, cells and fluids (gases, liquids and even colloids in suspensions). These operational methods depend on a variety of forces used in microfluidics, typically including magnetic, dielectrophoretic, optical, acoustic and inertial forces. By employing these forces, microfluidic devices can achieve high-efficiency fluid pumping, mixing and reactions, and also high-throughput cell and particle focusing, separation, trapping, extraction, detection and analysis. In this chapter, we introduce the forces used in microfluidic techniques and discuss their applications in microfluidic platforms commonly used in physical, chemical, biological and medical research.

5.2 Magnetic Force

Magnetic force, when utilized in microfluidics, usually applies magnetic particles, magnetic substrates or magnetic fields to the actuation of fluids. It is conventional in magnetic microfluidics to control

Microfluidics and Lab-on-a-chip
By Andreas Manz, Pavel Neužil, Jonathan S. O'Connor and Giuseppina Simone

Published by the Royal Society of Chemistry, www.rsc.org

fluids or droplets by using permanent magnets or electromagnets to drag magnetic particles inside fluids or droplets (Figure 5.1).[1] It is also referred to as magnetophoresis. Magnetic force is typically used for fluid actuation,[2] sample mixing and reactions[3] and cell and particle focusing,[4] separations[5] and detection.[6]

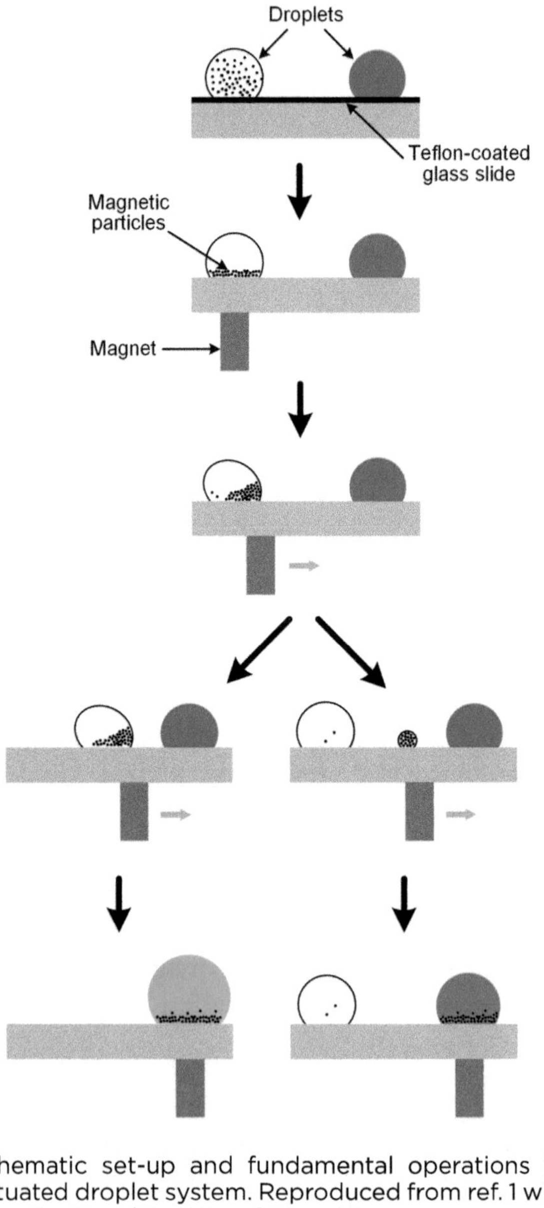

Figure 5.1 Schematic set-up and fundamental operations in a magnet-actuated droplet system. Reproduced from ref. 1 with permission from the Royal Society of Chemistry.

Magnetic forces can not only facilitate the mixing of fluids between droplets or in microchannels but can also improve fluid mixing inside a droplet. Instead of magnctic particles, there are several delicate and ingenious magnetic structures, such as a gyromixer,[7] which can be used inside a droplet to accelerate mixing and hence reactions.

Magnetic force has also been used to control the morphology of magnetically transformable nanostructures on a flexible substrate (Figure 5.2). Conventional substrates for magnetic microfluidics are smooth and coated with a hydrophobic thin film such as polyimide and polytetrafluoroethylene. It is widely known that surface roughness can change the properties of the surface, resulting in a different contact angle and making it hydrophobic or hydrophilic.[8] The roughness can be determined by fabricating nanostructures on the surface. Therefore, if the nanostructures are made of flexible and magnetic materials, the surface roughness can be tuned using magnetic force. As shown in Figure 5.2a, the droplet stays on the nanostructure surface with a large contact angle. When an external magnetic field is applied, depicted in Figure 5.2b, the magnetic nanostructures collapse, reducing the surface roughness and the apparent contact angle.[9]

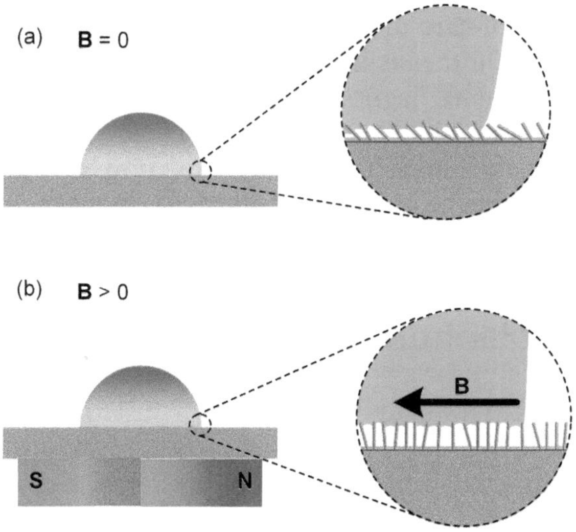

Figure 5.2 Schematic illustration of a water droplet on the surface of a nanostructured film under different magnetic field intensities (B). (a) Shows the droplet on a surface under no magnetic field, whilst (b) shows the change to the surface under a magnetic field, and the resultant change to the contact angle. Reproduced from ref. 9 with permission from American Chemical Society, Copyright 2011.

Compared with other microfluidic platforms, the unique advantage of magnetic microfluidics is the dual functionality of magnetic particles. In addition to serving as fluid or droplet actuators, magnetic particles can also serve as a functionalized solid substrate for molecule adsorption, which permits a wide range of applications in molecular diagnostics. Several researchers have demonstrated that such functionalized magnetic particles can control droplet motion and bind DNA molecules for solid-phase DNA extraction.[10,11] The chemical and biological function of magnetic particles makes magnetic microfluidics attractive for droplet-based bioassays. A typical example is the use of a microfluidic platform to detect the highly pathogenic avian influenza virus H5N1 on a throat swab sample.[12] In this device, a droplet containing superparamagnetic particles is actuated using magnetic force, resulting in sample mixing and reaction.

In summary, the dual functionality of magnetic particles, the flexible fluid control and the electrical 'power-free' operation of magnetic microfluidics give this system unique advantages, which can be deployed in two major applications. The first major application is using magnetic microfluidics for sample preparation, such as sample mixing, particle labelling and enrichment. Magnetic microfluidics-based sample preparation is often used to perform sample-to-answer analysis for point-of-care diagnostics. The second major application of magnetic microfluidics is in heterogeneous assays. It can be used to perform molecular diagnostics, modify surface properties and enhance chemical and biochemical reactions and also to enable cell and particle focusing and separation for detection. In addition, radioactive materials, other hazardous chemicals and reagents with a short shelf-life can be synthesized on demand.[13]

5.3 Dielectrophoretic Force

Dielectrophoresis (DEP), first utilized in a microfluidic system by Pohl,[14] is an effect where a particle is carried in an electric field as a result of its dielectric properties. Pohl explained this effect as 'the motion of suspensoid particles relative to that of the solvent resulting from polarization forces produced by an inhomogeneous electric field'. DEP is an analytical technique in which a polarizable particle suffers an attractive or repulsive force in a non-uniform electric field. A dipole approximation is used to describe the DEP force acting on a particle subjected to a non-uniform electric field. Multipole contributions, the perturbing effects arising from interactions with other cells and boundary surfaces and the influence of electrical double-layer polarizations play a role.[15] The applied electric field gradients for DEP applications are generated through microfabricated electrodes or the integration of microchannels in microfluidic devices filled with different liquid media.[16]

In recent years, DEP has been widely used in analytical applications, such as biosensors, cell therapeutics, drug discovery, medical diagnostics, microfluidics and lab-on-a-chip systems, to perform cell and particle manipulation, preconcentration, characterization and purification. Meanwhile, its use in the manipulation and analysis of subcellular analytes such as organelles and biomolecules such as nucleic acids and proteins, has also been reported.[17,18] The unique feature of DEP compared with other analytical techniques is that it is a label-free, specific, fast and potentially low-cost diagnostic technique. It provides tremendous analytical opportunities to characterize various bioanalytical and biomedical applications.[19]

Manipulation of both biological[20] and non-biological[21] analytes in a non-uniform electric field using DEP microdevices has attracted enormous interest in the last few decades. It enables researchers to study the physical and chemical properties of analytes in a non-invasive manner. Separation of nucleic acids (DNA and RNA) has been a central goal of analytical research. Several microdevices have been shown to allow selective and tuneable detection of DNA. A typical example is the ac insulator-based dielectrophoresis (iDEP) device.[22] Using this device, researchers can characterize the dynamics of sorting and achieve size-based trapping and sorting of DNA molecules. In addition to nucleic acids, DEP devices can also handle subcellular organelles and liposomes, microbes, bacteria, viruses, yeasts and microalgae. As an example, a new microfluidic chip for the continuous separation of microalgae cells based on ac dielectrophoresis was developed to increase the accuracy and efficiency of detection of microalgae cells in ballast water.[23] The microfluidic separation chip was designed by integrating 3D electrodes and an insulated triangular structure into the microchannel. This device was the first to separate two or more species of microalgae cells in a microfluidic chip by using negative and positive DEP forces simultaneously, offering a method with a number of advantages: simple operation, high efficiency, low cost, small size and with great potential for on-site pretreatment of ballast water.

In addition to its application to the manipulation of biomolecules, dielectrophoresis is a manipulation method for separating non-biological analytes, such as microparticles and carbon nanomaterials. Microparticles such as polystyrene beads, colloidal beads and other inorganic particles have served as model analytes for the initial characterization of DEP microdevices prior to analysing biological samples. Carbon nanomaterials such as carbon nanotubes, carbon nanofibres and graphene have been widely used due to their mechanical strength, thermal stability, and also their conductive and dielectric properties.

Recently, a study of surfactant and single-stranded DNA-wrapped single-walled carbon nanotubes (SWCNTs) suspended in aqueous solutions manipulated by iDEP was undertaken (Figure 5.3).[24] This method allowed the manipulation of SWCNTs with the help of arrays of insulating posts in a microfluidic device around which electric field gradients were created by the application of electric potential to the extremities of the device. Trapping of SWCNTs using low-frequency ac electric fields with an applied potential not exceeding 1000 V was demonstrated.

Typical DEP microdevices are two-dimensional (2D) due to the simplicity of their fabrication. Nevertheless, they can be ineffective because the fluid flowing at a distance from the electrodes is less affected by DEP, hence the particles may flow through the system undisturbed. This problem can be addressed by making 3D DEP microdevices.[25]

In summary, DEP is a powerful analytical technique for manipulating both biological and non-biological analytes. As it offers label-free, specific and fast assays, it has been widely used in microfluidics,

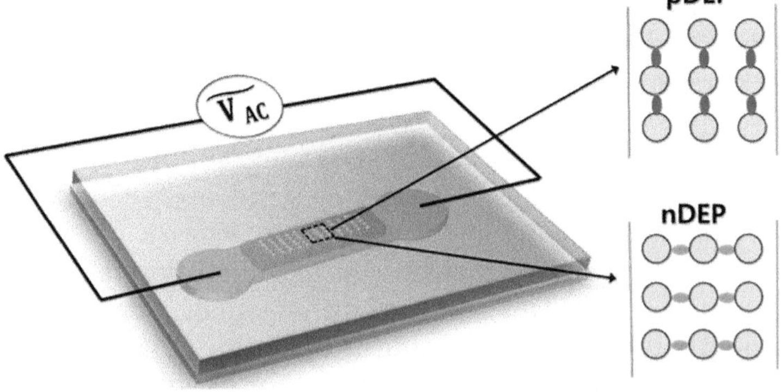

Figure 5.3 Selective trapping of SWCNTs between insulating posts in an iDEP device. Reproduced from ref. 24, https://doi.org/10.1021/acs.analchem.7b03105, with permission from American Chemical Society, Copyright 2017.

biosensors and lab-on-a-chip systems. The trend is now for DEP microdevices to characterize smaller and smaller analytes, from microbes such as bacteria and viruses to biomolecules such as DNA and proteins. However, fundamental research is still required to describe biomolecular DEP, critical for tailored applications. Further, being non-destructive in nature, DEP offers significant potential for the direct processing of biological fluids without sample preparation steps.

5.4 Optical Force

Optical force signifies that light incident on a surface gives rise to a force on that surface. It is usually described as the sum of two components: a 'scattering force', which pushes the particle along the propagation direction of the incident light, and a 'gradient force', which pulls the particle towards the highest intensity region due to the spatial intensity gradient.[26] Once the gradient force counterbalances the scattering force, an equilibrium state is established.

After the invention of optical tweezers by Ashkin and co-workers,[27,28] optical force became increasingly popular for manipulating objects. Optical tweezers are usually realized by using microscope lenses with high numerical apertures, which allow light to be focused as tightly as possible.[29] Because optical forces are feeble, usually measured in femto- to nanonewtons, they are only effective on microscale objects in a range from tens of nanometres to hundreds of micrometres. This range corresponds to the dimensions of cells and organelles so that optical tweezers are attractive for biological research, especially into inter- and intracellular processes. Optical tweezers have been widely used in chemical, physical, medical and biological applications. In physical applications, optical tweezers have the capability of manipulating objects in a non-destructive way, which allows the investigation of classical statistical mechanics, such as the measurement of molecular interactions.[30,31] In medical and biological applications, optical tweezers have been used to characterize the forces affected by molecular motors and to study the mechanical properties of single cells by evaluating the membrane elasticity. Moreover, they have also been exploited to estimate the viscoelastic properties of various samples, from single biomolecules, such as DNA, to aggregated protein fibres.[32,33] In addition, optical tweezers can also be utilized in *in vitro* fertilization or in microsurgery on cells for chromosome and gene modifications.[34]

In recent years, considerable efforts have been made in the development of integrated optofluidic devices that are able to handle biological samples. Optofluidics, which combines optical forces with microfluidics, has been developed for application in cell and particle manipulation, evanescent wave optical manipulation and in microfluidic components such as micropumps and microvalves. For instance, Figure 5.4a shows a generic single-cell manipulation tool that integrates optical tweezers and microfluidic chip techniques for handling the sorting of small cell populations with high accuracy.[35] In this arrangement, an array of holographic optical tweezers is used to create a microfluidic force sensor. The target cells can be handled precisely by optical tweezers to the final point in a non-invasive manner. The unique advantages of this sorter are its high recovery rate and purity in small-cell population sorting.

Optical tweezers can manipulate particles to control fluids. An optically driven micropump that employs viscous drag exerted on a spinning microrotor with left- and right-handed spiral blades on its rotational axis was developed using two-photon microfabrication (Figure 5.4b).[36] The twin spiral microrotor rotates at a high speed of over 500 rpm by focusing a laser beam. The high-speed microrotors

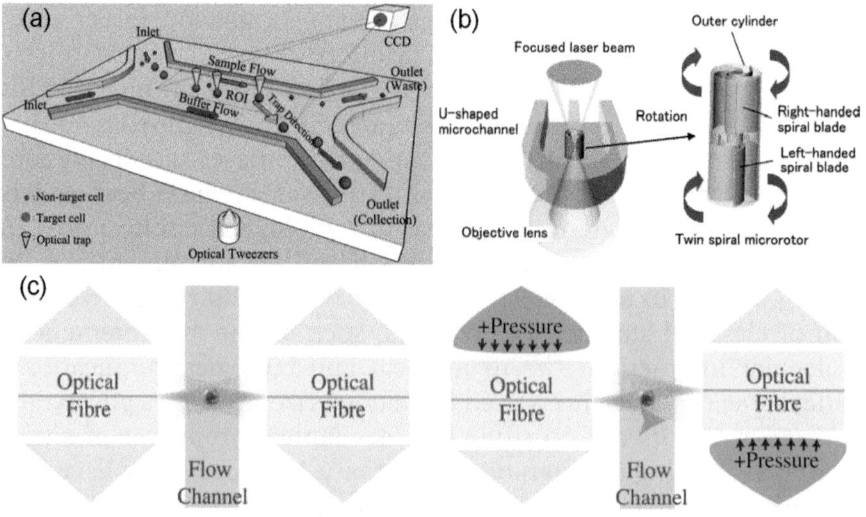

Figure 5.4 Optofluidic devices. (a) Reproduced from ref. 35 with permission from the Royal Society of Chemistry. (b) Reproduced from ref. 36 with permission from The Optical Society, Copyright 2009. (c) Reproduced from ref. 37 with permission from the Royal Society of Chemistry.

are trapped and rotated simultaneously by optical tweezers using a spatial light modulation technique. These viscous micropumps provide several advantages, including simplicity, ease of miniaturization, a readily controlled flow velocity, the production of a continuous flow and the safe transportation of biological samples such as cells.

In addition, Figure 5.4c shows a pneumatically actuated fibre-optic spanner integrated into a microfluidic lab-on-a-chip device for the controlled trapping and rotation of living cells.[37] The dynamic nature of the system allows interactive control over the rotation speed using the same optical power. The use of a multi-layered device makes it possible to rotate a cell both in the imaging plane and in the perpendicular plane, which allows tomographic imaging of the trapped living cell.

In summary, optical forces have been used in various applications, such as optical tweezers and optofluidics.

Compared with other manipulation methods, optical manipulation could be an ideal method for analysing rare cells or small sample populations as it has the advantages of precision measurement, single-cell specificity and the possibility of collecting the entire population or only part of it.

5.5 Acoustic Force

Acoustic force is usually induced by acoustic waves, which are generated by applying ac signals at a resonant frequency to an interdigitated transducer (IDT). The IDT is deposited on a piezoelectric substrate for surface acoustic waves (SAWs).[38] Acoustic waves can focus ultrasonic energy into specific points and produce acoustic nodes and anti-nodes to manipulate micro- and nanoparticles in a non-invasive manner.[39] Acoustics used in acoustofluidics can be divided into two types: travelling acoustic (Tacoustic) and standing acoustic (Sacoustic).[40] Tacoustic is a surface wave generated from a single IDT form by an interference of the original wave with a reflected wave producing a standing surface acoustic wave (SSAW).[41] Sacoustics are usually used to handle micro-objects and are also used in microfluidic actuation, whereas Tacoustics have been mainly used for the actuation of fluids and manipulation of micro-objects through the acoustic streaming flow and acoustic radiation force.[42] Acoustofluidics has been widely used for cell and particle manipulation in cytometry,[43] isolation of rare cells, cell focusing, washing, patterning and culture[44] and molecule manipulation for protein folding and enzymatic assays.[45]

Single microparticles, cells and organisms can be manipulated by acoustofluidic devices using surface acoustic waves. Figure 5.5a shows SAW-based acoustic tweezers. These constituted the first method to control precisely a single microparticle/cell/organism along an arbitrary path within a single-layer microfluidic channel in two dimensions.[46] The use of SAWs allows the device to utilize higher excitation frequencies, resulting in finer resolution in terms of particle manipulation compared with bulk acoustic waves. This technique is versatile and powerful. First, it can manipulate varieties of microparticles regardless of their electrical, magnetic or optical properties. Second, it can manipulate objects of different shapes and sizes, from nanometres to millimetres. Finally, it can also manipulate a single particle or groups of particles (*e.g.* tens of thousands of particles). The acoustic tweezers' versatility, biocompatibility and dexterity make them an excellent platform for a wide range of applications in the biological, chemical and physical sciences, including fundamental studies of the mechanical properties of micro- and nanoscale particles such as cells, DNA, proteins and other molecules.[47] Additionally, the ability to handle particles in large quantities at great speed could make this technique a powerful tool in many high-throughput assays such as cell sorting and separation.

Acoustofluidics can also be used for drug delivery (Figure 5.5b). The intracellular delivery of pharmaceutical agents to cardiac myoblasts by non-inertial contrast agent-free sonoporation has been demonstrated.[48] Real-time single-cell analysis and retrospective

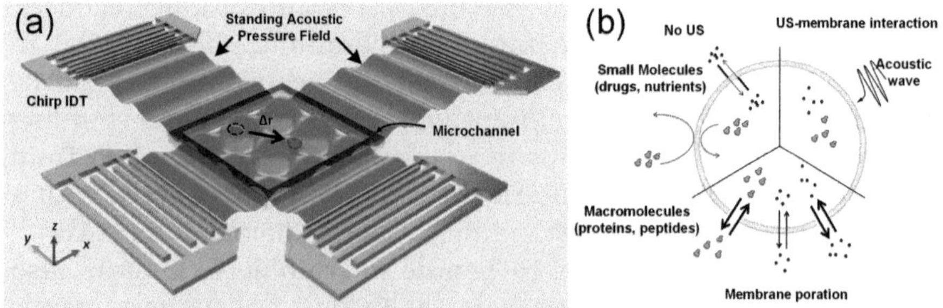

Figure 5.5 Acoustofluidic devices. (a) Reproduced from ref. 46 with permission from the authors, (b) Reproduced from ref. 48 with permission from AIP Publishing, Copyright 2011.

post-sonication analysis of insonated cardiac myoblasts was accomplished. This strategy provides an *in vitro* alternative to controversial *in vivo* models used for early-stage drug discovery, drug delivery programmes and toxicity measurements.

Compared with electrophoresis and magnetophoresis, acoustophoresis has the following advantages: biocompatibility, high-throughput performance, non-invasiveness, high selectivity and sensitivity, simple and low-cost fabrication, avoidance of ionic liquids or magnetic particles and flexibility in design. However, some problems remain. First, acoustofluidic devices always require a bulky function generator and amplifier, which limits the portability of acoustic-based platforms. Second, the energy dissipated in the device components leads to an increase in temperature above the critical threshold for living cells and biomolecules.[49]

5.6 Inertial Forces

The inertial force effect was first described by Segré and Silberberg,[50] who showed that randomly distributed particles at the inlet of a straight pipe migrated laterally to an annulus. This effect is also called the tubular pinch effect. As for a channel with square cross-section, an offset correction of the equilibrium position will occur so that the particles migrating in a square-section channel are located in the four equilibrium positions close to the midpoint of each channel wall due to the influence of the edge angle on the velocity and pressure distribution. In a high/low aspect ratio rectangular microchannel, the particles finally flow to two equilibrium positions close to the midpoint of the long channel wall (Figure 5.6).[51]

Many groups have attempted to explain the underlying mechanism. Theoretical and experimental analyses indicate that several forces influence this phenomenon. First, the main force is the drag force. When an object moves through a fluid or, relatively, when a fluid flows past an object, the drag force rises. In addition to the drag force, there is a dominant force in lateral particle migration, namely inertial lift force, including shear-gradient lift force F_S and wall-induced lift force F_W. The velocity profile of a Poiseuille flow in a microchannel appears to be parabolic, leading to a shear-induced inertial lift on the particle, which directs particles away from the channel centre. This is called the shear gradient-induced lift force. Further, there is a symmetric wake

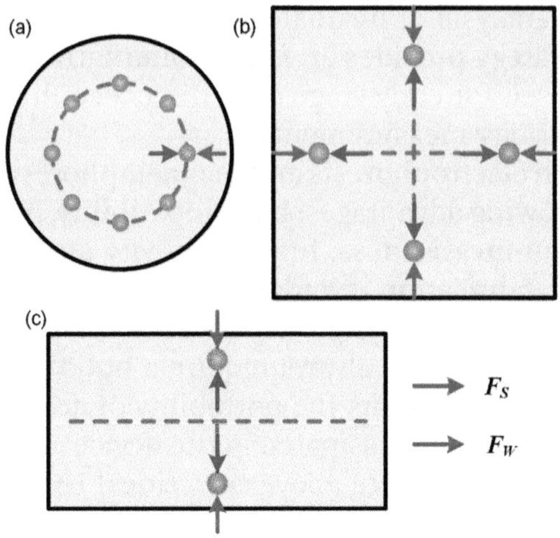

Figure 5.6 Equilibrium positions in (a) circular, (b) square and (c) rectangular channels based on shear-gradient lift force (F_S) and wall-induced lift force (F_W).

around the particles due to the rotation in the fluid,[52,53] as shown in Figure 5.7.

The third force is the Magnus force, which is also called rotation-induced lift force. In an in-viscous flow with uniform velocity, a stationary rigid sphere is rotating with a constant angular velocity. Assuming that there is no slip at the surface of the sphere, the fluid velocity at the bottom part is lower than the velocity at the upper part. According to the Bernoulli principle, the pressure is higher at the bottom part than at the upper part. As a result, a lift force appears to lift the sphere. This force is called the Magnus force. Magnus force results from the pressure difference induced by the streamline asymmetry due to the rotation of the sphere. The direction of the Magnus force is perpendicular to the plane defined by the vectors of the relative velocity and the axis of rotation.

Another force in inertial migration of particles is the Saffman force, also called slip–shear-induced lift force. The channel walls generate a fluid velocity gradient and a corresponding shear-induced particle rotation. The drag caused by walls makes the particle lag behind the fluid. This slip–shear motion will generate a lateral force on the particles called the Saffman force. The direction of the Saffman force is always towards the side where the magnitude of relative velocity is at a maximum.

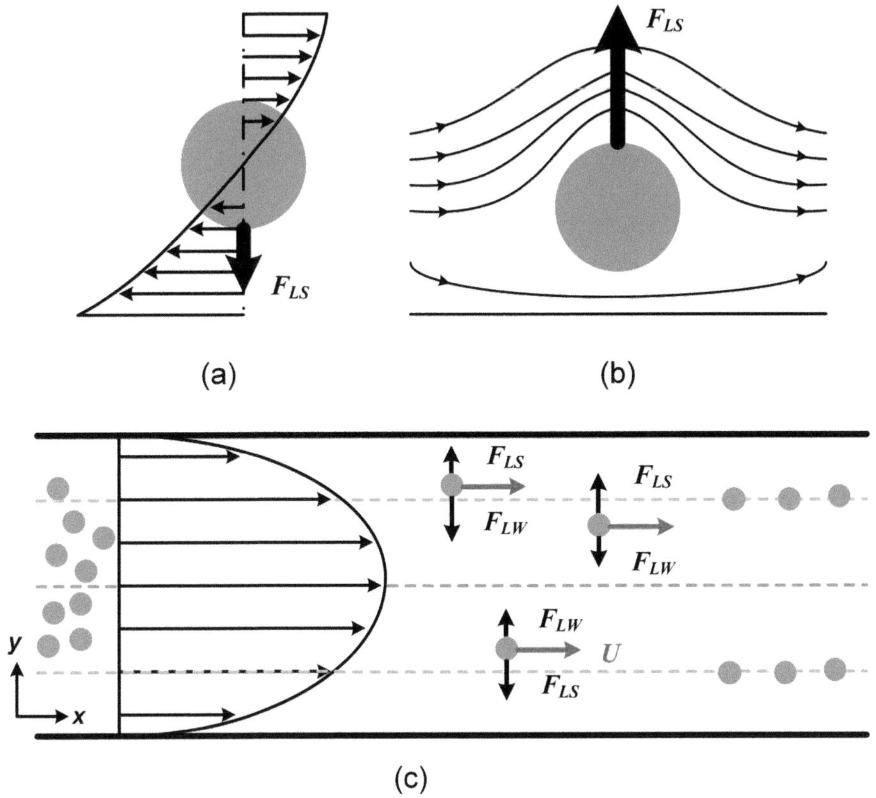

Figure 5.7 Flow field of (a) F_S and (b) F_W. (c) Flow field distribution in channel and inertial lift force schematic diagram. Reproduced from ref. 53, https://doi.org/10.3390/s18061762, under the terms of a CC BY 4.0 license, https://creativecommons.org/licenses/by/4.0/.

The fifth force is the deformability-induced lift force on deformable particles. Solid rigid particles are always used as a simple model in the study of the hydrodynamic behaviour of particles in a microchannel. However, in practice, particles such as cells and organelles are not rigid but deformable. This deformability will induce additional lift forces on the particles. The deformability-induced lift force is perpendicular to the main streamline and it is believed to be the effect of changing the shape of the particle and non-linearities caused by the matching of velocities and stresses at the deformable particle interface.[54]

The last force in inertial microfluidics is the Dean drag force, which occurs only in a channel where a curved fluid flow happens

(Figure 5.8).[55] The Dean drag force is induced by the Dean flow. The Dean flow is generated by the mismatch of the fluid momentum between fluid in the centre and the near-wall regions of a curved channel. The fluid near the channel centre has greater inertia than that close to the wall. Therefore, it flows outwards and induces the fluid near the channel wall to flow inwards based on the law of conservation of energy, eventually generating two symmetric and counter-rotating vortices, called the Dean flow. The direction of the Dean drag force is always along the Dean flow.

With the impact of these forces on particles, inertial microfluidics based on a secondary flow has become a powerful tool in cell and particle manipulation applications, such as focusing, enriching, sorting, separation and extraction. Secondary flow plays an important role in particle manipulation. It can be generated in two types of channels: (1) curved channels and (2) contraction–expansion array (CEA) channels. Secondary flow in a curved channel is also called the Dean flow. Fluid in a straight channel with CEA structures can also induce secondary flow, called geometry-induced secondary flow. In a straight channel with a series of steps on one side of the channel wall, locally curved fluids can be generated beside the step regions because of the change of the channel cross-section. These kinds of curved fluids can induce secondary flow-like fluid inside the curved channels. Therefore, this geometry-induced secondary flow is also called Dean-like flow. Secondary flow can induce additional drag force that can be used to modify the inertial migration of particles.

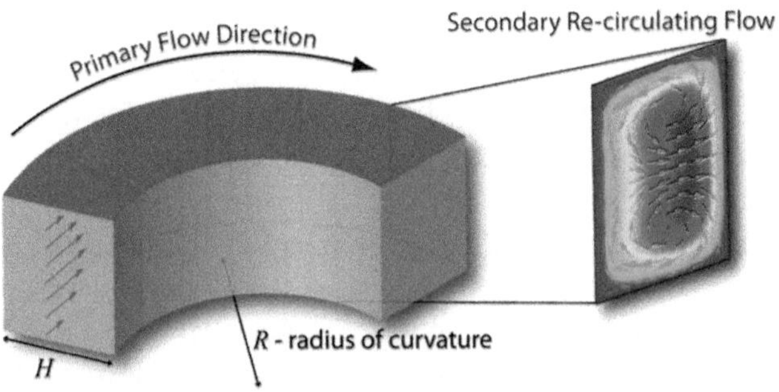

Figure 5.8 Dean flow. Reproduced from ref. 55 with permission from the Royal Society of Chemistry.

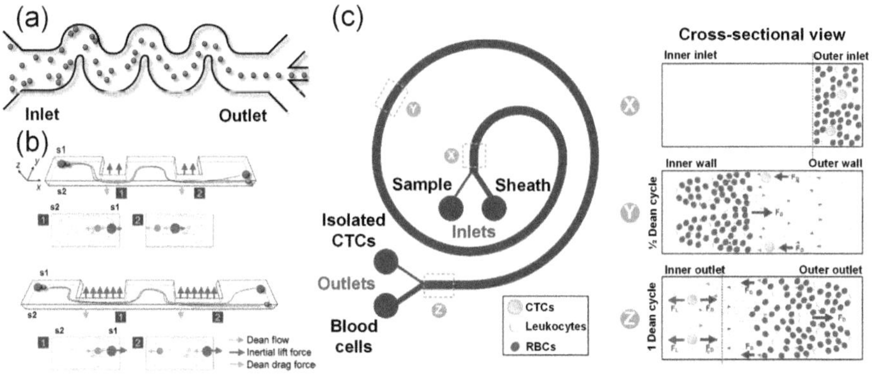

Figure 5.9 Inertial microfluidic devices. (a) Reproduced from ref. 56 with permission from National Academy of Sciences, USA, Copyright 2007. (b) Reproduced from ref. 61 with permission from Elsevier, Copyright 2014. (c) Reproduced from ref. 63 with permission from Springer Nature, Copyright 2013.

A number of excellent studies have investigated cell and particle focusing, separation and filtration based on inertial migration and secondary flows. Generally, they utilized different channel geometries, such as curved channels[56] (Figure 5.9a), CEA channels[57–61] (Figure 5.9b), spiral or double spiral channels[62,63] (Figure 5.9c) and serpentine channels.[64] The devices based on inertial migration and secondary flows usually provide high-throughput and effective separations of cancer cells with minimal damage, have the potential for improving the diagnosis of cancer and can contribute to circulating tumour cell studies and the development of point-of-care diagnostics.[65]

In summary, inertial forces used in inertial microfluidics provide a powerful tool for cell and particle manipulation. Inertial microfluidic devices have numerous advantages, including simple structures, high throughput and non-invasiveness. Compared with active methods such as dielectrophoresis, magnetophoresis and acoustophoresis, inertial microfluidics is independent of external fields such as electric, magnetic and acoustic fields. Moreover, it does not require complex equipment and supporting systems.

5.7 Centrifugal Forces

There are a large number of devices that use centrifugal force to move fluids inside microfluidic systems. In Figure 5.10, the sample is loaded into the system, which is subsequently rotated at elevated

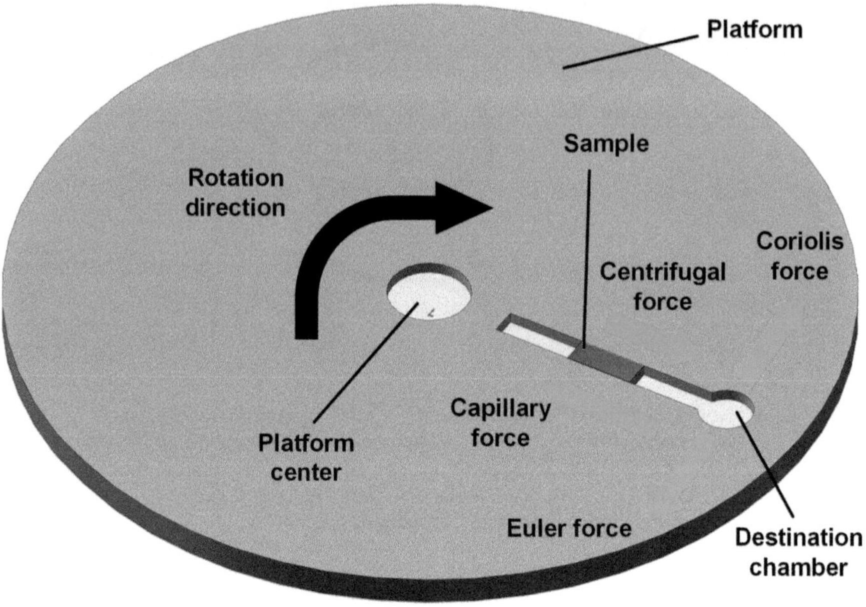

Figure 5.10 Centrifugal forces – principle. Reproduced from ref. 66 with permission from Springer Nature, Copyright 2017.

speed generating a centrifugal force, resulting in the fluids or particles moving inside the device.[66]

The resulting centrifugal force (F_{cen}) due to rotation speed in rad s^{-1} (ω), distance of the fluid from the rotation centre (r) and its specific mass (ρ) can be expressed as

$$F_{cen} = \rho\omega^2\bar{r} \tag{5.1}$$

and the corresponding pressure (P_{cen}) as

$$P_{cen} = \rho\omega^2\Delta r\bar{r} \tag{5.2}$$

Devices utilizing centrifugal forces are also sometimes called labs-on-a-disk. They are fabricated using different methods such as classical micromachining using lithography patterning and etching and also hot embossing, which is very popular. Hot embossing has the advantage of utilizing underused production facilities for DVDs such as the one near Salzburg, Austria, owned by SONY Corporation, which has been partially converted into bio-chip production. There are a number of groups developing this technology for various applications, *e.g.* the Swedish company Gyros used it for immunoassays at the beginning of this century.[67] Also, researchers at other institutions such as IMTEK were very active in this area, commercializing their Bio-Disk system[68] and developing a whole family of systems in subsequent years. It is possible to design and fabricate all systems required for microfluidics at the single substrate level, such as sample and reagent supply, reagent prestorage and release, liquid transport, valving and switching, metering and aliquoting, mixing, separation, droplet generation, detection and many others. As a result, centrifugal force-based devices have been shown to be able to perform blood plasma separation, cell lysis, nucleic acid purification, nucleic acid amplification, immunoassay, washing and blocking. The details of these achievements can be found in a comprehensive review by researchers at IMTEK.[69] According to eqn. (5.1) and (5.2), the amplitudes of both F_{cen} and P_{cen} are always positive since the disk rotating direction does not matter. Nevertheless, Madou's group developed a technique to create a force opposite to the centrifugal force inside the rotating disk.[70] The disk contained a chamber covered with a flexible part made of latex (Figure 5.11). The disk first rotated at high rpm, forcing the fluid to flow into the chamber with one wall made of latex. The centrifugal force was able to fill the chamber with the

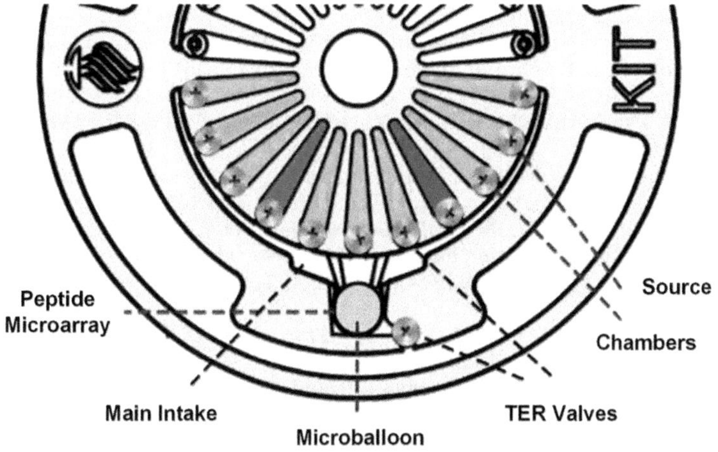

Figure 5.11 Centrifugal device with negative force. Reproduced from ref. 70 with permission from the Royal Society of Chemistry.

fluid until the flexible latex expanded and formed a balloon filled with the fluid. The rpm value was then lowered until the centrifugal forces dropped below the force formed by the flexible latex, forcing the fluid to flow in the opposite direction.

References

1. Z. Long, A. M. Shetty, M. J. Solomon and R. G. Larson, *Lab Chip*, 2009, **9**, 1567.
2. A. Hatch, A. E. Kamholz, G. Holman, P. Yager and K. F. Bohringer, *J. Microelectromech. Syst.*, 2001, **10**(2), 215.
3. K.-Y. Lien, J.-L. Lin, C.-Y. Liu, H.-Y. Leid and G.-B. Lee, *Lab Chip*, 2007, **7**, 868.
4. C. Liu, T. Stakenborg, S. Peeters and L. Lagae, *J. Appl. Phys.*, 2009, **105**, 102014.
5. N. Pamme and A. Manz, *Anal. Chem.*, 2004, **76**(24), 7250.
6. C. Wang, M. Yea, L. Cheng, R. Li, W. Zhu, Z. Shi, C. Fan, J. He, J. Liu and Z. Liu, *Biomaterials*, 2015, **54**, 55.
7. Y. Zhang and T.-H. Wang, *Microfluid. Nanofluid.*, 2012, **12**, 787.
8. R. N. Wenzel, *J. Phys. Colloid Chem.*, 1949, **53**, 1466.
9. Q. Zhou, W. D. Ristenpart and P. Stroeve, *Langmuir*, 2011, **27**, 11747–11751.
10. Y. Zhang, S. Park, K. Liu, J. Tsuan, S. Yang and T. H. Wang, *Lab Chip*, 2011, **11**, 398.
11. Y. Zhang and T.-H. Wang, *Adv. Mater.*, 2013, **25**, 2903.
12. J. Pipper, M. Inoue, L. F. P. Ng, P. Neuzil, Y. Zhang and L. Novak, *Nat. Med.*, 2007, **13**, 1259.
13. Y. Zhang and N. T. Nguyen, *Lab Chip*, 2017, **17**, 994.
14. H. A. Pohl, *J. Appl. Phys.*, 1951, **22**, 869.
15. R. Pethig, *Biomicrofluidics*, 2010, **4**, 022811.
16. N. Abd Rahman, F. Ibrahim and B. Yafouz, *Sensors*, 2017, **17**, 449.
17. M. Viefhues and R. Eichhorn, *Electrophoresis*, 2017, **38**, 1483.
18. A. Nakano and A. Ros, *Electrophoresis*, 2013, **34**, 1085.
19. D. Kim, M. Sonker and A. Ros, *Anal. Chem.*, 2019, **91**, 277.
20. K.-T. Liao and C.-F. Chou, *J. Am. Chem. Soc.*, 2012, **134**, 8742.
21. H. Ding, J. Shao, Y. Ding, W. Liu, H. Tian and X. Li, *ACS Appl. Mater. Interfaces*, 2015, 7, 12713.
22. P. V. Jones, G. L. Salmon and A. Ros, *Anal. Chem.*, 2017, **89**, 1531.
23. Y. Wang, J. Wang, X. Wu, Z. Jiang and W. Wang, *Electrophoresis*, 2019, **40**, 969.
24. M. T. Rabbani, C. F. Schmidt and A. Ros, *Anal. Chem.*, 2017, **89**, 13235.
25. C. Iliescu, L. Yu, G. Xu and F. E. H. Tay, *J. Microelectromech. Syst.*, 2006, **15**, 1506.
26. P. Paie, T. Zandrini, R. M. Vazquez, R. Osellame and F. Bragheri, *Micromachines*, 2018, **9**, 200.
27. A. Ashkin, J. M. Dziedzic, J. E. Bjorkholm and S. Chu, *Opt. Lett.*, 1986, **11**, 288.
28. A. Ashkin, *IEEE J. Sel. Top. Quantum Electron.*, 2000, **6**, 841.
29. K. C. Neuman and S. M. Block, *Rev. Sci. Instrum.*, 2004, **75**, 2787.
30. J. C. Crocker and D. G. Grier, *Phys. Rev. Lett.*, 1994, **73**, 352.
31. Y. N. Ohshima, H. Sakagami, K. Okumoto, A. Tokoyoda, T. Igarashi, K. B. Shintaku, S. Toride, H. Sekino, K. Kabuto and I. Nishio, *Phys. Rev. Lett.*, 1997, **78**, 3963.
32. K. Svoboda, P. P. Mitra and S. M. Block, *Proc. Natl. Acad. Sci. U. S. A.*, 1994, **91**, 11782.
33. C. Bustamante, S. B. Smith, J. Liphardt and D. Smith, *Curr. Opin. Struct. Biol.*, 2000, **10**, 279.
34. M. Waleed, S.-U. Hwang, J.-D. Kim, I. Shabbir, S.-M. Shin and Y.-G. Lee, *Biomed. Opt. Express*, 2013, **4**, 1533.

35. X. Wang, S. Chen, M. Kong, Z. Wang, K. D. Costa, R. A. Li and D. Sun, *Lab Chip*, 2011, **11**, 3656.
36. S. Maruo, A. Takaura and Y. Saito, *Opt. Express*, 2009, **17**, 18525.
37. T. Kolb, S. Albert, M. Haug and G. Whyte, *Lab Chip*, 2014, **14**, 1186.
38. J. Friend and L. Y. Yeo, *Rev. Mod. Phys.*, 2011, **83**, 647.
39. A. E. Christakou, B. Vanherberghen, M. Wiklund, M. Ohlin, M. A. Khorshidi, T. Frisk, B. Önfelt and N. Kadri, *Integr. Biol.*, 2013, **5**, 712.
40. X. Ding, P. Li, S.-C. S. Lin, Z. S. Stratton, N. Nama, F. Guo, D. Slotcavage, X. Mao, J. Shi, F. Costanzo and T. J. Huang, *Lab Chip*, 2013, **13**, 3626.
41. T. Leong, L. Johansson, P. Juliano, S. L. McArthur and R. Manasseh, *Ind. Eng. Chem. Res.*, 2013, **52**, 16555.
42. G. Destgeer and H. J. Sung, *Lab Chip*, 2015, **15**, 2722.
43. L. Schmid, D. A. Weitz and T. Franke, *Lab Chip*, 2014, **14**, 3710.
44. Y. Jia, B. Wang, Z. Zhang and Q. Lin, *Sens. Actuators, A*, 2015, **231**, 1.
45. M. Wiklund, A. E. Christakou, M. Ohlin, I. Iranmanesh, T. Frisk, B. Vanherberghen and B. Önfelt, *Micromachines*, 2014, **5**, 27.
46. X. Ding, S.-C. S. Lin, B. Kiraly, H. Yue, S. Li, I.-K. Chiang, J. Shi, S. J. Benkovic and T. J. Huang, *Proc. Natl. Acad. Sci. U. S. A.*, 2012, **109**, 11105.
47. X. Ding, Z. Peng, S.-C. S. Lin, M. Geri, S. Li, P. Li, Y. Chen, M. Dao, S. Suresh and T. J. Huang, *Proc. Natl. Acad. Sci. U. S. A.*, 2014, **111**, 12992.
48. D. Carugo, D. N. Ankrett, P. Glynne-Jones, L. Capretto, R. J. Boltryk, X. Zhang, P. A. Townsend and M. Hill, *Biomicrofluidics*, 2011, **5**, 044108.
49. A. Barani, H. Paktinat, M. Janmaleki, A. Mohammadi, P. Mosaddegh, A. Fadaei-Tehrani and A. Sanati-Nezhad, *Biosens. Bioelectron.*, 2016, **85**, 714.
50. G. Segré and A. Silberberg, *Nature*, 1961, **189**, 209.
51. A. A. S. Bhagat, S. S. Kuntaegowdanahalli and I. Papautsky, *Microfluid. Nanofluid.*, 2008, **7**, 217.
52. J. Zhang, S. Yan, D. Yuan, G. Alici, N. T. Nguyen, M. Ebrahimi Warkiani and W. Li, *Lab Chip*, 2016, **16**, 10.
53. Y. X. Gou, Y. X. Jia, P. Wang and C. K. Sun, *Sensors*, 2018, **18**, 1762.
54. H. Amini, W. Lee and D. Di Carlo, *Lab Chip*, 2014, **14**, 2739.
55. D. Di Carlo, *Lab Chip*, 2009, **9**, 3038.
56. D. Di Carlo, D. Irimia, R. G. Tompkins and M. Toner, *Proc. Natl. Acad. Sci. U. S. A.*, 2007, **104**, 18892.
57. J. Zhang, M. Li, W. H. Li and G. Alici, *J. Micromech. Microeng.*, 2013, **23**, 085023.
58. Z. Wu, Y. Chen, M. Wang and A. J. Chung, *Lab Chip*, 2016, **16**, 532.
59. A. J. Chung, D. R. Gossett and D. Di Carlo, *Small*, 2013, **9**, 685.
60. A. J. Chung, D. Pulido, J. C. Oka, H. Amini, M. Masaeli and D. Di Carlo, *Lab Chip*, 2013, **13**, 2942.
61. M. G. Lee, J. H. Shin, S. Choi and J.-K. Park, *Sens. Actuators, B*, 2014, **190**, 311.
62. L. Wu, G. Guan, H. W. Hou, A. A. Bhagat and J. Han, *Anal. Chem.*, 2012, **84**, 9324.
63. H. W. Hou, M. E. Warkiani, B. L. Khoo, Z. R. Li, R. A. Soo, D. S.-W. Tan, W.-T. Lim, J. Han, A. A. S. Bhagat and C. T. Lim, *Sci. Rep.*, 2013, **3**, 1259.
64. J. Zhang, W. H. Li, M. Li, G. Alici and N. T. Nguyen, *Microfluid. Nanofluid.*, 2014, **17**, 305.
65. M. G. Lee, J. H. Shin, C. Y. Bae, S. Choi and J. K. Park, *Anal. Chem.*, 2013, **85**, 6213.
66. W. Al-Faqheri, T. H. G. Thio, M. A. Qasaimeh, A. Dietzel, M. Madou and A. a. Al-Halhouli, *Microfluid. Nanofluid.*, 2017, **21**, 102.
67. H. Song, J. M. Rosano, Y. Wang, C. J. Garson, B. Prabhakarpandian, K. Pant, G. J. Klarmann, A. Perantoni, L. M. Alvarez and E. Lai, *Lab Chip*, 2015, **15**, 1320.
68. J. Ducrée, S. Haeberle, S. Lutz, S. Pausch, F. v. Stetten and R. Zengerle, *J. Micromech. Microeng.*, 2007, **17**, S103.

69. O. Strohmeier, M. Keller, F. Schwemmer, S. Zehnle, D. Mark, F. von Stetten, R. Zengerle and N. Paust, *Chem. Soc. Rev.*, 2015, **44**, 6187.
70. M. M. Aeinehvand, L. Weber, M. Jiménez, A. Palermo, M. Bauer, F. F. Loeffler, F. Ibrahim, F. Breitling, J. Korvink, M. Madou, D. Mager and S. O. Martínez-Chapa, *Lab Chip*, 2019, **19**, 1090.

6 Flow Control

6.1 Introduction

Controlled and repeatable sample injection and precise dispensing of reagents are crucial for the reliable operation of microfluidic devices. Different designs have been proposed, based on electroosmosis and pneumatic actuation. Combined systems including, for example, electrokinetic injection and isotachophoresis (ITP) preconcentration can be used for preconcentration and injection. Dielectric elastomer actuators can be integrated with the capillary electrophoresis (CE) for varying volume injection. An overview of the methods of injection and sample pretreatment is reported in the following sections.

6.2 Hydrodynamic Flow Control

In microfluidics, laminar flow and diffusion predominantly influence the transport of solute molecules in a liquid. Therefore, pressure offers the possibility of controlling the fluid behaviour in microfluidic systems, in combination with the design of the microfluidic channel. Channels with a V, H or Y configuration can be designed for controlled mixing, filtering and focusing of the sample. Figure 6.1 illustrates two simple channel layouts, both with four ports. The pressure applied to any one of them controls the flow, as in an equivalent electrical circuit with ohmic resistors. For example, in Figure 6.1a, if a vacuum is applied to port 1, equal flow rates can be expected from ports 2, 3 and 4 towards the centre and the sum of those flows is the flow rate through port 1.

Microfluidics and Lab-on-a-chip
By Andreas Manz, Pavel Neužil, Jonathan S. O'Connor and Giuseppina Simone

Published by the Royal Society of Chemistry, www.rsc.org

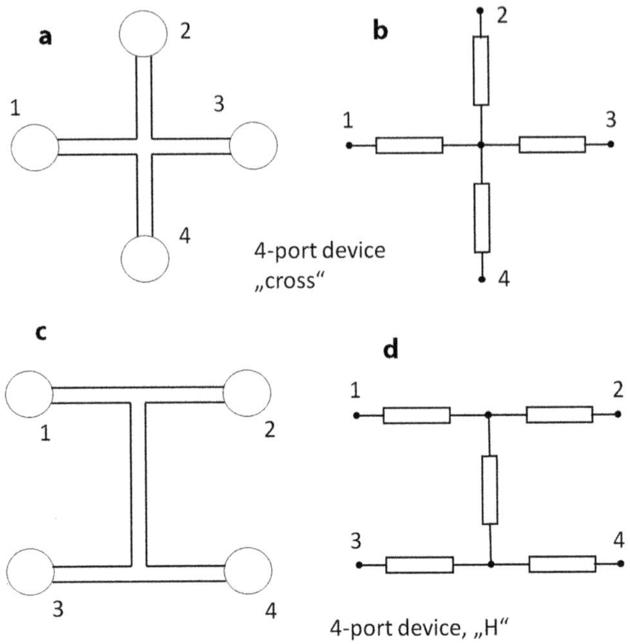

Figure 6.1 Schematic channel layouts and electrical equivalent circuits. (a), (b) Cross intersection of four channels; (c), (d) H-shaped arrangement of channels.

6.2.1 V-shaped Channel

Figure 6.2a shows a typical example of flow in a V-shaped channel, where two different streams meet.

As discussed in Chapter 1, while two fluids may be miscible, diffusion is the prevalent mechanism of mixing in microfluidics. The characteristic diffusion time is a function of the diffusion coefficient, so the two streams flow independently for a designed length.

In Chapter 1, we observed that the time of diffusion is defined by eqn (1.18), which we now want to use for predicting the mixing length. For the geometry reported in Figure 6.2a, with a defined microfluidic channel and known fluids, it is possible to predict the length of the channel required for these fluids to mix. Figure 6.2b shows the relationship between the channel width and the length required for full mixing. Because of the square root dependence in the Einstein–Smoluchowski equation, eqn (1.18), reducing the diffusion distance is essential for rapid mixing. For instance, by reducing the width of the channel, the diffusion distance is shortened and therefore faster mixing between the two phases occurs. However, to counteract the increase in fluidic resistance that this causes, short channel segments and deeper channels are used.

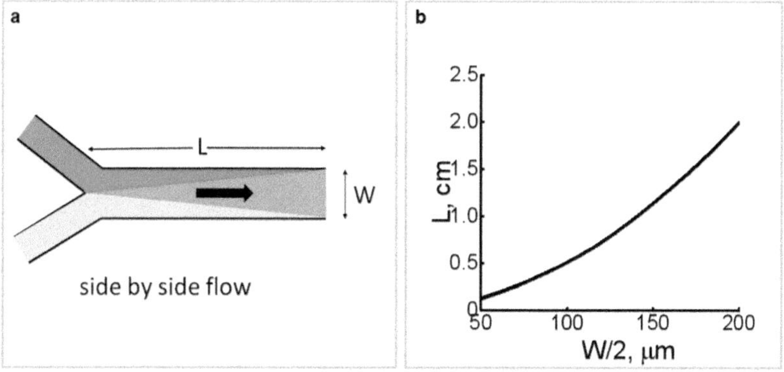

Figure 6.2 V-shaped channel. (a) Schematic of a V-shaped channel layout. (b) Relationship between the width of the channel and the length required to mix the two fluids for a given linear flow rate.

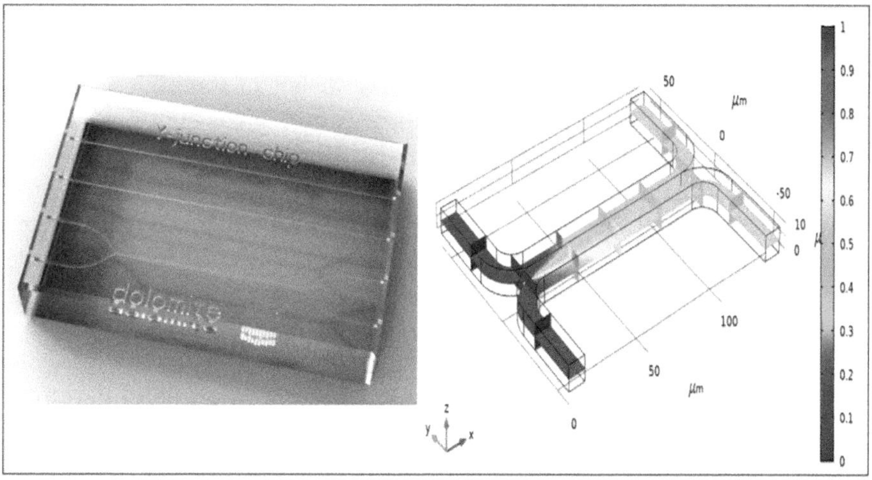

Figure 6.3 H-shaped channel. Left: a chip (Dolomite Microfluidics). Right: an example simulation of a concentration gradient (Comsol). Reprinted with permission, courtesy of Dolomite Microfluidics Inc.

6.2.2 H-shaped Channel

It is possible to control the fluid behaviour by using an H-shaped filter, as shown in Figure 6.3. The two fluids injected at the two inlets have different characteristics, the fluid in the top channel (blue colour) consists of a mixture of two different molecular species with two different coefficients of diffusion; the fluid in the bottom channel (red colour) is a pure compound.

While flowing along the channel, the component of the top stream with the highest diffusion coefficient will diffuse into the bottom stream faster than the slower component, meaning that the chip acts as a filter or an extractor.

6.2.3 Y-shaped Channel with Three Inlet Channels

The working principle of Y-shaped channels is independent of the geometric arrangement at the inlet whether merging at an acute angle or with right-angle geometry. Figure 6.4a shows the arrangement of the fluids at the inlet; the target liquid flows from the middle channel and the functional liquids flow from the two side channels. The control of the flow is achieved by controlling the flow rates.

To focus or narrow the centre flow to a greater or lesser extent, the laminating flow rates can be increased or decreased[1] (Figure 6.4b). The top set of flow rates has the fastest functional liquid flow rate. This decreases down the three flow rates shown.

The concentration of the middle stream can be increased or decreased by controlling the flow rates. This geometry is of interest for two-dimensional focusing; notably cellular focusing, which consists of the flow of aligned single cells flowing one after another at the centre of the channel.

The perfect symmetrical geometry of the microchannel permits focusing in the middle stream, with this condition maintained while the side flow rates are equal. Any change to one of the side flow rates unbalances the flow inside the microfluidic channel, destroying the focusing effect.

6.2.4 Combination of V-, H- and Y-shaped Channels

The combination of all the fluidic elements allows the manipulation of concentrated solutions to produce different concentrations at the outlets.

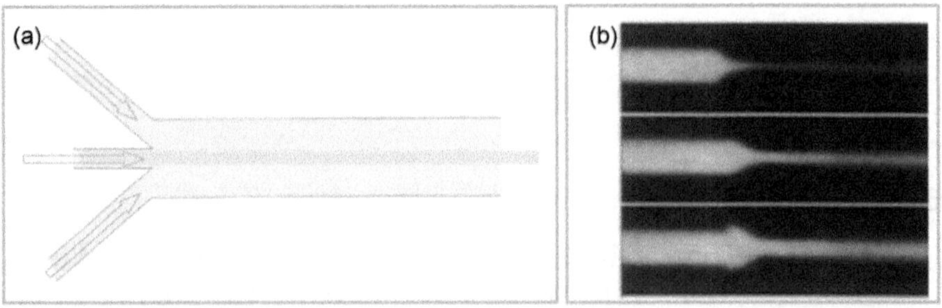

Figure 6.4 Y-shaped channel. (a) Inlet configuration of the three channels. (b) Lamination of the three fluids according to three different flow rate conditions. Reproduce from ref. 1 with permission from American Chemical Society, Copyright 1997.

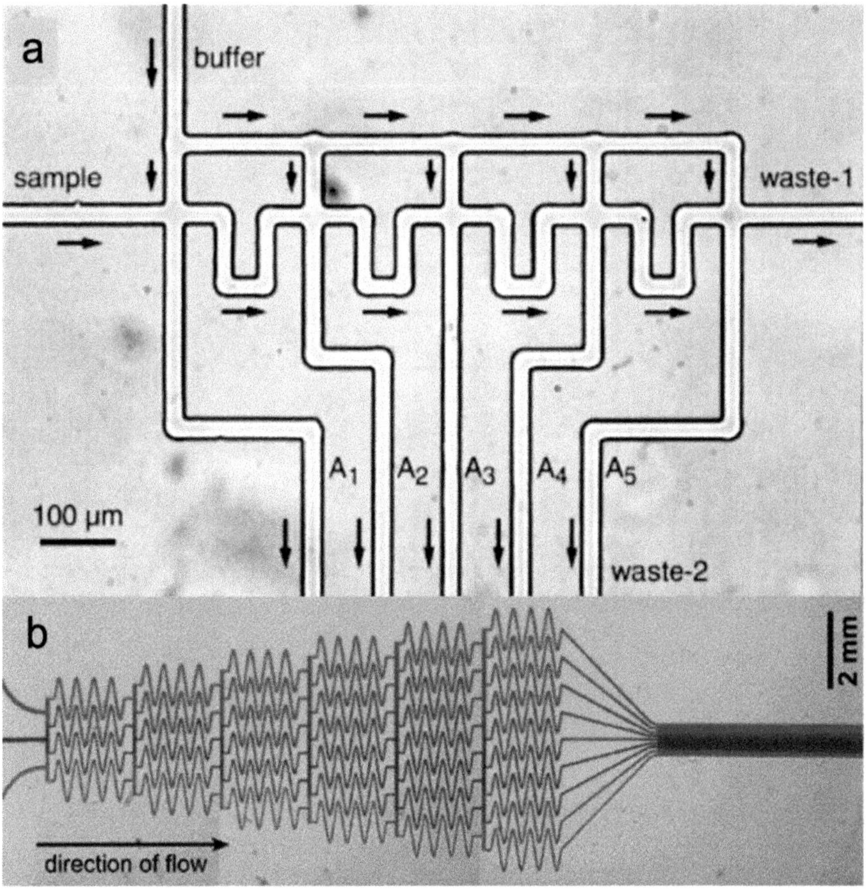

Figure 6.5 (a) Serial mixer/diluter. (b) Complex systems of mixer/diluter. (a) Reproduced from ref. 2 with permission from American Chemical Society, Copyright 1999. (b) Reproduced from ref. 3 with permission from American Chemical Society, Copyright 2001.

Figure 6.5a shows a serial mixer used for diluting a given compound in solution.[2] The concentrated sample and buffer enter the flow manifold at the top left. While flowing through the channel, the sample and the buffer are mixed iteratively, resulting in a cascading dilution of the solution. The mixing channel is bent only to allow the contact between the two liquids and provide an appropriate contact time. Part of the diluted solution is collected at the exit, part is mixed with more of the fresh buffer for further dilution and again part of the diluted solution is collected at the exit and part of it is diluted again. At the outlets A_1 to A_5, several dilutions can be obtained for further experiments. A more complicated geometry, called a 'Christmas tree' design, was fabricated for the dilution of three components based on

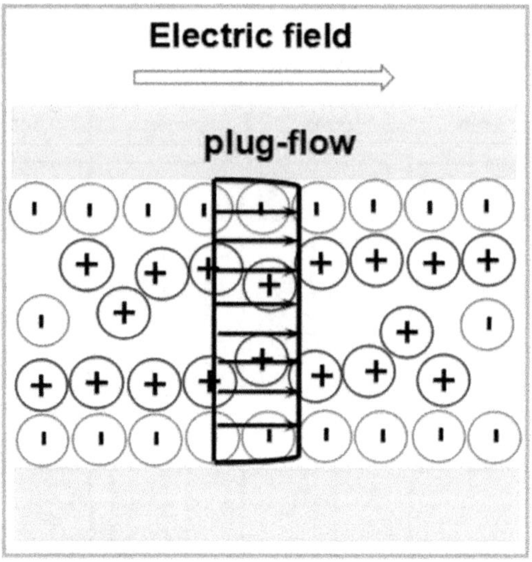

Figure 6.6 Plug-like flow in a microfluidic channel generated by electroosmotic forces.

the same working principle, shown in Figure 6.5b.[3] Several applications can benefit from this design, *e.g.* separation techniques requiring defined pH gradients across the channel width.

6.3 Electroosmotic Flow Control

Figure 6.6 shows a schematic of the working principle of electroosmotic flow (EOF), which underpins electroosmotic pumping. Here, the principle is introduced to explain two important operations, gated and pinched injections, for microfluidic manipulation.

EOF pumping is the most commonly used pump operation based on electrokinetic phenomena. Positively charged species move towards the negative polarity of an applied electric field along the negatively charged surface, due to the exposed hydroxyl groups. In microfluidics, the electrical double layer of the surface is far more dominant than in bulk flows. The greater prominence of the double layer means that the interactions with the surface produce a shear force on the surrounding fluid and induce a flow. Cations have an electrophoretic velocity in the same direction as the EOF, whereas neutral species have no electrophoretic velocity and as such flow at the same rate as the EOF, and anions have an electrophoretic velocity opposed to the electroosmotic flow and hence are the slowest species. The main advantage of using EOF pumping is that it works in small channels without a need for applied pressure.[4]

The first example involving electrokinetic fluid handling is the dispensing or injection of a given volume of sample in a reproducible manner. The volume of the sample can be given by the geometry of the channel arrangement or it may be larger, caused by diffusion of the solute, but if the fluid control is appropriate the volume is still reproducible. The control of the fluid volume given by geometry is called 'metering'. Quantitative analysis of a sample or bioassay applications require the injection of a well-defined volume of fluid. Injection makes use of a cross channel or a double T arrangement in combination with electrokinetic fluid manipulation.

Figure 6.7a shows the schematic layout of a microfluidic chip that can be used for electrokinetic manipulation. The offset cross or double V provides an injection area of a certain distance, thereby increasing the volume to be injected compared with cross injection intersections. Figure 6.7b is an equivalent electrical circuit. This has the double role of serving like a schematic for understanding the fluidic resistance (see Chapter 1) and also as the equivalent circuit to describe the electrical behaviour of the fluidic network. There are five nodes, four for the inlets and outlets and one at the intersection, and also four resistors representing the impedance in the four channel segments. The voltage is applied at the four terminal nodes and only the voltage at the inner node can be calculated once the individual impedances are measured. A model of equations based on Ohm's law and the Kirchhoff rules allows one to calculate the voltages required for a specific pathway of the current through the channel network and for similitude of the fluidic flow pathway.

Figure 6.8 shows an example of voltages used to perform a gated injection; (1) is the sample channel, (2) is the sample waste channel,

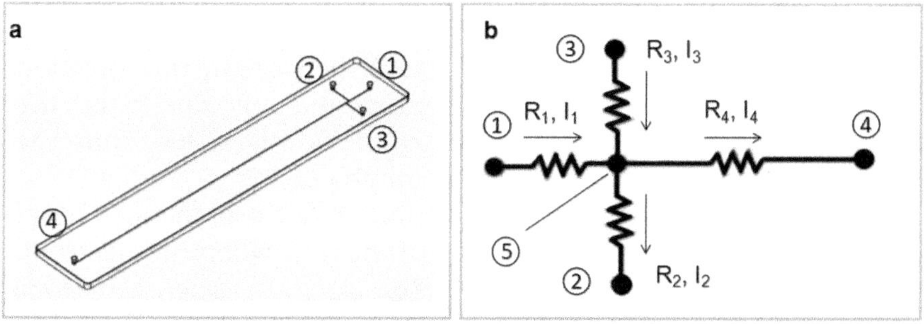

Figure 6.7 (a) Schematic of an offset cross-shaped channel chip (microfluidic ChipShop). (b) Equivalent electric circuit: R = resistance, I = intensity. Reproduced with permission, courtesy of microfluidic ChipShop GmbH.

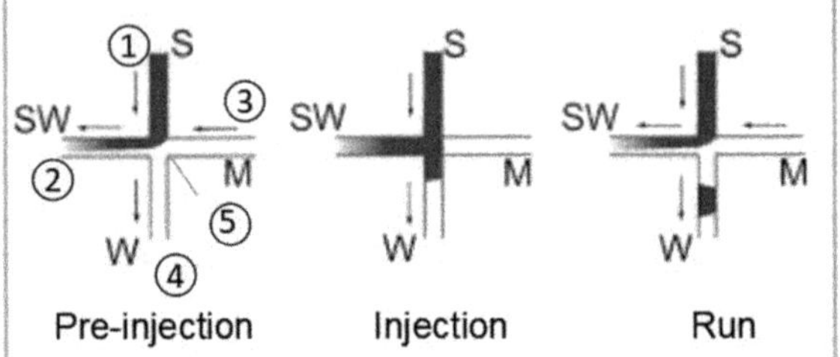

	Pre-injection	Injection	Run
V_1	0.7 HV	0.7 HV	0.7 HV
V_2	0.3 HV	float	0.3 HV
V_3	1.0 HV	1.0 HV	1.0 HV
V_4	0.0	float	0.0

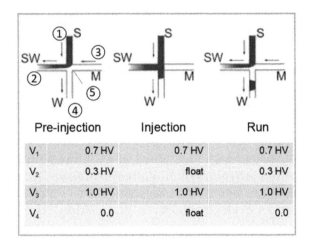

Figure 6.8 Gated electroosmotic injection. Schematic and operating conditions. The table provides an example of the voltage setting for the three operations. HV is the applied voltage, S (1) is the sample channel, SW (2) is the excess sample channel, M (3) is the buffer channel and W (4) is the exit channel leading to further downstream processing of the injected sample volume.

(3) is the buffer channel, (4) is the run channel, which may lead to another part of the device or a detector, and (5) is the cross junction. Before the injection, (2) and (4) are set to the lowest voltage, so the electroosmotic flow is directed to them. The sample and buffer reservoirs (1) and (3) are set at a higher positive potential, so that the sample and buffer flow from (1) to (2) and from (3) to (4), respectively. The potential of (3) is higher than that at (1), so the flow of sample is directed into (2). At the time of the injection, the voltages at exits (2) and (4) are briefly disconnected (floating), to allow them to assume the same potential as the intersection (5). This allows a reproducible volume of the sample to flow into the main channel (4), forming the sample plug. When the voltage is established again, the injection plug is isolated from the crossing potentials and only experiences the potential difference of (1) and (4), hence the flow is directed into the main channel (4) for further downstream processing.

This operation is called 'gated injection' or 'electrokinetic valving' and the injected volume is a function of the time when the electrokinetic valve is open.

As the gated injection is EOF controlled, it must be taken into account that the operation is biased. As mentioned before, different charged species have different velocities. This means that positively charged ions are injected to a larger extent than neutral species and neutral species to a larger extent than negatively charged ions.

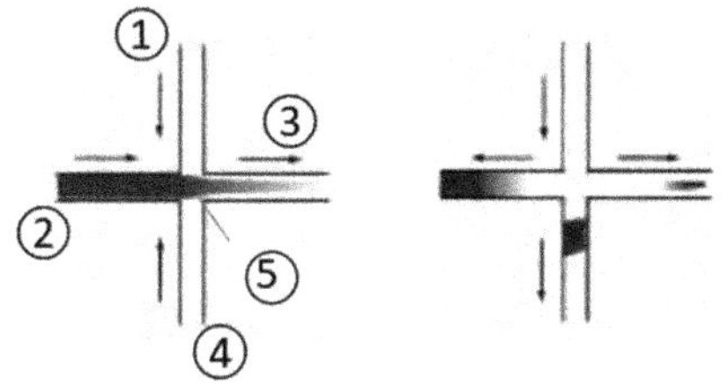

V_1	0.3 HV	1.0 HV
V_2	1.0 HV	0.2 HV
V_3	0.0	0.2 HV
V_4	0.3 HV	0.0

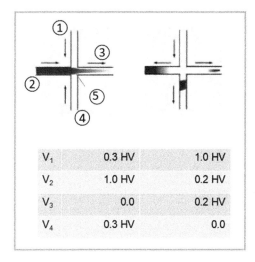

Figure 6.9 Pinched injection. Schematic and operating conditions for electroosmotic flow. The table displays an example of the voltage setting for the three operations. HV is the applied voltage, (2) is the sample channel, (3) the excess sample channel, (1) is the buffer channel and (4) is the exit channel, leading to further downstream processing of the injected sample volume.

The same layout can be used in a 'pinched injection' configuration. Figure 6.9 shows the working principle; the sample reservoir is channel 2. Prior to injection, there is continuous sample pumping across the intersection to the waste reservoir, channel 3. The voltages are set so that the EOF passes from 1, 2 and 4 towards 3. Once an established flow is created and 5 is 'loaded' with sample, the voltages switch so that the EOF flows from 1 towards 2, 3 and 4. This flushes the sample plug at 5 downwards while also pushing the remaining sample in channel 2 and 3 back to the respective reservoirs.

Pinched injection allows a predefined and a fixed volume of sample. The bias effect can be avoided if the loading step is given sufficient time to allow the concentration composition of the sample solution to be present at the intersection.

References

1. S. C. Jacobson and J. M. Ramsey, *Anal. Chem.*, 1997, **69**(16), 3212.
2. S. C. Jacobson, T. E. McKnight and J. M. Ramsey, *Anal. Chem.*, 1999, **71**(20), 4455.
3. S. K. W. Dertinger, D. T. Chiu, N. L. Jeon and G. M. Whitesides, *Anal. Chem.*, 2001, 73(6), 1240.
4. S. L. Barker, D. Ross, M. J. Tarlov, M. Gaitan and L. E. Locascio, *Anal. Chem.*, 2000, **72**(24), 5925.

7 Valving and Pumping

7.1 Introduction

Before micropumping is fully explored, larger external pumping methods should be discussed as these are still very commonly used. The simplest method to pump a fluid through a lab-on-a-chip (LOC) system is by using an external piston pump, *e.g.* a positive displacement syringe pump. Here, a syringe is fitted into the pump and connected by tubing to the chip. The pump functions by driving the syringe plunger forwards and backwards using a metal plate connected to a motor. LOC systems use syringe pumps because of the ease of control, high-throughput capabilities and highly precise flow. Sophisticated commercial syringe pumps are available that can produce minimum flow rates in the picolitre per minute range. Syringe pumps are not always ideal as the fluid volume is limited by the syringe used, which needs to be refilled by either removing the syringe and manually refilling it or by programming the pump to draw fluid into the syringe. This method requires a switch valve to redirect the flow to a refill reservoir. Also, if a cell-laden fluid is within the syringe, then it is possible for the cells to form a sediment, reducing the homogeneity of the fluid. Finally, there is no pressure control with syringe pumps. The flow rate is entered and the resultant pressure is not directly monitored. Sophisticated pumps can detect dramatic pressure changes and halt the pumping to avoid damage, but this is not an indication of the pressure within a microfluidic device. Excess pressure can easily damage cells or even rupture the chip.

Microfluidics and Lab-on-a-chip
By Andreas Manz, Pavel Neužil, Jonathan S. O'Connor and Giuseppina Simone

Published by the Royal Society of Chemistry, www.rsc.org

Pressure-driven pumps are an alternative solution. Here, pressurized air is used to drive fluid through the system at a constant pressure. These pumps are advantageous because the mechanical parts are only involved in controlling the air. Additionally, the pumps are versatile, capable of stable flows and can replicate *in situ* pressures, *e.g.* the cycle of a heartbeat. However, flow sensors also have to be implemented, otherwise there is no way to determine the flow rate. Nanolitre injections are also not possible due to the lack of flow-rate control. It is also worth noting that pressure-driven pumps require more components than a syringe pump.

Peristaltic pumps are very common in microfluidics and are discussed in greater detail in Section 7.4.

Micropumps allow for the accurate handling of the microflow in microfluidic devices. In standard applications, micropumps are coupled with other microfluidic components – mixers, separators, filters, *etc.* – to ensure a very high degree of control. Specific pumps are more suitable for different fluids or experiments, and the differences between the pumps are explained in the following.

One general classification, based on actuators, includes mechanical (piezoelectric, electrostatic, pneumatic, thermopneumatic) or non-mechanical (electrokinetic, surface tension driven) micropumps.

The conceptual design of the micropumps is based on the theory of microfluidics (see Chapter 1). In particular, by scaling down the characteristic length, the surface forces (friction and surface tension) scale down in proportion to the square of the length scale but the body or volume forces (*e.g.* inertial and gravitational forces) scale down in proportion to the cube of the characteristic length. This means that the surface forces are 10 times more prevalent than the body forces. On the micrometre scale, the Navier–Stokes equation results in linear and predictable Stokes flow (or creeping flow). Also, the Stokes equation [eqn (1.6)] is equal to zero, hence the pressure and drag forces upon the fluid cancel each other out with relatively little inertia.

Given the conditions and with the Navier–Stokes relation being a linear equation, all the flows in the creeping flow limit are kinematically reversible. This means that if a fluid is displaced by some forward actuation stroke, such as a moving piston in a micropump, and the reverse stroke is in exactly the opposite direction, then all fluid will return to its original position at the end of the cycle, hence there is zero net fluid motion. Therefore, one of the challenges in designing micropumps is ensuring flow asymmetry in order to produce a net

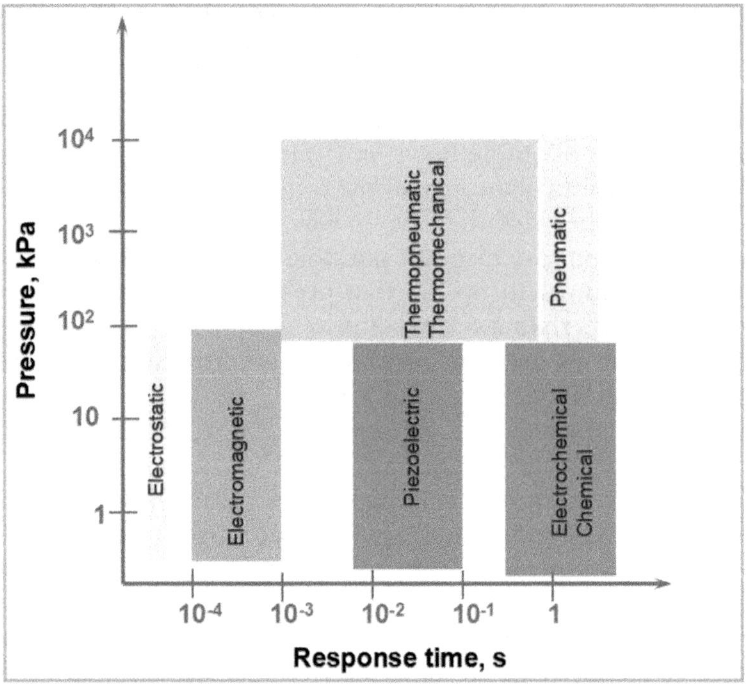

Figure 7.1 Categorization and operating windows of micropumps and microvalves.

pumping motion. The pressure profile of the Poiseuille flow can be modelled using the equations in Table 1.3.

Figure 7.1 shows the categorization of all the existing pumping methods based on the operational pressure and response time. For most applications, low-pressure liquid pumps are used. High pressures or high voltages are used only for molecular separations with high efficiency, such as in chromatography and electrophoresis.

7.2 Actuation Sources for Mechanical Micropumps

Mechanical actuation consists in coupling the mechanical deformation of a moving boundary to increase or decrease the fluid pressure to drive flow. Actuators can be classified as external or integrated, but all are based on the same working principle, to actuate a mechanical stroke cycle.

An example of a common positive displacement pump is a membrane pump, as shown in Figure 7.2a,[1,2] consisting of a closed fluidic pump chamber with a flexible diaphragm on the top side. The actuation stroke is when the diaphragm is depressed, hence the pressure increases and the volume decreases. To reach equilibrium, the volume then passes to the right of the chamber to reduce the pressure. Meanwhile, the passive stroke is when the diaphragm is raised, increasing the volume and decreasing the pressure, thus drawing in fluid from the left to reach an equilibrium state. This configuration uses valves at the inlet and outlet to stop any unwanted reverse flow. Figure 7.2b shows the pressure–volume cycle during the pump's operation between actuation and passive strokes. The ratio between the stroke-volume change and the initial volume of the chamber represents the compression ratio, CR:

$$CR = \frac{\Delta V}{V_0} \tag{7.1}$$

where V_0 is the initial volume and ΔV is the stroke-volume change caused by pumping.

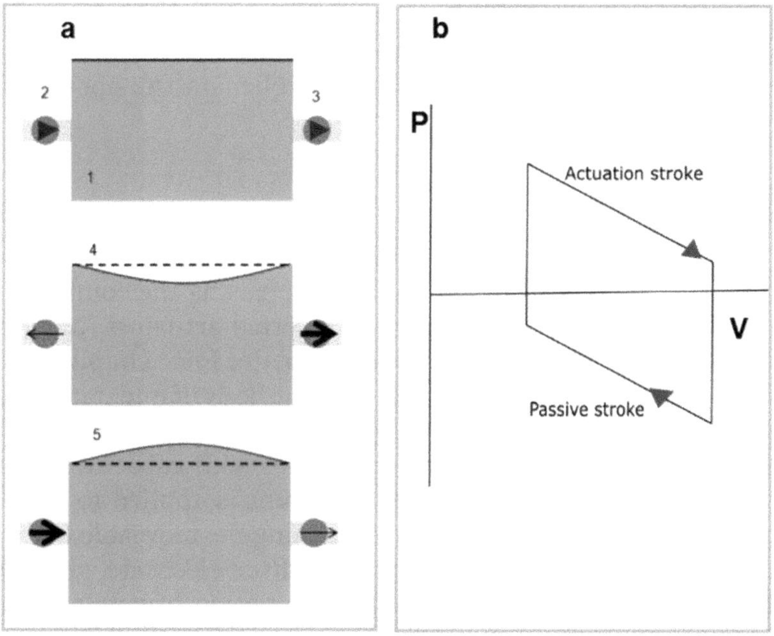

Figure 7.2 (a) 1, Pump chamber with diaphragm; 2, inlet valve; 3, outlet valve; 4, pump phase; 5, supply phases. (b) Pressure (*P*)—volume (*V*) diagram showing the strokes.

7.2.1 External Actuators

Actuators are required to move the diaphragm. External actuators are components that are independent of a microfluidic device but require coupling to the integrated micropump. External actuators can produce larger forces and stroke displacement but are normally larger than microfluidic components and increase the total size of the micropump. External actuators may utilize electromagnetic, piezoelectric or pneumatic actuation.

Most electromagnetic actuators work by energizing a solenoid coil, which produces a magnetic field gradient that produces a force on a ferromagnetic plunger.[3] The plunger then compresses a membrane but, upon de-energization of the coil, the membrane and plunger return to the resting position. The strokes can be controlled by tailoring the coil current and the number of wire turns. Piezoelectric actuators function by converting applied electronic currents to a mechanical translation, thus exerting a force upon a connected membrane, in a very similar cycle to that with the electromagnetic actuator.[4] Finally, pneumatic actuation uses a high-pressure gas or a vapour source coupled to the pump membrane. Gas or vapour can be forced into a pneumatic chamber, thus increasing the volume contained so that the chamber exerts pressure on the membrane. The response speeds of pneumatic systems are slower than those of the previous two because the system relies on the opening and closing of valves.

7.2.2 Integrated Actuators

Integrated actuators are micromachined as a pump component during the microfabrication process, therefore increasing the complexity of the process. However, compared with external actuators, integrated actuators offer faster response times and better force coupling to the pump at the cost of lower efficiency. Two methods of integrated actuation are called electrostatic and thermopneumatic actuation.[5] An electrostatic actuator consists of two electrodes with capacitive charge, one fixed and one moveable. Once a current is applied to the electrodes, an electric field is established, causing the moveable electrode to move closer to or further away from the fixed electrode, depending on the polarity.[6] The thermopneumatic actuation principle is based on the thermal expansion of a fluid or gas within a sealed actuation chamber. The fluid or gas expands during heating, applying pressure upon the membrane, which deforms and generates large forces and stroke volumes, but requires relatively large amounts of energy.[7] Micropumps with integrated actuators are positive displacement pumps.[8,9]

7.3 Fluid Rectification

Rectification is the act of converting an alternating flow into a one-directional flow. One method to force a one-way flow is to integrate check valves into the pump (Figure 7.3). Check valves have a smaller flow resistance in the forward flow direction than in the reverse direction, thus producing a net forward pumping flow.[10]

In supply mode, the membrane is displaced upwards and creates a negative pressure inside the pump chamber, which sucks fluid in. The direction of the inlet check valve allows the fluid to enter the chamber with little resistance while at the same time, at the exit check valve, the fluid must move against the check valve, where there is a greater resistance in this direction, thus restricting the re-entering flow (low leakage backflow). In pump mode, the membrane is depressed, creating a positive pressure inside the chamber, 'opening' the outlet check valve and 'closing' the inlet check valve. In one pump stroke, a fluid volume ΔV is displaced.

The fixed-geometry rectification micropumps are the simplest alternative to positive displacement pumps.[11] Check valves are easier to integrate into the fabrication at the inlet and outlet channels with variable cross-section. Flow rectification is established due to the valves' physical channel structure adding a geometric flow resistance asymmetry that rectifies the flow, typically by using the diffuser or nozzle configuration. Because of the reversible actuation stroke,

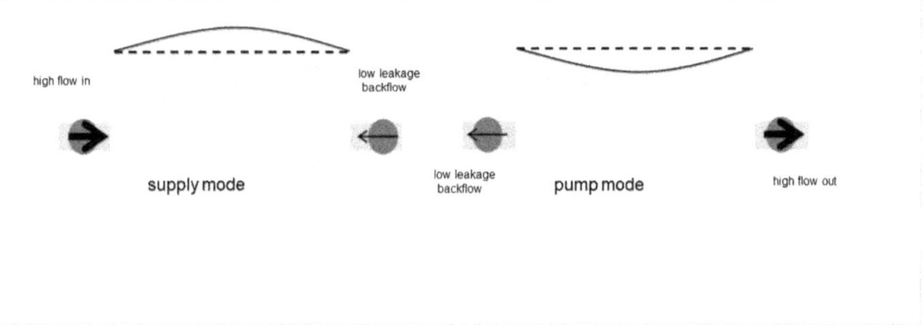

Figure 7.3 Working principle of positive displacement pumps integrating check valves at the inlet and outlet.

the fixed-geometry valve acts as both a diffuser and a nozzle element, depending on the flow direction (Figure 7.4a). The diverging/converging geometry changes the trend of the pressure and velocity inside the channels. During the positive stroke, the fluid is drawn through the expansion along the diffuser, causing the fluid to decelerate with increases in pressure; when the stroke is negative, the fluid flows through the nozzle, where the velocity increases and the pressure decreases (Figure 7.4b).

An analogous explanation can be applied to explain the pressure losses. Through the diffuser, during the positive stroke, the pressure losses are lower than those of the backflow, that is, the flow through the nozzle during the negative stroke (Figure 7.4c). A turbulent boundary layer between a central jet and the sidewall can be generated as a secondary effect of the differences mentioned above.

A limitation of fixed-valve micropumps is related to the strong inertia that exists in very small channels, which impacts significantly on the amount of leakage. The efficiency is measured by the ratio of the forward and backward pressures:

$$\eta = \frac{P_{\text{forward}}}{P_{\text{backward}}} \tag{7.2}$$

A characteristic application of the working principle of the fixed-geometry rectification micropump is using the Tesla valve, a bifurcating loop that connects two parts of a microchannel (Figure 8.5). In the

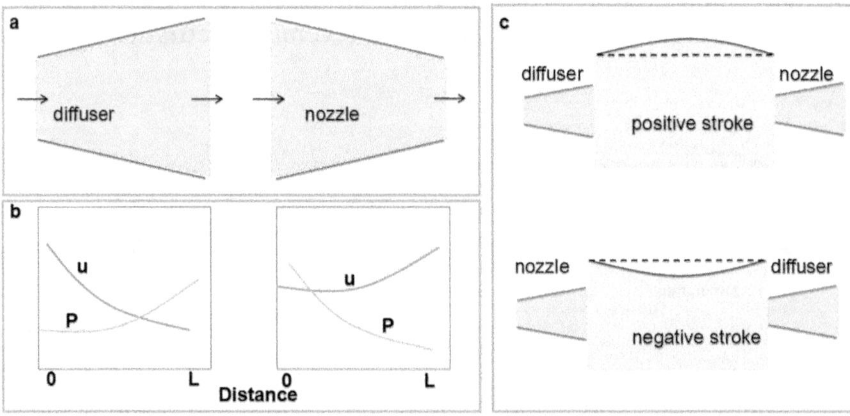

Figure 7.4 Fixed-geometry rectification micropump. (a) Diffuser/nozzle geometry. (b) Pressure and velocity trend according to the length of the asymmetric channel. (c) Fixed-geometry rectification micropumps.

forward direction, the fluid flows through the main microchannel. In the backward direction, the bifurcating geometry of the loop region causes the fluid to flow in the loop region instead of through the main channel. This extra flow through the loop region increases the pressure drop in the backflow mode, reducing the velocity.

7.4 Peristaltic Pumps

Peristaltic pumps are simple and the most widely used pumps available for microfluidic applications. External peristaltic pumps generate flow by compressing a flexible tube that is coiled around a rotary motor fitted with rollers. The rollers pinch the tube during rotation, causing the fluid within to be driven forwards. External peristaltic pumps are very easy to use, cheap, capable of high throughput and can pump large volumes very rapidly. These benefits make peristaltic pumps very appealing, with one common use being the circulation of cooling liquids. However, the flow generated pulsates due to the gradual compression and relaxation of the tubing. Therefore, the flow is not constant, which can cause detrimental shear effects and also an inability to produce complex uniform flow profiles, *e.g.* ramps and pulses. Last, the response times of external peristaltic pumps are slower and more difficult to determine than in the previously mentioned methods.

Integrated peristaltic pumps function by the peristaltic movement produced by a series of actuation membranes, which displace the volume in the desired flow direction. The membranes work *via* volume displacement with valving, so that the channel is opened or blocked to try to reach the ideal compression ratio of 1, where the complete volume of the chamber is transferred per stroke.[12] The simplest integrated peristaltic pump works by using at least three actuation membranes in series to obtain a non-reversible pump stroke. The flow direction is determined by the sequence of membrane actuation, either forwards, as shown in Figure 7.5, or backwards. At the end of this sequence, the net effect is that one stroke volume, ΔV, has been displaced through the device. The flow rate through this pump depends on the stroke volume and the frequency of the membrane sequence.

7.5 Non-mechanical Micropumps

Non-mechanical micropumps convert non-mechanical energy into kinetic energy of the fluid and can be based upon electrokinetic pumping, electrowetting and surface tension gradients.

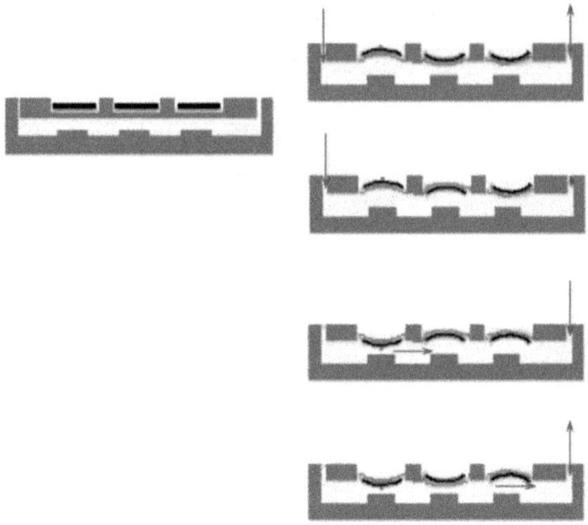

Figure 7.5 Four-phase actuation sequence of a peristaltic micropump. The membrane can be in one of two positions where the 0 position is the non-actuated (closed) membrane position and the 1 position is the open position. Initially, all three membranes are in the closed position (0,0,0). 1, One membrane rises, pulling a stroke volume, *V*, into the number one position (1,0,0). 2, The number two membrane is actuated to displace the *V* forwards one position and draw an additional *V* (1,1,0). 3, Membrane 3 is actuated, pulling one *V* into the outlet (0,1,1). At the same time, membrane 1 relaxes, causing backflow of one stroke volume (0,1,0). 4, Membrane 2 relaxes, displacing the fluid forwards (0,0,1), and finally membrane 3 is relaxed (0,0,0) and fluid is displaced forwards.

7.5.1 Electrokinetic Pumping

Electroosmotic flow (EOF) pumping is the most common pumping method that is based on the electrokinetic phenomena, as introduced in Chapter 6. Charge movement produces a shear on the surrounding fluid, resulting in bulk fluid flow. The main advantage is that EOF pumping can work in small channels without a need for high pressures.[13,14]

Most of the surfaces used for the fabrication of microfluidic devices can develop a charge. The surface of insulating materials such as glass or some plastics becomes negatively charged when in contact with an aqueous solution. For example, glass presents silanol groups (–Si–OH) at the surface that can be deprotonated ($–Si–O^-$), so it carries a negative charge. This is due to surface hydrogens being removed by the solution, the amount depending on the pH or the presence of electrolytes to also draw off hydrogen. The charged surface attracts the counter-charges from the electrolyte, which couple with the surface charges and form a strongly associated charge layer known as the Stern layer or fixed layer. The thickness of the Stern layer ranges between 1 μm in pure water down to 0.3 nm in 1 M NaOH solution. The Stern layer does not completely shield the surface charges, hence a weakly associated diffuse layer of counter-charges forms adjacent to the Stern layer; this second layer is called the diffuse layer. The Stern layer is relatively immobile and stable, whereas the charges in the diffuse layer are free to move (Figure 7.6a). The potential of the shear plane between the fixed Stern layer and the diffuse layer is called the

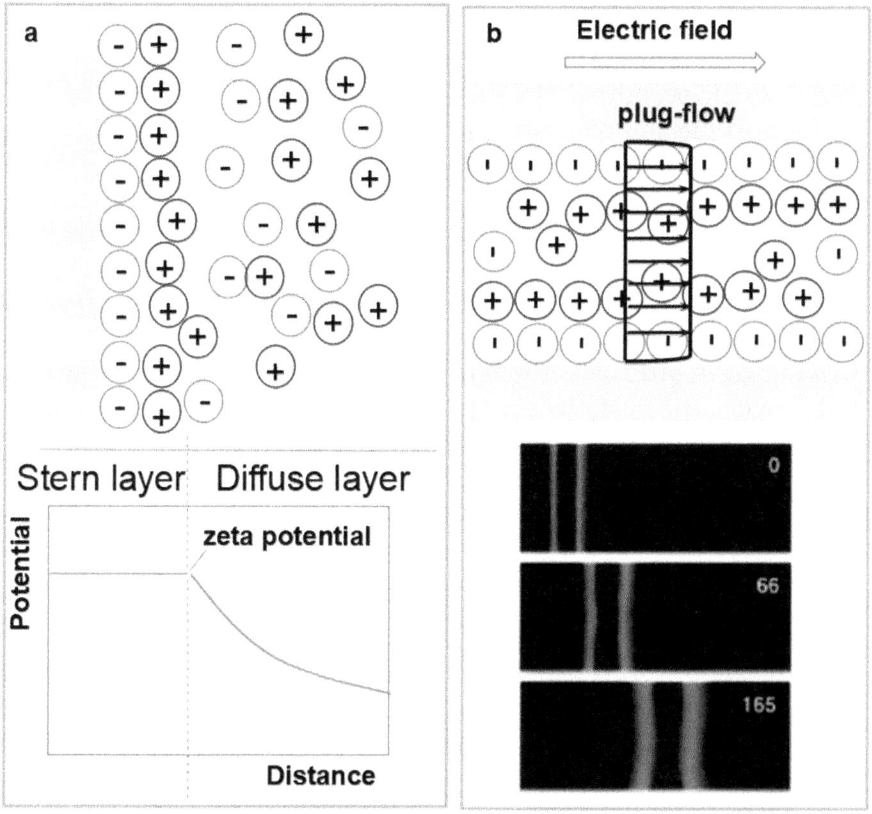

Figure 7.6 (a) Schematic of the double layer and trend of the potential according to the distance. (b) Plug-like flow in a microfluidic channel. Top: the charges in the diffuse layer and at the centre of the channel form the plug flow. Bottom: fluorescence micrograph of small volume elements in the electroosmotic plug flow from 0 to 165 s. Reproduced from ref. 15 with permission from American Chemical Society, Copyright 1998.

zeta potential (ζ). The combination of the two layers forms the electrical double layer (EDL). The arrangement of the charges in the EDL determines the electric potential. The total length of the surface electrostatic effect is called the Debye length. The EDL is modelled by a Boltzmann distribution, where the potential close to the wall, φ, is approximately given by

$$\varphi = \varphi_w \left[1 - \exp\left(\frac{-\xi}{\lambda_D} \right) \right] \tag{7.3}$$

where φ_w is the wall potential, ξ is the coordinate perpendicular to the wall and λ_D is the electrical double layer thickness (Debye length) given by

$$\lambda_D = \sqrt{\frac{\varepsilon kT}{2F^2 z^2 c}} \tag{7.4}$$

where ε is the permittivity of the medium, k is Boltzmann's constant, T is the absolute temperature, F is the Faraday constant, z is the charge number and c is the concentration of the species.

Because the microfluidic channel is made with insulating materials, a tangential electric field may be applied down the length of the channel to produce an electroosmotic flow as the positively charged ions move towards the negatively charged cathode and *vice versa*. In the bulk region of the microfluidic channel, the charge neutrality is maintained in order that there is no net charge movement. However, under an electric field, the positive ions in the diffuse layer move more because the surface charges allow rapid 'hopping' within the electric field. This causes the ions to generate a shear force upon the fluid (Figure 7.6b).[15] In small microfluidics channels, these surface forces are far more dominant than in the bulk, hence the observed flow.

In the absence of a pressure gradient, the flow equation determines that the Coulomb force and the viscous dissipation are balanced and that the Coulomb force can be expressed by the following equation:

$$\mu \nabla^2 u = \rho E = \varepsilon E \nabla^2 \varphi \tag{7.5}$$

where E is the electric field and u the velocity. Combining eqn (7.3) and (7.5), the solution of the differential equation provides the expression for the velocity profile:

$$u = \frac{\zeta \lambda_D \varepsilon E}{\mu} \left[1 - \exp\left(\frac{-\xi}{\lambda_D} \right) \right] \tag{7.6}$$

For situations where a small Debye length is present, the flow determines the plug shape. It is also possible to define the electroosmotic mobility:

$$\mu_{eo} = \frac{\zeta\varepsilon}{4\pi\mu} \tag{7.7}$$

where ε is the dielectric constant and μ the viscosity, and the profile of velocity is then expressed by

$$u = \mu_{eo}E \tag{7.8}$$

EOF-driven flow has the advantage of driving the flow with a small surface force. This results in the flow presenting a flat solvent front, with slanted edges due to the surface drag. Flat fronts have much higher resolution movement than with pressure-driven flows that form parabolic solvents fronts. This allows for more accurate flow programming with less time needed to compensate for the uneven distribution. The main disadvantage is the intricate dependence on the solution pH, dielectric constants and constituent concentrations, where every change influences the zeta potential and thus the flow.

7.5.2 Electrowetting

As emphasized above, surface forces are very important at the microscale and the surface tension can be used as an effective actuator for micropump applications.

Surface tension-driven pumps actuate along the fluid–fluid interface, generating a movement that produces a shear driving force, driving the flow. An application is provided by electrowetting.[16] From the thermodynamic perspective, electrowetting relies on the change in the liquid–solid surface tension of an electrode surface, by charging the electrode, and hence the electrical double layer. The surface tension at the liquid–solid interface is given by the Gibbs free energy including the chemical and electric charge terms. The variation of the surface tension can be related to the capacitance and voltage of the electrode by Lippmann's equation:

$$\gamma = \gamma_0 - \frac{1}{2}CV^2 \tag{7.9}$$

where γ_0 is the surface tension when no potential is applied and C is the capacitance per unit area charged to voltage V. The water contact angle may model as a function of the bias using Young's equation [eqn (7.10)] and Lippmann–Young's equation [eqn (7.11)]:

$$\lambda_{SA} = \gamma_{SG} - \gamma_{AG}\cos\theta \tag{7.10}$$

$$\cos\theta = \cos\theta_0 + \frac{CV^2}{2\gamma_{AG}} \tag{7.11}$$

where γ_{SA} is the surface tension between the solid and aqueous phases, γ_{SG} is the surface tension between the solid and gas phases, γ_{AG} is the tension between the aqueous and gas phases, θ is the contact angle and θ_0 is the contact angle with no bias voltage. The pressure generated by the change in the surface tension in a channel of diameter d is given by the following equation:

$$\Delta P = \frac{\gamma_{AG}}{d(\cos\theta - \cos\theta_0)} \tag{7.12}$$

and inserting eqn (7.10) and (7.11) in eqn (7.12), the final expression

$$\Delta P = \frac{\varepsilon\varepsilon_r V^2}{2\lambda d} \tag{7.13}$$

is obtained, where ε is the aqueous dielectric constant of the insulator, ε_r is the relative permittivity, λ is the thickness of the dielectric and d is the thickness of the channel. Hence, as the electrode is biased, a double layer initially charges at the gas–liquid interface. This provides a force so that the aqueous layer will displace the gaseous phase to move to the surface of the electrode to reduce the increase in surface tension.

7.5.3 Marangoni Pumps

The surface tension can still be used for moving fluids by other means, such as imposing temperature or concentration gradients to generate a non-uniform gradient of surface tension at a liquid–gas interface. This is called the Marangoni effect. A gradient of surface tension at the interface generates a force parallel to the interface and directly towards the larger surface tension, which produces a shear force, pumping the fluid because of the momentum transport.[17]

In Figure 7.7, temperature is used as the propelling force by generating gradients of surface tension, as increasing the temperature causes the surface tension to decrease.[18] Three heaters are used to produce the gradient of temperature: the two lateral ones for heating

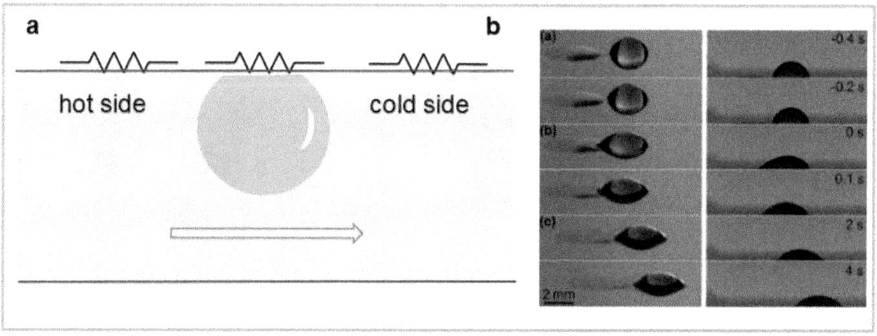

Figure 7.7 (a) Schematic of a Marangoni micropump. (b) Droplets moving by the thermocapillary effect. Reproduced from ref. 18 with permission from AIP Publishing, Copyright 2011.

and cooling the fluid and a central one. The central heater by resistive dissipation produces a vapour bubble, with a gradient of temperature around it. The droplet flows by the thermocapillary effect from the hot to the cold side.

7.6 Microvalves

Microvalves allow the fluid to pass in one direction and suppress it in the other. Microvalves can be classified as passive or active (*e.g.* check valves) if they need activation.

The requirements for a good valve include zero leakage and dead volume, zero response time and compatibility with fluids. These requirements are ambitious and it is difficult to satisfy all of them.

7.6.1 Passive Valves

The capillary burst microvalve (Figure 7.8a) exploits the high-energy air–liquid interface for opening/closing the channels. The working principle is based on the flow of fluid in a capillary channel being halted at the valve by the air–liquid interface surface tension. Microvalves consist of very small channels where the fluidic resistance is so high, it blocks the flow until a high pressure is applied; the required pressure can be enhanced by hydrophobic channel coatings. Once overcome, flow continues through the microvalve. However, once the channel is wetted, the fluid can flow to the next section without control.[19,20]

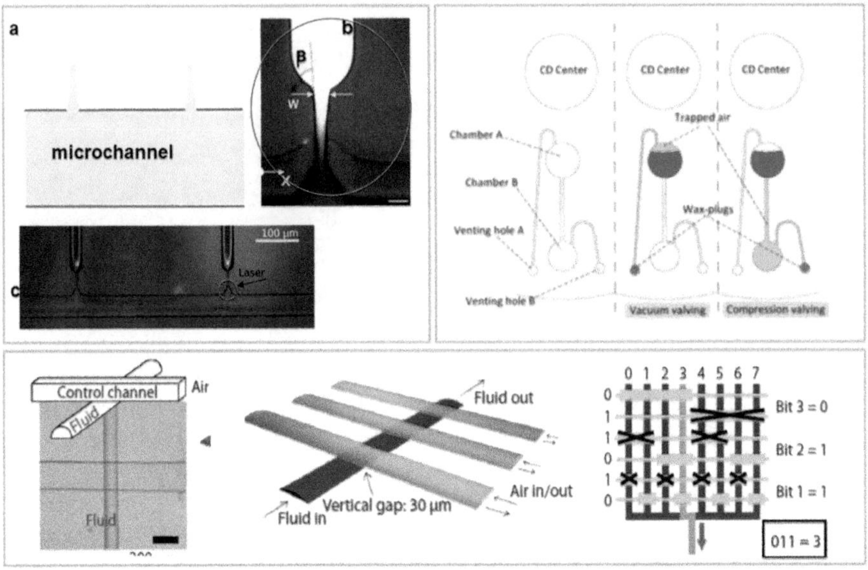

Figure 7.8 Passive valves. (a) Burst capillary valve; (b) centrifugal valve; (c) pinch valve: from left to right, schematic of the pinch between the control and the fluid channel, working principle and array design. (a) Reproduced from ref. 19 with permission from AIP Publishing, Copyright 2017.

Another example is the centrifugal microvalve (Figure 7.8b).[21] The fluid moves due to the rotation of the microfluidic platform, thus by centrifugal forces. Similarly to the previous example, the fluid passes through channel constrictions; the constrictions essentially act as centrifugal microvalves that control the flow into the micrometric chamber, while the CD-like platform rotates.[22] The required unconventional spinning system represents the main drawback of this valve, as it is more difficult to integrate into existing set-ups.

Another interesting example of passive micrometric valves is the pinch valve, which stops the fluid flow by creating a blockage inside the channel by applying a mechanical force; the removal of the force guarantees that the fluid flows again along the channel. This application requires that the channel is deformable, hence the common use in elastomeric chips. A typical application is shown in Figure 7.8c,[23,24] where the channel is pinched by a second air- or water-filled channel that crosses the main channel. The pinching channel is located on top of the fluid-carrying channel and separated by a thin gap. The main advantage of pinch valves is that many valves can operate simultaneously by pressurizing the control channels in parallel or in sequence (or both). Fabrication poses an issue for the pinch valve, because the main fluid microchannel roof has to have a rounded cross-section, otherwise the valve does not seal well when the channel is pinched. Depending on the size, the valves can operate at frequencies up to hundreds of hertz.

The main channel and the control channel need a significant overlap for optimum performance. The control channel must be able to create a complete stroke along the fluid channel and for this reason a narrow control channel might not be suitable. On the other hand, wide control channels can block more of the fluid channel, permitting the microfluidic multiplexer. Metering constitutes one of the main limitations of such valves, as the volume between two valves cannot easily be determined, and this represents a problem for those applications where the exact volume value is required, *e.g.* quantitative (bio)chemical assays. Despite some limitations also due to the methods applicable for the samples flowing inside the channel, these micrometric valves revolutionized the field of microfluidics.

7.6.2 Active Valves

Active valves require an external actuator. The valves can be designed as normally open or normally closed according to the rest position. The actuation sources are the same as we have discussed for the micropumps, but are selected depending on the pressure limits, the way in which the stroke is generated, the response time and the reliability.

The earliest implementation of valving in a microfluidic channel was the electrokinetic valve. The fundamental concept is based on electroosmosis. Figure 7.9a shows a typical application with a T-type channel intersection. By applying a voltage between the two reservoirs R1 and R2, the liquid flows in the same direction as the current flow. Two mechanisms occur when the current between the two reservoirs R1 and R2 is maintained and the flow to the side channel is hindered (*e.g.* hydrophobic coating of the channel or filling it with a nanoporous material). First, the fluid still wants to flow from the first to the second reservoir following the current flow because of the applied voltage. Second, the impossibility of moving towards the reservoir R2 builds up the pressure at the intersection, which produces a flow of the fluid in the direction with the lowest resistance towards R3; thus the fluid divides into one part moving to the field-free region by hydrodynamic pumping and a second part following the initial part. The ratio of the

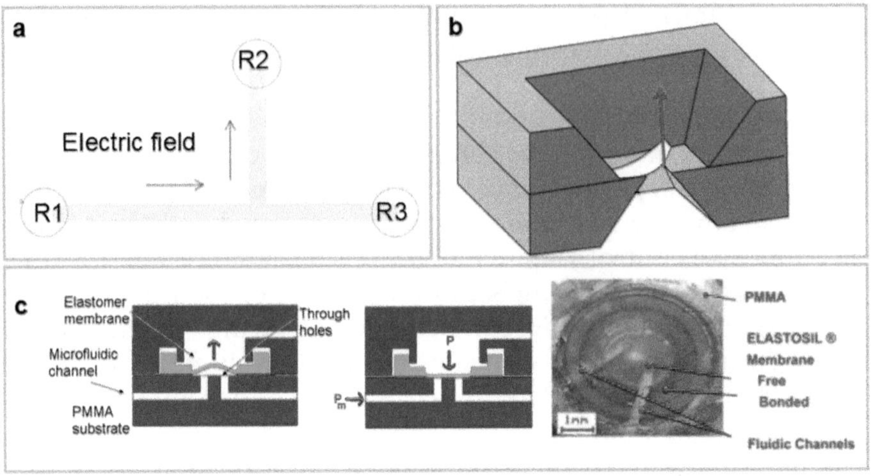

Figure 7.9 Active valves. (a) Electroosmotic valve with hydrodynamic flow; (b) cantilever valve; (c) schematic of a monolithic doormat valve and real-world application. Reproduced from ref. 29 with permission from Springer Nature, Copyright 2009.

two flows depends on the ratio of the flow resistances. The electrokinetic valve is widely used in electrophoresis but has some limitations due to a strong dependence on the material used for the channels and the properties of the buffers (*e.g.* pH, concentration). Further, the high voltage necessary for operation makes the method expensive, unsafe and not suitable for biological applications. The possibility of coupling the electroosmotic pump with hydrodynamic pumping can extend the use of the electroosmotic pump to a wider set of applications, including biological applications, where a high voltage can be lethal for the samples.[25]

Cantilever valves are the simplest example of active valves. A strip opens and closes the channel by the effect of an applied pressure (Figure 7.9b); although the response time is fast, leakages are common because of the difficulty in pushing the strip to adhere reversibly to the channel surface. The silicon cantilever valves integrated in silicon micropumps need to be actuated at frequencies higher than the resonance frequency for operation.[26] In this case, the valve can move water at flow rates between 50 and 2000 μL min^{-1} and a forward and backward pressure of a few kPa. Various static and dynamic silicon flaps have been investigated.[27,28]

Check valves are mechanical devices that allow fluid to flow in only one direction. These consist of a flap impeding the reverse flow by forming a seal with the channel. An increase in the back-pressure improves the sealing, and for reopening the valve the downstream pressure must be made lower than the upstream pressure. Check valves can be integrated with fixed-volume pumps and allow the rectification of the fluid. Important limitations are the difficulty of metering microlitre volumes and the ease of contamination.

Based on the same principle, active valves can also be made with materials that are sensitive to stimuli from the external environment (such as gels) and that can shrink (open the channel) or swell (close the channel) in response to changes.

A sandwich-like elastomeric valve, which consists of an elastomeric membrane pad sandwiched between the control channel and the fluidic channel (Figure 7.9c),[29] has also been implemented. This particular valve is called a doormat valve. By applying a negative pressure to the control channel, the membrane deflects downwards, opening the valve and allowing the two sides of the microchannel to communicate under the wall. Application of an overpressure to the control line deflects the membrane upwards and the channel is closed. The fabrication of the doormat valves is relatively simple and the metering is limited simply by the resolution of the fabrication process of the monolithic architecture.

References

1. G. T. A. Kovacs, *Micromachined Transducers Sourcebook*, McGraw Hill, New York, 1998.
2. W. L. McCabe, J. C. Smith and P. Harriot, *Unit Operations of Chemical Engineering*, McGraw - Hill, New York, 7th edn, 2005.
3. M. T. Ke, J. H. Zhong and C. Y. Lee, *Sensors*, 2012, **12**(10), 13075.
4. H. T. G. van Lintel and F. C. M. Van De Pol, *Sens. Actuators, A*, 1989, **15**, 153.
5. K. Abi-Samra, L. Clime, L. Kong, R. Gorkin, T. H. Kim, Y. K. Cho and M. Madou, *Microfluid. Nanofluid.*, 2011, **11**(5), 643.
6. L. Zheng, Y. Wu, X. Chen, A. Yu, L. Xu, Y. Liu, H. Li and Z. L. Wang, *Adv. Funct. Mater.*, 2017, **27**, 1606408.
7. B. Yang, B. Wang and W. K. Schomburg, *J. Micromech. Microeng.*, 2010, **20**, 095024.
8. A. Terray, J. Oakey and D. W. Marr, *Science*, 2002, **296**(5574), 1841.
9. D. L. Greisch, J. B. Chemelli, inventors; Eastman Kodak Co, assignee, Micro dispensing positive displacement pump, *US Pat.*, US 5,593,290, 1997.
10. A. A. Deshmukh, D. Liepmann and A. P. Pisano, Characterization of a micromixing, pumping and valving system, in *Transducers' 01 Eurosensors XV*, Springer, Berlin, Heidelberg, 2001, pp. 922–925.
11. J. Loverich, I. Kanno and H. Kotera, *Microfluid. Nanofluid.*, 2007, **3**(4), 427.
12. J. Xie, J. Shih, Q. Lin, B. Yang and Y. C. Tai, *Lab-on-a-Chip*, 2004, **4**(5), 495.
13. A. Manz, C. S. Effenhauser, N. Burggraf, D. J. Harrison, K. Seiler and K. Fluri, *J. Micromech. Microeng.*, 1994, **4**(4), 257.
14. X. Wang, C. Cheng, S. Wang and S. Liu, *Microfluid. Nanofluid.*, 2009, **6**(2), 145.
15. P. H. Paul, M. G. Garguilo and D. J. Rakestraw, *Anal. Chem.*, 1998, **70**(13), 2459.

16. M. G. Pollack, R. B. Fair and A. D. Shenderov, *Appl. Phys. Lett.*, 2000, 77(11), 1725.
17. A. S. Basu and Y. B. Gianchandani, *J. Micromech. Microeng.*, 2008, **18**(11), 115031.
18. Y. Zhao, F. Liu and C. H. Chen, *Appl. Phys. Lett.*, 2011, **99**(10), 104101.
19. J. Eriksen, B. Bilenberg, A. Kristensen and R. Marie, *Rev. Sci. Instrum.*, 2017, **88**(4), 045101.
20. H. Cho, H. Y. Kim, J. Y. Kang and T. S. Kim, *J. Colloid Interface Sci.*, 2007, **306**(2), 379.
21. R. Gorkin, J. Park, J. Siegrist, M. Amasia, B. S. Lee, J. M. Park, J. Kim, H. Kim, M. Madou and Y. K. Cho, *Lab Chip*, 2010, **10**(14), 1758.
22. J. Ducrée, S. Haeberle, S. Lutz, S. Pausch, F. Von Stetten and R. Zengerle, *J. Micromech. Microeng.*, 2007, **17**(7), S103.
23. M. A. Unger, H. P. Chou, T. Thorsen, A. Scherer and S. R. Quake, *Science*, 2000, **288**(5463), 113.
24. A. K. Au, H. Lai, B. R. Utela and A. Folch, *Micromachines*, 2011, **2**(2), 179.
25. T. E. McKnight, C. T. Culbertson, S. C. Jacobson and J. M. Ramsey, *Anal. Chem.*, 2001, **73**(16), 4045.
26. G. H. Feng and E. S. Kim, *J. Micromech. Microeng.*, 2004, **14**(4), 429.
27. E. H. Yang, S. W. Han and S. S. Yang, *Sens. Actuators, A*, 1996, **57**, 75.
28. L. Smith and B. Hok, *Proceedings of the 6th International Conference Solid-State Sensors and Actuators (Transducers '91)*, San Fransisco, CA, USA, 1991, p. 1049.
29. G. Simone, G. Perozziello, G. Sardella, I. Disegna, S. Tori, N. Manaresi and G. Medoro, *Microsyst. Technol.*, 2010, **16**(7), 1269.

8 Mixing

8.1 Introduction

Mixing represents one of the fundamental standard operations in microfluidics with the aim of homogenizing different fluids. Typical applications that require mixing are related to chemical and biochemical reactions and biological manipulations.[1–3] Each application requires a specific characteristic time for mixing and reactions. However, the passive mechanism of mixing in microfluidics depends solely on diffusion, owing to the scale of the velocities of the fluids and the size of the channels. The theory of mixing by diffusion was presented in Chapter 1. Mixing by diffusion is acceptable when the characteristic channel dimension of the reactants is of the order of few micrometres and, above all, if the flow rate is slow enough to allow the reaction to take place within the length of the channel (see Figure 8.1).[4,5] For this reason, biochemical assays or chemical reactions that involve a short characteristic reaction time can be performed in microfluidics. Several strategies have been applied to speed up the mixing, either by reducing the diffusion distances, triggering the movement of the fluid in directions different from the flow, or by integrating active mixers to use external energy for inducing turbulence.

Passive mixers are simple but the mixing efficiency is strongly coupled to flow rate and geometry.[6] Moreover, passive mixers are typically suitable for low-viscosity fluids containing diffusive species, such as colloidal particles in water. Elastic instabilities have been

Microfluidics and Lab-on-a-chip
By Andreas Manz, Pavel Neužil, Jonathan S. O'Connor and Giuseppina Simone

Published by the Royal Society of Chemistry, www.rsc.org

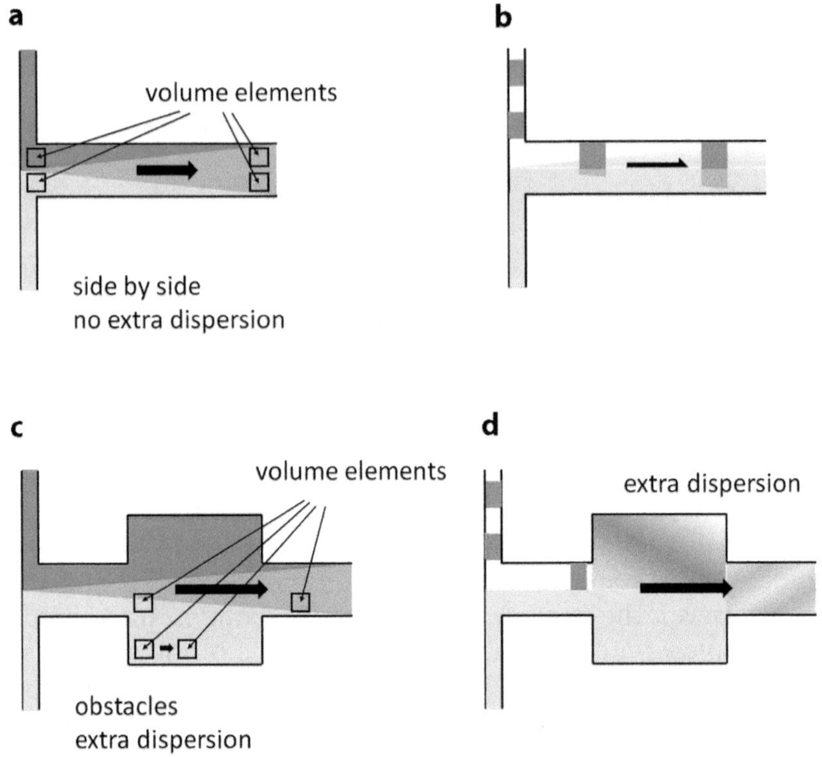

Figure 8.1 Schematic of laminar flow mixing. (a) Side-by-side, mixing only by diffusion, no extra dispersion; (b) sequential sample injection is possible in this case; (c) introducing 'obstacles' may increase the speed of mixing, as diffusion and flow disturbance are superimposed, extra dispersion in the direction of flow; (d) sequential sample handling is not possible in this case.

shown to drive the mixing of weakly viscoelastic polymer solutions in microfluidic devices. There is growing interest in the continuous mixing of strongly viscoelastic materials, *i.e.* yield stress fluids, in microchannels, which until now has only been demonstrated at the macroscale.[7]

8.2 Mixing by Diffusion and Passive Mixing Methods

In Chapter 1, we observed that the statistical movement of a single molecule in a fluid can be described as a random walk. The movement is characterized by the 1-dimensional Einstein–Smoluchowski equation:

$$x = \sqrt{2Dt} \tag{8.1}$$

which expresses the relationship between the diffusion distance, x, the diffusion constant, D, and time, t, so that as time progresses the average diffusion distance changes with the square root of time; alternatively, if a given diffusion distance halves, the diffusion time is reduced by a factor of 4; if the distance is reduced 10-fold, the time is reduced 100-fold. Simply put, reducing the channel width or increasing the depth, hence increasing the contact area, reduces the time for complete diffusion. However, there are some operational problems in applying one of the two solutions suggested above. With narrowing of the channels, the fluidic resistance increases and channels of high aspect ratio need special machining technology for fabrication. Figure 8.1a shows a side-by-side example, illustrating that the two miscible fluids will mix downstream after a given length. In this case, there is no extra dispersion except the one by the flow itself. Assuming plug flow, as is the case with electroosmotic flow, two given volume elements will not change position with respect to each other. If samples are injected in sequence, they will retain that sequence (see Figure 8.1b).

To speed up mixing, several strategies have been pursued, such as generating vortices and turbulence in the main flow, disturbing the streamlines of the laminar flow. Figure 8.1c shows such a case schematically, where an 'obstacle' is introduced to disturb the flow. In this case, there is extra dispersion, which means that two given sample volumes do not maintain their relative position and sequentially injected samples will also mix with each other (Figure 8.1d).

One strategy is to split the main channel into an array of smaller channels, circumventing the high resistance of just one small channel while increasing the contact area of the different fluids (Figure 8.2).[8] In this configuration, the small channels merge again once complete mixing has occurred. If the two fluids are labelled with different colours, the mixing can be easily evaluated by using the colour as an indicator (usually the lighter colour in use) of the extent of mixing. Normally, for this configuration, the mixing at the walls is slower, because it represents the greatest diffusion distance required.

The role of the dimensionless Péclet number was also discussed in Chapter 1. This provides a measure of the diffusive mixing, which, given the correct geometry and the fluid dynamics, could be sufficient for efficient and complete mixing. A high value of the Péclet number indicates that the diffusion mechanism is insufficient for complete mixing and the design of the microchannel needs to be modified to increase the diffusion velocity or triggering chaotic advection and Dean vortexes. Another typical configuration consists of grooves with a certain depth fabricated into the bottom of the channel. For pressure-driven flow, the grooves introduce anisotropic flow resistances into an otherwise isotropic system. The resistance applied to the fluid is less when it flows along the ridges and valleys than when flowing perpendicular to them. The difference in the resistance induces corkscrew-like fluidic streamlines and consequent enhanced mixing of the fluids. An arrangement of different types of grooves can produce chaotic stirring. These are fundamental ideas that have been implemented in designs with the purpose of developing effective passive mixers.

8.2.1 Lamination and Intersecting Channels

The lamination flow configuration can be realized by different types of feed arrangements; examples are shown in Figures 8.2 and 8.3a.[9] Such a laminated feed stream passes into an inverse bifurcation structure and into a subsequent folded delay-loop channel where mixing

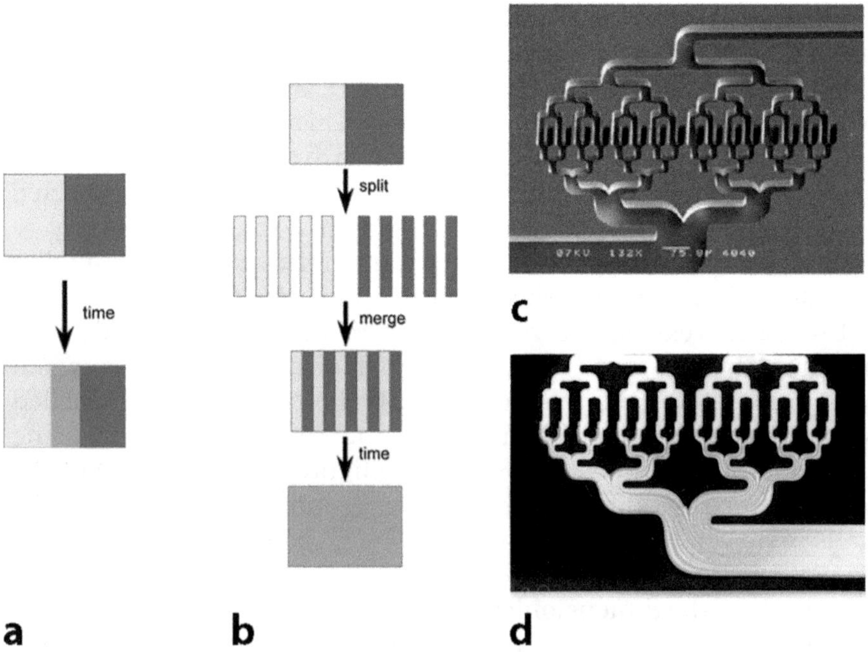

Figure 8.2 Principle of lamination. (a) A single pair of streams start mixing over time at the interface; (b) schematic of the split and merge; (c) and (d) show a silicon microchip with fluid inlets on two levels (of which the lower one is not visible) and outlet on one level.

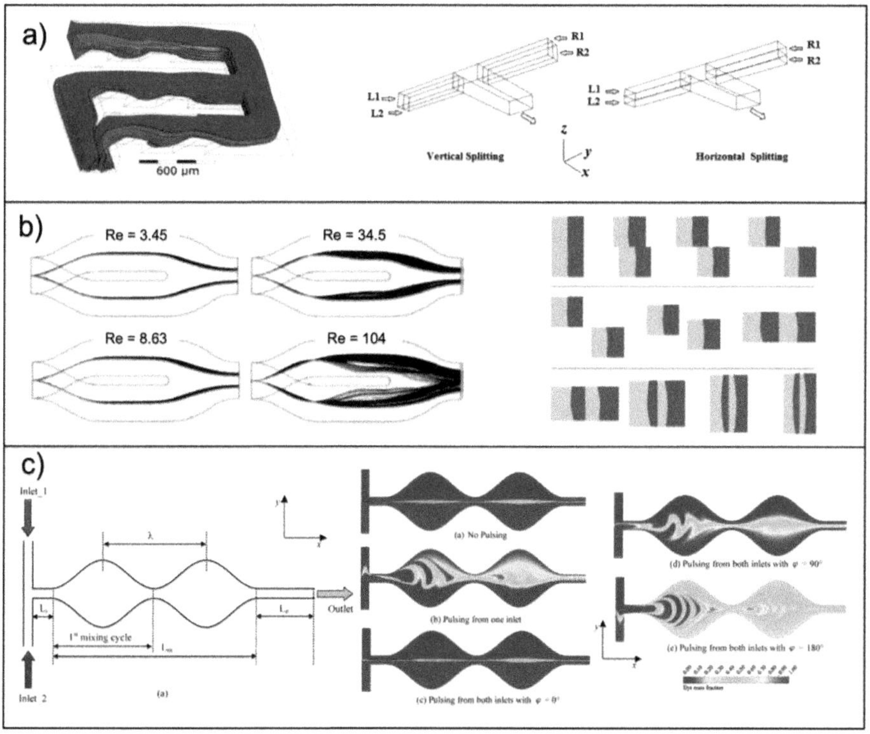

Figure 8.3 Passive mixing with a modified mechanism of diffusion. (a) Lamination and intersecting channels; (b) split-and-recombine; (c) convergent-divergent. (a) Reproduced from ref. 9a with permission from Springer Nature, Copyright 2010 and ref. 9b with permission from John Wiley and Sons, Copyright 2012.

takes place. The basic concept is that the flow in each channel can be rotated through a further angle before being recombined, with serial recombination and splitting. However, the simplest lamination does not include the split/recombination, but it is characterized by two perpendicular channels that are connected and the main mechanism of transport is diffusion (see also Section 6.2 in Chapter 6). The advantages of such simple designs are the reduction of the diffusive path, by using thinner channels, and the change in flow direction by 90°, with the transposing fluids, to generate pressure against one another to aid the diffusion, with a significant contribution of Taylor dispersion. This configuration permits the doubling of the interfacial contact area between two fluids. The benefit of geometric focusing can be easily seen by looking at the Fourier number [eqn (1.25)], which in the case of a straight rectangular channel comprising two fluid lamellae is given by

$$\mathrm{Fo} = 4\left(\frac{hL}{w}\right)\left(\frac{D}{Q}\right) \tag{8.2}$$

where h, L and w denote the geometric characteristics of the channel height, length and width, respectively, D is the diffusion coefficient and Q is the volumetric flow rate. A decrease in the channel width leads to an increase in Fo, implying advanced diffusion. Then, the diffusion distances are reduced by compressing the fluid lamellae to a few micrometres, corresponding to liquid mixing in the milliseconds range. Here, it is important to highlight the effect of the value of the diffusion coefficient D on the efficiency of the mechanism of mixing. It has been shown that under low Reynolds number conditions, water–ethanol mixing is superior to water–water mixing, because the diffusion coefficient of ethanol in water is higher than that of water in water. Given this condition, diffusion is an effective driving force for mass transfer at the interface.

T-configuration mixers can be combined with splitting channel mixers to further improve the mixing for lamination by the formation of vortex pairs.

Split-and-recombine mixers, also named intersecting channels, are used to create sequentially multi-laminating patterns, different from the parallel approach of the interdigital feeds. There are three essential requirements, splitting, rearrangement and recombination. Side intersecting channels are used to split, rearrange and combine component streams to enhance mixing, by creating a bimodal width distribution. The better efficiency of the mixing is the consequence of the combined effect of unbalanced collisions of the fluid streams and the formation of Dean vortices. In a typical design, the channels running parallel to the flow direction have a width of a few micrometres, whereas those intersecting the parallel channel network have a width five times that of the narrow channels and most importantly they are aligned at a bent angle (*e.g.* 45°) (Figure 8.3b). The main and the bent streams are then conveyed in a wider channel for mixing. Better mixing can be achieved by changing the cross-section of the microchannel where plugs made of different fluids pass.

8.2.2 Convergent–Divergent Channels

Split-and-recombine mixers can also consist of a special configuration called convergent–divergent channels. The convergent–divergent geometry uses a main channel that is separated into two sub-channels and then recombined at periodic intervals along the microchannel length (Figure 8.3c).[10,11] Two discrete liquid samples are fed into the mixing chamber laminarly at a constant flow rate. This results in perfectly laminar parallel flow and very little mixing. However, as soon as the flow from one inlet is pulsed, the pressure also pulses and the asymmetry distorts the laminar interface. Consequently, the constantly changing flow lines within the liquid plug cause a self-folding effect within the plug, yielding a significant improvement in mixing performance. When using such a design, the performance of mixing is enhanced even at low Reynolds numbers. By increasing the Reynolds number, the mixing performances improve owing to the greater centrifugal forces, inducing Dean-like vortex structures (which is explained in Section 8.2.3). The mixing performance is further enhanced by the interaction between the Dean vortex structures and newly formed separation vortices in the divergent areas of the microchannel. The decrease in the aspect ratio improves the mixing index and reduces the pressure drop due to the greater centrifugal force exerted in the central area of the channel.

8.2.3 Chaotic Mixing

8.2.3.1 Serpentine Channels

Serpentine channels are a popular design strategy used in passive micromixers. A channel with repeated curved sections and turns generates cross-sectional flows and induces chaotic advection capable of enhancing the mixing (Figure 8.4a).[12]

The geometry of the channel forces the fluid into a curved trajectory, thus experiencing centrifugal forces. The flow field that develops inside such a system is consistent with the formation of transversal vortices, also known as Dean flows. The important dimensionless group for flow description in the curved-channel mixer is the Dean number, De:

$$\mathrm{De} = \mathrm{Re}\sqrt{\frac{d}{R}} \tag{8.3}$$

where Re is the Reynolds number, d is the channel diameter and R is the radius of the curvature of the channel. By altering d and R, it is possible to tune De and the mixing efficiency. The serpentine channel in Figure 8.4a demonstrates the development of four vortices, inducing large internal interfacial areas, for De > 140. When the fluid velocity is high, the mixing in simple serpentine/spiral designs is rapid and the formation of secondary Dean vortices has been observed, leading to a transition to a chaotic advection regime. This regime corresponds to Reynolds numbers that are too high in practical microfluidic settings.

Different clamping angles can be used for connecting or combining channels, hence there is a large degree of freedom for designs. During the split-and-recombine mixing of viscoelastic fluids, the act of mixing takes place by very small fluctuations in the flow streamlines that are not apparent in the microstructure. These provide a geometrically simple way of promoting advective transport in microchannels using serpentine- or spiral-shaped designs. The smooth and non-structured channel surfaces lack any high local shear, which is attractive for biological applications. Such work is targeted at handling biomolecules and cells, which can be easily damaged or ruptured by shear forces. Additionally, serpentine channels are easier to fabricate than 3D surface structures, such as in Figure 8.4a.

8.2.3.2 Three-dimensional Serpentine Structures

Three-dimensional serpentine micromixers have been designed and fabricated to induce a chaotic mixing effect. A typical configuration is shown in Figure 8.4b; the mixer consists of an interconnected

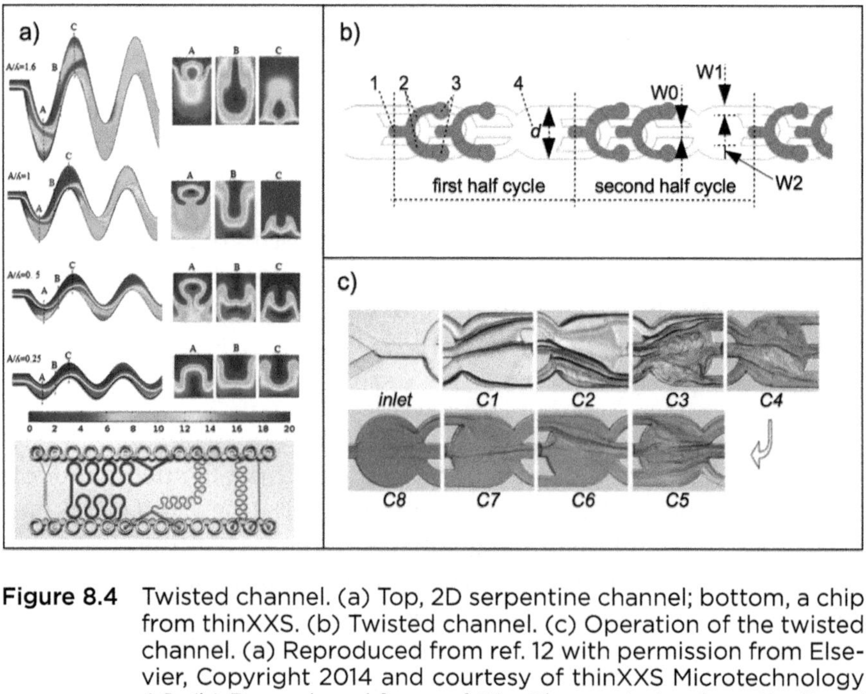

Figure 8.4 Twisted channel. (a) Top, 2D serpentine channel; bottom, a chip from thinXXS. (b) Twisted channel. (c) Operation of the twisted channel. (a) Reproduced from ref. 12 with permission from Elsevier, Copyright 2014 and courtesy of thinXXS Microtechnology AG. (b) Reproduced from ref. 13 with permission from the Royal Society of Chemistry.

multi-channel network through which the bulk fluid volumes are divided into smaller ones and chaotically reorganized.

Characteristic of these serpentine mixers, the mixing effect is enhanced as Re increases. The earliest implementation was limited to using silicon micromachining and only overlapping U-shaped turns were possible. Subsequently, with the advent of more sophisticated methods of microfabrication, it has been possible to investigate C and L turns or sigmoid arrangements, and in fact small changes of these geometries can generate a significant change in the efficiency of mixing. The general system works by injecting multiple fluid streams into an expansion chamber where viscous flow instabilities are triggered. As shown in Figure 8.4c, the flow and mixing process can be divided into three main stages. The first stage is injection into chamber 1 (C1) and flow into C2. The second stage is turbulence generation and mixing within C3 and C4 and the third stage is laminar flow restoration within C5 and C6.[13] From the inlet to C2, the flow remains fairly stable. Here, the less viscous liquid is confined by the more viscous liquid and hence thin fluid streams are formed. Flow instability is initiated at the bottom of C2, causing the flow to transition into an unstable state, which becomes stronger as the flow moves downstream. The flow fully develops turbulence in C3. The turbulent fluid motion between C2 and C3 nonetheless results in a significantly improved mixing performance. The flow then slowly 'calms' between the C4 and C8 mixer units. After C8, the flow is restored to a steady laminar flow. As mentioned above, a three-dimensionally crossing manifold micromixer permits rapid mixing in a short channel length (<1 mm).

8.2.3.3 Tesla Structures with Coandă Effect

Tesla mixers are an asymmetric structure of twisted channels that offers more fluidic resistance in one direction than in the other, therefore acting as a homogenizer (Figure 8.5).[14,15] The Tesla structure is a modification of the Coandă design and effect, also known as boundary layer attachment, discovered by the Romanian aircraft inventor Henri Coandă. He found that a stream of fluid flowing next to a convex surface tends to stay attached to the surface, contrary to the expectation that, by inertia, the fluid would detach from the surface and follow the tangent. In the Tesla mixer, the Coandă effect occurs at the walls of the end of the loops. The design is characterized by staggered curved channels that permit enhanced mixing by Dean vortex structures that are induced in the curved tapered channels. As mentioned previously, the presence of the vortexes inside the fluid enhances the diffusion mechanism and reduces the diffusion time. The important parameters are the ratio of the diffuser gap to the channel width and the ratio of the curved gap to the channel width (w/W in Figure 8.5); however, the latter ratio has a greater effect on the performance of the mixer. The optimization results showed that the mixing and pressure drop characteristics were both very sensitive to the geometric parameters.

8.2.3.4 Embedded Barriers and Staggered Herringbone Structures

The simplest method of obtaining chaotic advection is to insert obstacles into the mixing channel, which breaks the streamlines, disrupting the laminar flow, and enhances mixing. The complete mathematical definition of chaotic mixing is outside the scope of this book; however, the curious reader can acquire further understanding from a paper by Khakhar *et al.*[16]

Figure 8.6a shows an example of how microfluidic channels with embedded barriers represent a solution for promoting chaotic mixing. First, all the species are fed into the micromixer channel

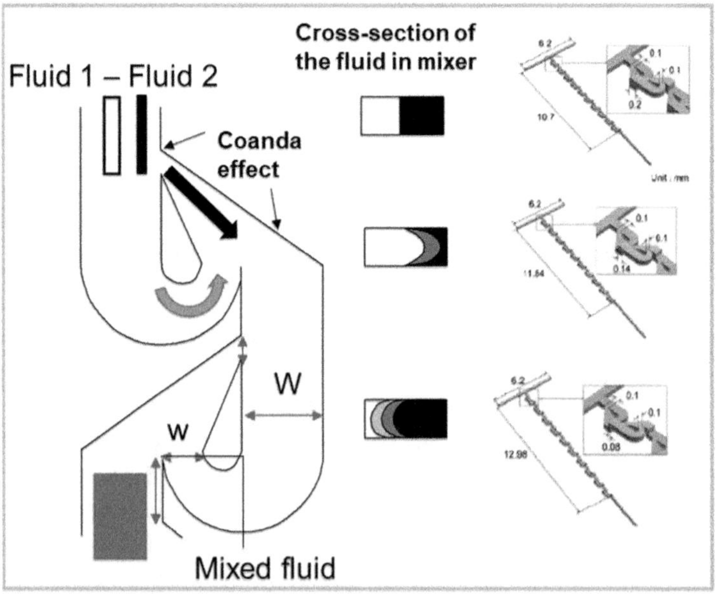

Figure 8.5 Tesla mixer and principle of the Coandă effect, and a 3D application. Reproduced from ref. 15 with permission from Elsevier, Copyright 2015.

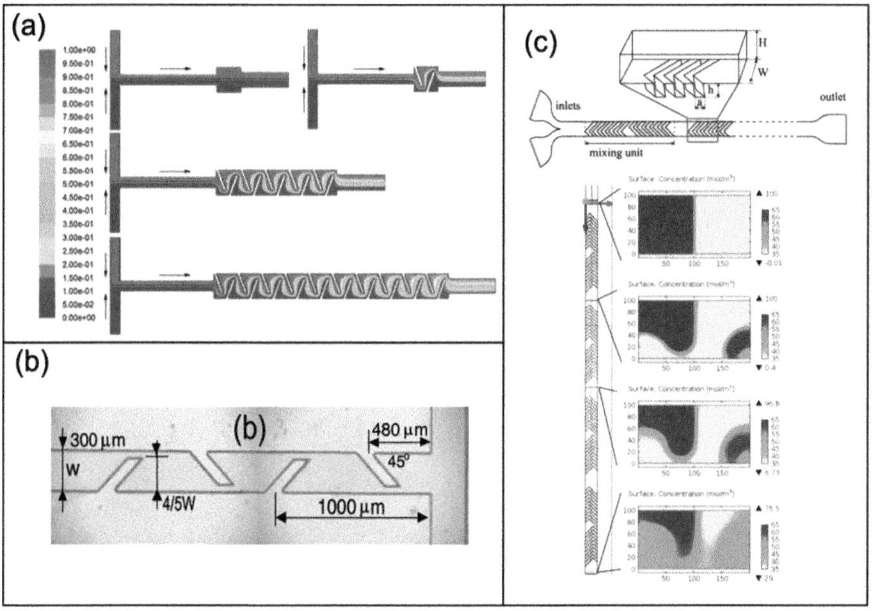

Figure 8.6 (a) Passive mixing obtained by staggering the fluid with increasingly larger obstacles, producing better mixing efficiencies. (b) An example. (c) Staggered herringbone structures. (a) Reproduced from ref. 17 with permission from Elsevier, Copyright 2012. (b) Reproduced from ref. 19 with permission from the Royal Society of Chemistry. (c) Reproduced from ref. 20, https://doi.org/10.3390/mi6010136, under the terms of a CC BY 4.0 license, https://creativecommons.org/licenses/by/4.0/.

from the two inlet channels. These then flow down the mixing channel, where the barriers produce stretching of the interfacial contact area between the flows of the two species, which in turn enhances mixing. Variation of the obstacle size also affects the mixing efficiency.[17] The design includes shifted trapezoidal blades for disturbing the laminar flow; here, several mixing principles are involved, *i.e.* vortices, transversal flows and chaotic advection are combined to achieve a high mixing efficiency even at Reynolds numbers lower than 100.

To improve the mixing performance in a chaotic configuration, staggered herringbone structures can be incorporated on the microchannel floor of micromixers, as described by Whitesides' group in 2002.[18] The principle is based on the ongoing folding of the initial lamellae after passing the cycles defined by the herringbone pattern (Figure 8.6c).[19,20] The incompressible fluid starts from a rotating (helical) motion, then it enters a channel that has slanted grooves on its floor. The swirling patterns are more efficient with an asymmetric herringbone groove design, the centre of which is shifted periodically. The fluid that enters and flows into the floor grooves transmits its momentum to the rest of the fluid in the channel, causing it to rotate; the slanted grooves produce a recirculation.

The performance of such a mixer is modulated by controlling the aspect ratio of the mixing channel, the ratio of the groove depth to the channel height, the ratio of the groove width to the groove pitch, the asymmetry factor of the groove, the angle of the groove and the numbers of grooves per half channel. More specifically, the mixing performance increases with increase in groove width.

8.3 Mixing Yield Stress Fluids

So far we have considered Newtonian fluids, but many biological fluids are non-Newtonian, *e.g.* blood. The main difference between Newtonian and non-Newtonian fluids concerns the relationship between shear stress and viscosity (see Chapter 1). The rheological behaviour of a yield stress material is described by the following non-linear equation:

$$\tau = \mu(\gamma)\gamma = \tau_v + K\gamma^b \tag{8.4}$$

with yield stress τ_v and consistency index K. To analyse the behaviour of non-Newtonian fluids, it is convenient to introduce the Bingham number, Bi (which measures the relative importance of the yield stress), which is given by

$$\text{Bi} = \frac{\tau_v}{\mu(\gamma)\gamma} \tag{8.5}$$

At low Reynolds number, γ is small, so the viscosity is a function of the shear stress, and therefore the shear rate for a passive mixer is

$$\gamma = \frac{U}{d} \tag{8.6}$$

At low shear rates, where Bi approaches 1, most of the material moves through the nozzle as a solid plug and convective mixing is inhibited, whereas at higher shear rates, where Bi approaches 0, viscous stresses fluidize the material, allowing mixing caused by chaotic motion. In contrast, in a passive mixer two opposite phenomena occur and the yield stress of the materials is restricted by two competing constraints, low Pe (low Q, sufficient mixing time) and Bi < 1 (high Q, sufficient fluidization).

8.4 Active Mixing

Active mixers, in which external energy in some form induces chaotic behaviour, are necessary for faster and more efficient mixing. Active mixing can be triggered by using piezo actuators,[21–23] ultrasound waves,[24,25] electric fields[26–28] and magnetic stirrer particles/beads.[29–31] The presence of the additional velocity components brings chaotic behaviour and enhanced mixing within a shorter time frame. However, such systems require far more design considerations and intensive fabrication, which may nullify the advantages.

Mixers including both passive and active components have been designed, for example the combination of electroosmotic flow with a constricting T-channel.[32] An electric field was applied between the inlets, causing oscillation of the two fluids through a constricting channel, increasing the interfacial area, and hence increasing the diffusion and the resultant mixing. Another example is the combination of flexible pillars functionalized with magnetic material that can be tuned to vary the mixing.[33]

8.5 Mixing Efficiency

We have discussed the difficulties facing mixing efficiency and the implementation of passive and active mixing to overcome these difficulties. All of these methods use some technique to evaluate the efficiency of mixing. Currently, the most commonly utilized procedure is mixing two differently coloured liquids together and analysing the product. In the simplest case, we assume a 1 : 1 mixing ratio, hence the final liquid should only be the colour of the product, *e.g.* yellow and blue to produce green.[34] This principle has been implemented countless times in simulations and experimentally,[35,36] with Figures 8.3, 8.4 and 8.6 giving examples of colour mapping of to show the mixing efficiency. Fluorescent intensity, with either confocal or regular fluorescence microscopy, variants are also very common, where the dilution or degradation of fluorescent materials is measured.[37–39] Additionally, Fourier transform infrared (FT-IR) spectroscopy,[40] microparticle movement,[41] Raman spectroscopy[42] and all the other detection methods covered in this book have been developed and implemented to determine the mixing efficiency. The segregation index has been reported as a measure of the efficiency, often discussed as a function of the pressure drop,[43,44] but it is more common to report the efficiency directly *via* the prior colour mapping method.

References

1. C. Y. Lee, C. L. Chang, Y. N. Wang and L. M. Fu, *Int. J. Mol.*, 2011, **12**(5), 3263.
2. P. S. Dittrich and A. Manz, *Nat. Rev. Drug Discovery*, 2006, **5**(3), 210.
3. B. H. Weigl, R. L. Bardell and C. R. Cabrera, *Adv. Drug Delivery Rev.*, 2003, **55**(3), 349.
4. S. Haeberle and R. Zengerle, *Lab Chip*, 2007, 7(9), 1094.
5. H. A. Stone, A. D. Stroock and A. Ajdari, *Annu. Rev. Fluid. Mech.*, 2004, **36**(1), 381.
6. H. M. Xia, S. Y. Wan, C. Shu and Y. T. Chew, *Lab Chip*, 2005, **5**(7), 748.
7. T. J. Ober, D. Foresti and J. A. Lewis, *Proc. Natl. Acad. Sci. U. S. A.*, 2015, **112**(40), 12293.
8. F. G. Bessoth, A. J. de Mello and A. Manz, *Anal. Commun.*, 1999, **36**, 213.
9. (a) T. Tofteberg, M. Skolimowski, E. Andreassen and O. Geschke, *Microfluid. Nanofluid.*, 2010, **8**, 209; (b) M. Roudgar, E. Brunazzi, C. Galletti and R. Maur, *Chem. Eng. Technol.*, 2012, **35**(7), 1291.
10. A. Afzal and K. Y. Kim, *Chem. Eng. J.*, 2012, **203**(1), 182.
11. A. Afzal and K. Y. Kim, *Sens. Actuators, B*, 2015, **211**, 198.
12. M. K. Parsa, F. Hormozi and D. Jafari, *Comput. Fluids*, 2014, **105**, 82.
13. H. M. Xia, Z. P. Wang, Y. X. Koh and K. T. May, *Lab Chip*, 2010, **10**, 1712.
14. S. Hossain, M. A. Ansari, A. Husain and K. Y. Kim, *Chem. Eng. J.*, 2010, **158**, 305.
15. A. S. Yang, F. C. Chuang, C. K. Chen, M. H. Lee, S. W. Chen, T. L. Su and Y. C. Yang, *Chem. Eng. J.*, 2015, **263**, 444.
16. D. V. Khakhar, H. Rising and J. M. Ottino, *J. Fluid Mech.*, 1986, **172**, 419.
17. Y. Fang, Y. Ye, R. Shen, P. Zhu, R. Guo, Y. Hu and L. Wu, *Chem. Eng. J.*, 2012, **187**, 306.
18. A. D. Stroock, S. Dertinger, A. Ajdari, I. Mezic, H. Stone and G. Whitesides, *Science*, 2002, **295**(5555), 647.
19. M. Wang, Z. Wang, M. Zhang, W. Guo, N. Li, Y. Deng and Q. Shi, *J. Mater. Chem. B*, 2017, **5**, 9114.
20. E. L. Toth, E. Holczer, K. Ivan and P. Fürjes, *Micromachines*, 2014, **6**(1), 136.
21. S. O. Catarino, L. R. Silva, P. M. Mendes, J. M. Miranda, S. Lanceros-Mendez and G. Minas, *Sens. Actuators, B*, 2014, **205**, 206.
22. H. Yu, J. W. Kwon and E. S. Kim, *J. Microelectromech. Syst.*, 2006, **15**(4), 1015.
23. T. Mashimo, *IEEE Trans. Ultrason. Ferroelectr. Freq. Control*, 2013, **60**(10), 2098.
24. M. Bezagu, S. Arseniyadis, J. Cossy, O. Couture, M. Tanter, F. Montic and P. Tabeling, *Lab Chip*, 2015, **15**, 2025.
25. S. S. Guo, S. T. Lau, K. H. Lam, Y. L. Deng, X. Z. Zhao, Y. Chen and H. L. W. Chan, *J. Appl. Phys.*, 2008, **103**, 094701.
26. I. Glasgow, J. Battona and N. Aubrya, *Lab Chip*, 2004, **4**, 558.
27. C. K. Harnett, J. Templeton, K. A. Dunphy-Guzman, Y. M. Senousy and M. P. Kanouff, *Lab Chip*, 2008, **8**, 565.
28. L. -M. Fu, R. -J. Yang, C. -H. Lin and Y. -S. Chien, *Electrophoresis*, 2005, **5**, 1814.
29. L.-H. Lu, K. S. Ryu and C. Liu, *J. Microelectromech. Syst.*, 2002, **11**(5), 462.
30. A. Rida and M. A. M. Gijs, *Anal. Chem.*, 2004, **76**(21), 6239.

31. M. Ballard, D. Owen, Z. Grant Mills, P. J. Hesketh and A. Alexeev, *Microfluid. Nanofluid.*, 2016, **20**(6), 88.
32. C. Y. Lim, Y. C. Lam and C. Yang, *Biomicrofluidics*, 2010, **4**(1), 14101.
33. B. Zhou, W. Xu, A. A. Syed, Y. Chau, L. Chen, B. Chew, O. Yassine, X. Wu, Y. Gao, J. Zhang, X. Xiao, J. Kosel, X.-X. Zhang, Z. Yao and W. Wen, *Lab Chip*, 2015, **15**, 2125.
34. C.-H. D. Tsai and X.-Y. Lin, *Micromachines*, 2019, **10**, 583.
35. S. Sivashankar, S. Agambayev, Y. Mashraei, E. Q. Li, S. T. Thoroddsen and K. N. Salama, *Biomicrofluidics*, 2016, **10**, 034120.
36. R. A. Taheri, V. Goodarzi and A. Allahverdi, *Micromachines*, 2019, **10**(11), 786.
37. L. Chen, G. Wang, C. Lim, G. H. Seong, J. Choo, E. K. Lee, S. H. Kang and J. M. Song, *Microfluid. Nanofluid.*, 2009, **7**, 267.
38. T. Park, M. Lee, J. Choo, Y. S. Kim, E. K. Lee, D. J. Kim and S.-H. Lee, *Appl. Spectrosc.*, 2014, **58**(10), 1172.
39. T. Daniel, S. White and J. Lewis, *Nat. Mater.*, 2003, **2**, 265.
40. E. Chung, R. Gao, J. Ko, N. Choi, D. W. Lim, E. K. Lee, S. I. Chang and J. Choo, *Lab Chip*, 2013, **13**, 260.
41. S.-S. Hsieh, J.-W. Lin and J.-H. Chen, *Int. J. Heat Fluid Flow*, 2013, **44**, 130.
42. J. Jahn, O. Žukovskaja, X.-S. Zheng, K. Weber, T. W. Bocklitz, D. Cialla-May and J. Popp, *Analyst*, 2017, **142**, 1022.
43. Y. Lin, X. Yu, Z. Wang, S.-T. Tu and Z. Wanga, *Chem. Eng.*, 2011, **171**(1), 291.
44. Y. Men, V. Hessel, P. Löb, H. Löwe, B. Werner and T. Baier, *Chem. Eng. Res. Des.*, 2007, **85**(A5), 605.

9 Droplet Formation and Manipulation

9.1 Introduction

Droplets are made by a well-controlled volume of liquid moving in a fluidic channel while being transported by a carrier outer fluid or driven by external forces.[1,2] The opportunity to control the volume, the composition and the motion of the droplets makes microfluidic droplet formation interesting for numerous applications in chemistry and biology.[3,4] In fact, droplets allow the manipulation of discrete fluid packets on the microscale, acting as microreactors, that provide numerous benefits for conducting biological and chemical assays. These uses of microdroplets exploit the advantages of allowing high-throughput and multiplex experiments, with more complex systems allowing the control of a single droplet, which is incredibly useful for single-cell analysis.

In this chapter, the theory of wettability is introduced, leading into the theory behind droplet formation and manipulation with real-world examples.

Today, microdroplet generators are integrated into several applications, with examples related to polymerase chain reaction (PCR),[5] enzyme assays,[6,7] particle synthesis[8] and single-cell genomics.[9]

Microfluidics and Lab-on-a-chip
By Andreas Manz, Pavel Neužil, Jonathan S. O'Connor and Giuseppina Simone

Published by the Royal Society of Chemistry, www.rsc.org

9.2 Multiphase Microflows: Droplet and Plug Flow in Microchannels

9.2.1 Theory of Wettability

Before analysing multiphase microflow in microfluidic channels, it is opportune to introduce and recall some fundamentals of the theory of wettability.

Referring to Figure 9.1a, the surface tension (force per unit length or energy per unit area) is associated with the interface between each of three phases: the solid substrate (s), the liquid droplet (l) and the surrounding ambient phase, which for simplicity is often denoted vapour (v). Each interface exerts a force on the three-phase contact line and the balance between the interfacial surface tensions, *i.e.* γ_{sv} (solid–vapour), γ_{sl} (solid–liquid) and γ_{lv} (liquid–vapour), must be in equilibrium according to Young's law:

$$\gamma_{sv} - \gamma_{sl} - \gamma_{lv}\cos\theta_s = 0 \tag{9.1}$$

Young's law predicts the value of the static contact angle θ_s as a function of the surface energy of the different materials. The static contact angle is not uniquely defined and it can be comprised between two values, the first obtained by slowing to a stop at an advancing front, $\theta_{s,a}$, and another value (smaller) obtained by slowing to a stop at a receding front, $\theta_{s,r}$.

9.2.2 Dynamic Contact Angle

The contact angle formed between a flowing liquid front (advancing or receding, Figure 9.1b) and a solid surface is not constant but reflects the balance between capillary forces and viscous forces. As observed in Chapter 1, the capillary number is a scale of the ratio between the drag force of the flow on a plug and the capillary forces.

However, the relationship between the capillary number (Ca) and the contact angle accounts for the static and dynamic angles according to Tanner's law:

$$\theta_d^3 - \theta_s^3 \approx \mathrm{Ca} \tag{9.2}$$

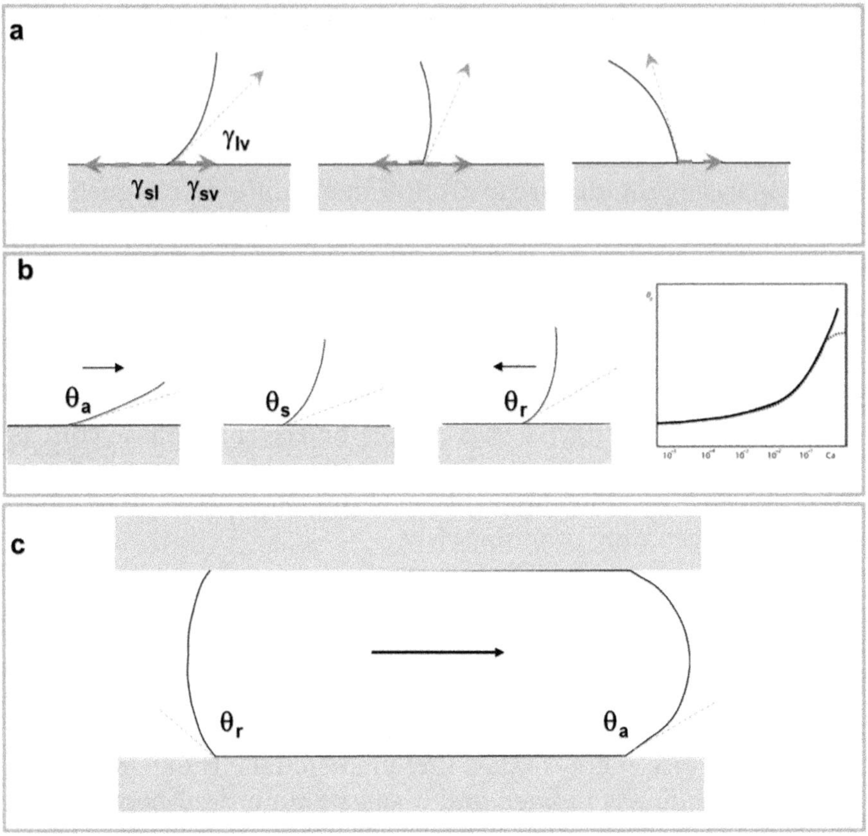

Figure 9.1 (a) Left: example of advancing (a), static (s) and receding (r) angles formed between a flowing liquid front and a solid surface; Right: relationship between the dynamic contact angle and Ca. (b) Surface tension associated with the interface between each of the three phases: the solid substrate (s), the liquid droplet (l) and the surrounding ambient phase, which for simplicity is often denoted vapour (v). Each interface exerts a force on the three-phase contact line and the balance between the interfacial surface tensions γ_{sv} (solid–vapour), γ_{sl} (solid–liquid) and γ_{lv} (liquid–vapour) must be in equilibrium according to Young's law. (c) In a droplet, the advancing and receding angles contribute to the capillary pressure drop.

where θ_d and θ_s are the dynamic and static contact angles, respectively. Linearization of Tanner's law yields

$$\theta_d = \left(\theta_s^3 + A\mathrm{Ca}\right)^{\frac{1}{3}} = \theta_s\left(1 + \frac{1}{3}\frac{A\mathrm{Ca}}{\theta_s^3}\right) \tag{9.3}$$

The difference between the static and dynamic angles is determined by the sign of Ca (Figure 9.1b); the values of the advancing and receding contact angles are then given by

$$\theta_{\mathrm{a}} = \theta_{\mathrm{s}} + \frac{1}{3}\frac{A|\mathrm{Ca}|}{\theta_{\mathrm{s}}^{2}} \tag{9.4}$$

$$\theta_{\mathrm{d}} = \theta_{\mathrm{s}} - \frac{1}{3}\frac{A|\mathrm{Ca}|}{\theta_{\mathrm{s}}^{2}} \tag{9.5}$$

Therefore, an advancing contact angle is always larger than the static contact angle and a receding contact angle is always smaller than the static contact angle.

9.2.3 Microflow Blocked by Plugs

A plug of fluid travelling inside a capillary experiences a total pressure drop given by the motion inside the capillary, inducing a friction pressure drop and a second contribution given by a capillary pressure drop.

The capillary pressure drop depends on the interface, either advancing or receding contributions denoting a positive or negative in the function of the contact angles, respectively. The capillary pressure drop is derived directly from the Laplace law, which relates the pressure difference at a spherical interface of radius of curvature a by the following relationship:

$$\Delta P = \frac{2\gamma}{a} \tag{9.6}$$

Further, the contact angle is related to both the channel radius, R, and a by

$$\cos\theta = -\frac{R}{a} \tag{9.7}$$

Combining the above two expressions, the advancing and receding angle expression yields

$$\Delta P_{\mathrm{a}} = -\frac{2\gamma}{R}\cos\theta_{\mathrm{a}} \tag{9.8}$$

and

$$\Delta P_{\mathrm{a}} = \frac{2\gamma}{R}\cos\theta_{\mathrm{r}} \tag{9.9}$$

The capillary pressure drop is due to the difference in the capillary forces between the advancing and receding fronts, shown by the two different contact angles of advancing and receding, θ_a and θ_r, respectively. If θ_a is larger than π/2, there is a positive pressure drop associated with the advancing interface. If θ_r is smaller than π/2, the receding front contributes positively to the pressure drop and negatively in the opposite case. The advancing and receding angles can be opportunely modified by varying the surface tension. The pressure drop due to the friction on the solid walls is given by the modified Hagen–Poiseuille expression

$$\Delta P_{\text{drag}} = -\frac{8U}{R^2}\left(\mu_1 L_1 - \mu_2 L_2\right) \tag{9.10}$$

where the subscripts 1 and 2 relate the liquid plug to the surrounding carrier fluid, U is the average liquid velocity and L_1 and L_2 are the total length of contact of liquid 1 and liquid 2 with the solid wall ($L_1 + L_2 = L$, total length of the tube). Each interface, advancing and receding, contributes (positively or negatively in the function of the contact angles) to the capillary pressure drop (Figure 9.1c).

As a rule of thumb, the wettability of the channel surface defines which of the two liquid phases becomes the droplet (see Figure 9.2). Hydrophilic surfaces are good for oil-in-water droplets, whereas hydrophobic surfaces yield water-in-oil droplets.

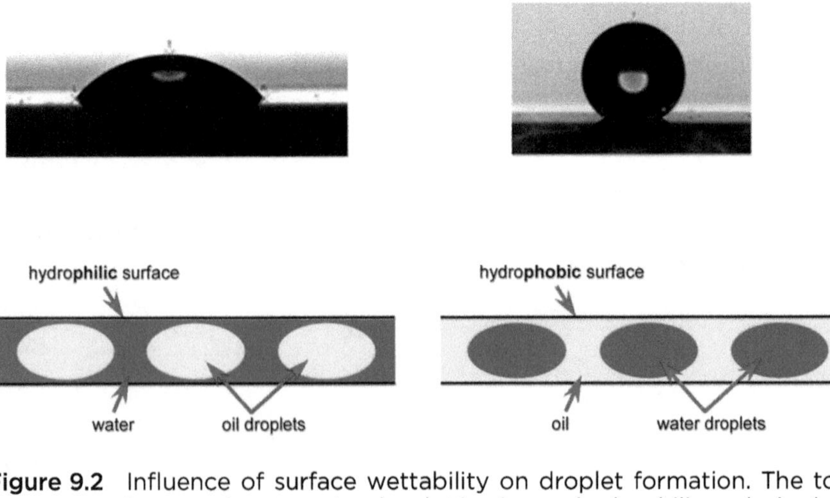

Figure 9.2 Influence of surface wettability on droplet formation. The top images show a water droplet in air on a hydrophilic and a hydrophobic surface. The schematic below shows which type of droplet is more likely to be formed.

9.3 Microdroplets Fluidic Channel

The formation of monodispersed droplets in a continuous microflow is a phenomenon referred to as 'dropletting'. The droplets can be produced in concentric capillaries (as in Figures 9.4 and 9.7) and on-chip in T-junctions (two-, three- or five-port intersections) (see Figure 9.3).

In such flow focusing devices, four types of modes can be distinguished. The first mode, also called a type of instability because the flow breaks up into droplets, is called squeezing (Figure 9.4). This occurs in T-junctions at a low velocity (low capillary and Weber numbers). The second type, dripping, is obtained in flow focusing and T-junctions with a higher flow velocity, where a flowing liquid is reduced to a filament under the action of the other fluid; because of the conflicting surface tension forces, the filament is not stable. Then, the filament breaks down into droplets at some distance from the channel entrance. Upon this condition, there are two regimes of flow: dripping, where the droplets are immediately formed, and jetting,

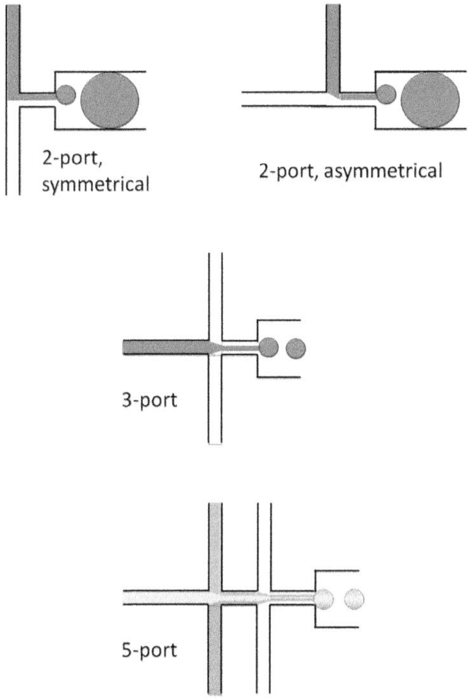

Figure 9.3 Examples of on-chip channel layouts for droplet formation.

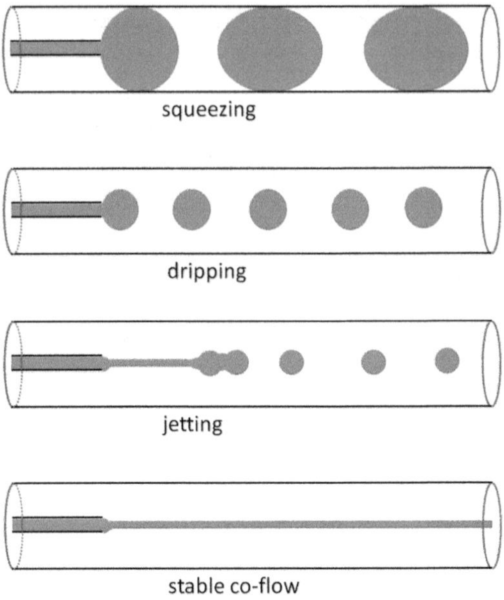

Figure 9.4 Examples of formation of droplets. The flow instability of the two phases breaks up in different ways.

where a laminar flow is still visible downstream before breaking up into droplets. The ultimate case is a stable co-flow, which is not of interest for obtaining droplets.

9.3.1 T-junctions

T-junctions (or, in the case of one channel, an L-junction) in microfluidic channels consist of two channels of the same depth merging at an angle of 90°. Within this geometry, the droplet detachment is a function of the flow velocity and the capillary number. The size of the droplets is determined solely by the ratio of the volumetric flow rates of the two immiscible fluids. With the condition of a rectangular cross-section,

$$\frac{L}{a} = 1 + w\frac{Q_{\text{drop}}}{Q_{\text{outer}}} \tag{9.11}$$

where L is the length of the fluid segment, w is a dimensionless parameter, and Q_{drop} and Q_{outer} are the flow rates of the droplet and the outer fluid, respectively. The differences in viscosity and surface energies cause surface instabilities, which in turn determine the droplet detachment method. Detachment can change depending on Ca and whether Ca is (a) $<10^{-2}$ or (b) $>10^{-2}$.

Usually, the droplets are formed in a squeezing regime when Ca $<10^{-2}$. Figure 9.5 shows the steps involved in droplet formation. First, the liquid penetrates the main channel where the carrier fluid flows, forms a blob and develops a neck, which elongates and becomes thinner. The forming droplet has the same size as the channel. However, droplets can be smaller than the size of the channel if the incoming rate of liquid is modulated in frequency in a synchronized regime. The droplet size is inversely proportional to the frequency. The ratio of the viscosities of the inner and outer fluids influences the dynamics of formation and also the shape of the neck. As the capillary and Weber numbers are low, the interfacial forces dominate shear stress, so the pressure drop across the droplet is responsible for the break-up of the neck and droplet formation.

9.3.2 Mixing Inside Droplets Formed in a T-junction

T-junction designs offer the possibility of merging two different fluids for mixing within a single droplet. Mixing by diffusion alone would be a slow mixing mechanism. However, droplets with the same size as the channel show internal stirring. This is due to the

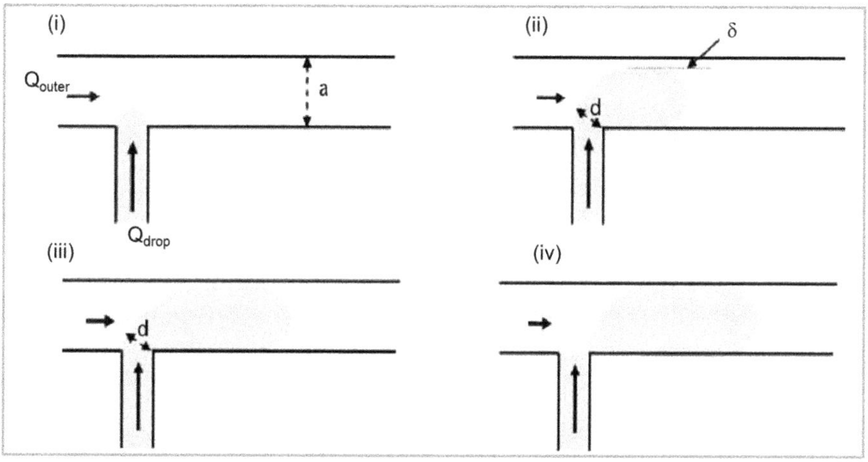

Figure 9.5 T-junction: steps involved in droplet formation at low flow velocity. (i) After the side stream Q_{drop} enters the main channel. (ii) Blob formation, which has approximately the size of a main channel width. (iii) If the discontinuous liquid has a flow rate higher than that of the continuous liquid ($Q_{drop} > Q_{outer}$), the droplet elongates in the main channel, but if $Q_{drop} < Q_{outer}$ then there is no elongation and the separation is immediate. (iv) The droplet detaches.

friction between the droplet and the channel wall, combined with the phase boundary which represents a barrier to flow. The internal streamlines undergo recirculation inside the plug. The dissymmetry of the recirculation flow in the turns induces a reorientation of the fluid domains. This reorientation is essential for the mixing of liquids in the plug, as increases in the number of striations increases the mixing efficiency.

It is worth noting that the striation thickness in a straight channel, s_t, is given by

$$s_t = s_{t0}\frac{L}{x} \tag{9.12}$$

where L is the length of the plug (droplet), x is the distance travelled and s_{t0} is the initial striation thickness. In the winding channel, we have

$$s_t(n) = s_{t0} \times 2^{-n} \tag{9.13}$$

where n is the number of fold, stretch and reorient steps.

The dynamics of mixing in the straight and meandering channel has been modelled as shown in Figure 9.6a.[10] Two recirculation patterns form inside the plug, one on the left and one on the right. The striation thickness increases proportionally to the length of the plug and inversely to the distance that the droplet travels in the range of time under consideration. Each time the plug has travelled a double distance of droplet length, the striation thickness is halved. An example is shown in Figure 9.6b.[11] The mixing inside the droplets takes place by diffusion of one species into the other through diffusion. This is optimized by a mechanism of folding and stretching between the filets of the diffusing species, the diffusion time t is given by

$$t = \frac{s_t^2}{2D} \tag{9.14}$$

To obtain complete mixing and a uniform concentration, a much longer channel is required.

It has been shown that the efficiency of the mixing inside the droplets can be improved by including winding or snaking channels. The dynamics is shown in Figure 9.6b.

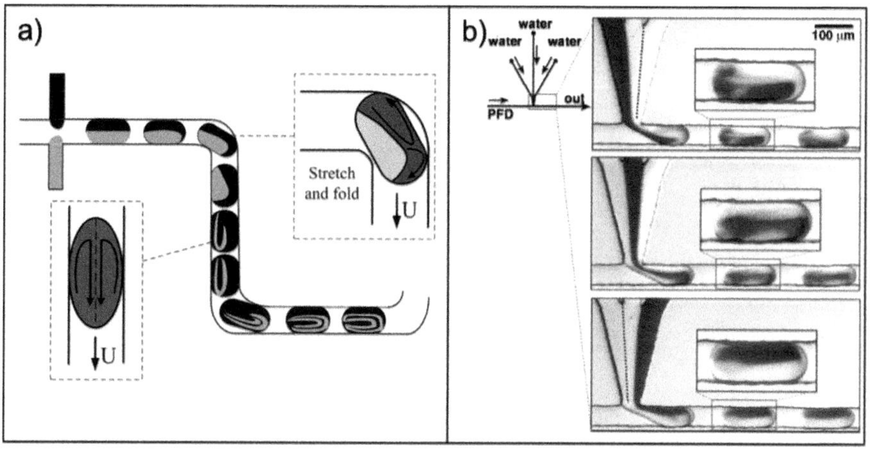

Figure 9.6 (a) Dynamics of the mixing in a droplet travelling in a channel. (b) Micrograph of a three-component droplet showing the mixing pattern. (a) Reproduced from ref. 10 with permission from the Royal Society of Chemistry. (b) Reproduced from ref. 11 with permission from American Chemical Society, Copyright 2003.

9.3.3 Micro Flow Focusing Devices

The free jet breaks up somewhat randomly under the effect of the Plateau–Rayleigh instability and produces polydisperse droplets. A microflow focusing device is capable of making these instabilities occur at regular controlled moments to generate monodisperse droplets. The principle of a flow focusing device, shown in Figure 9.7, can be described as follows: (i) the dispersed phase liquid enters the orifice, forming an obstacle to the flows coming from each of the sides. (ii) The side flow pinches the inner flow and a liquid blob is formed, which is linked to the incoming dispersed phase by an elongating thread. (iii) The thread thins and becomes unstable due to the Plateau–Rayleigh instabilities. (iv) The thread breaks and droplets are formed. Many parameters characterize the flow: the actuation parameters, Q_i and Q_e or P_i and P_e; the fluid characteristics, μ_i, μ_e, γ, ρ_i and ρ_e, which denote, respectively, the viscosity of the dispersed and continuous phases, the surface tension between the two liquids and the density of the two liquids in addition to the inner and outer phase flow rates and pressures. Also, the geometry and the flow resistances of the channels (R_i and R_e) also influence the dynamics of the droplets in a flow focusing channel. The actuation of the flow can be controlled by pressure or flow rates by keeping Q_i/Q_e or P_i/P_e

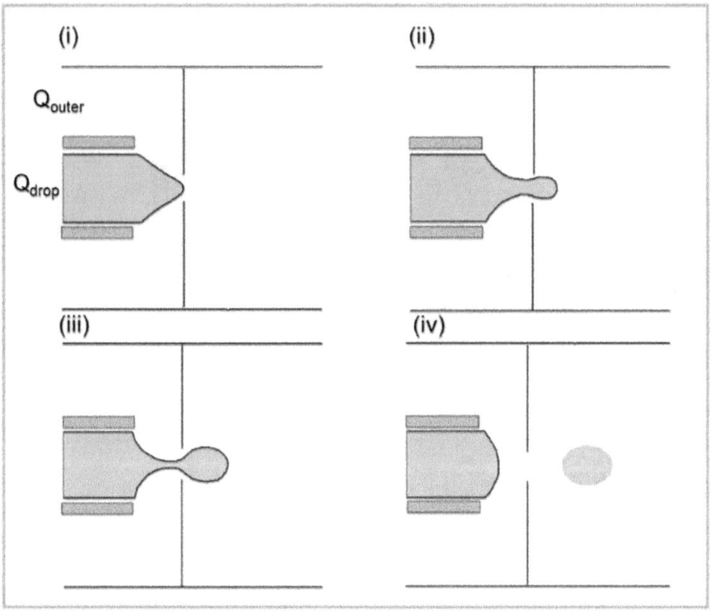

Figure 9.7 (i, ii) Flow focusing: steps of droplet formation at low flow velocity. (iii) The inner stream forms a blob, which grows inside the outer fluid channel. (iv) The pinching forces are too great, causing instability at the filament to 'cut the neck' sharply and so the droplet detaches.

constant. A constant ratio of the flow rates implies that the capillary number ratio depends on the flow rates and viscosities ($Ca_i/Ca_e = \mu_i Q_i/\mu_e Q_e$), and a constant ratio of pressures provides a constant ratio of the capillary numbers.

Dripping and jetting are the main flow regimes observed in flow focusing devices. In the dripping regime, the flow rates are small enough and the droplet forms immediately after the nozzle, whereas in the jetting regime, a thread or filament stretches far into the outlet channel. In the first case, the drops are larger. The transition to the jetting regime takes place when the Weber number is lower than the critical Weber number if $P_e << P_i$. The frequency at which the droplets are generated depends on the ratio of the flow rate of the inner phase and the volume of the droplet. The size of the droplets depends on the flow conditions, the geometry of the focusing region, the viscosities of the fluids and the surface tension between the fluids. For a given geometry and viscosity ratio, the size of the droplets depends on the ratio of the applied pressures P_i and P_e or the ratio of the applied flow rates Q_i and Q_e. The droplet volume decreases with decrease in flow rates and pressure ratios.

9.4 Electrowetting-on-dielectric

There are two means of actuation: acoustic and electric. In this chapter, we deal only with electric actuation, called electrowetting, because it is the principle of most digital microfluidics systems. For a long time, the application of this principle to aqueous electrolytes was limited by the decomposition of water, electrolysis, beyond a few hundred millivolts. Later, the electrolysis was overcome by the introduction of a thin insulating layer to separate the conductive liquid from the metal electrode. This allowed for systems where insulating, also referred to as dielectric, droplets can be manipulated upon electrodes, called electrowetting-on-dielectric (EWOD).

As can be seen in Figure 9.8a, the droplets reside on a horizontal planar substrate and are usually composed of aqueous solutions with a typical droplet diameter of 1 mm or less. The medium surrounding the droplet can be either air or another immiscible liquid such as silicone oil, or the aqueous droplet is enclosed in a thin layer of oil surrounded by air.[12]

In the presence of an electric field, electric charges gather at the interface between conductive and non-conductive materials. These gathering electric charges exert a force on the interface and deform the interface when possible. This effect is of particular interest when electric forces are exerted on a liquid–gas interface at the vicinity of the contact line with a solid, resulting in a change of the contact angle; an example is shown in Figure 9.8b.

9.4.1 Theory of Electrowetting

The relative importance of the gravitational force to the surface tension force can be expressed by the Bond number (Bo), expressed as

$$\mathrm{Bo} = \frac{\Delta\rho g R^2}{\gamma_{\mathrm{lv}}} \tag{9.15}$$

where $\Delta\rho$ is the difference in density between the droplet and the surrounding medium, g is the acceleration due to gravity, R is the radius of curvature of the droplet and γ_{lv} is the liquid–vapour surface tension. In electrowetting applications, gravitational forces are small in comparison with the surface tension force. In the absence of an external electric field, the shape of the droplet is determined

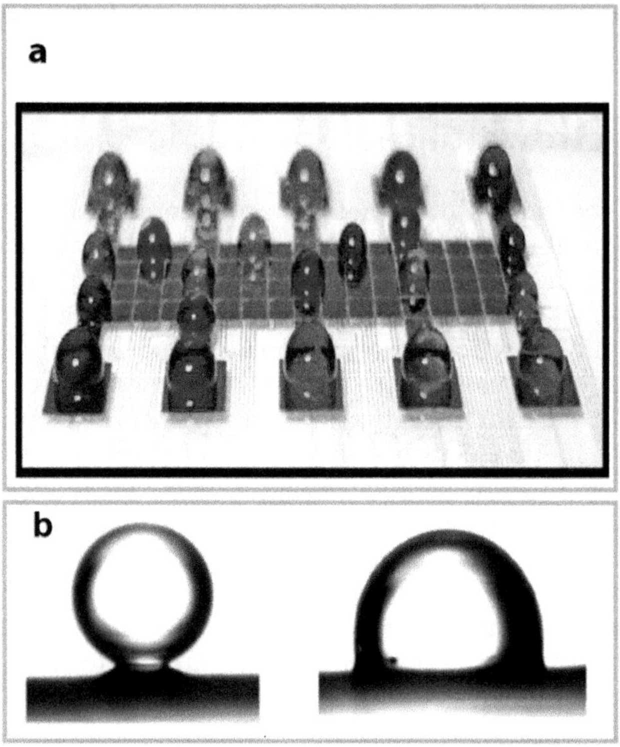

Figure 9.8 (a) Digital microfluidics system electrically actuated (electrowetting). (b) Droplet before and after being exposed to an electric field causing electrowetting. Reproduced from ref. 12 with permission from John Wiley and Sons, Copyright © 2010 WILEY-VCH Verlag GmbH & Co. KGaA, Weinheim.

solely by the surface tension, as discussed earlier. The solid–liquid interface behaves as a parallel-plate capacitor, with a capacitance per unit area, C, and a potential difference, V, between the liquid and the substrate. Once an electric field is applied, the shape of the droplet changes, as modelled in Figure 9.9. Energy stored in the droplet, which acts as a capacitor, is $CV^2/2$ and the work is CV^2. Application of the voltage causes a new force, which induces variations of the contact angle of equilibrium according to the following expression:

$$\lambda_{sv} - \left(\gamma_{sl} - \frac{CV^2}{2}\right) - \gamma_{lv}\cos\theta_{sW} = 0 \tag{9.16}$$

where θ_{sW} is the contact angle after the application of the electrowetting voltage. By combining eqn (9.5) and (9.15), the Berge–Lippmann–Young equation is derived, which correlates the angle of contact with the angle of contact in the presence of the electric field:

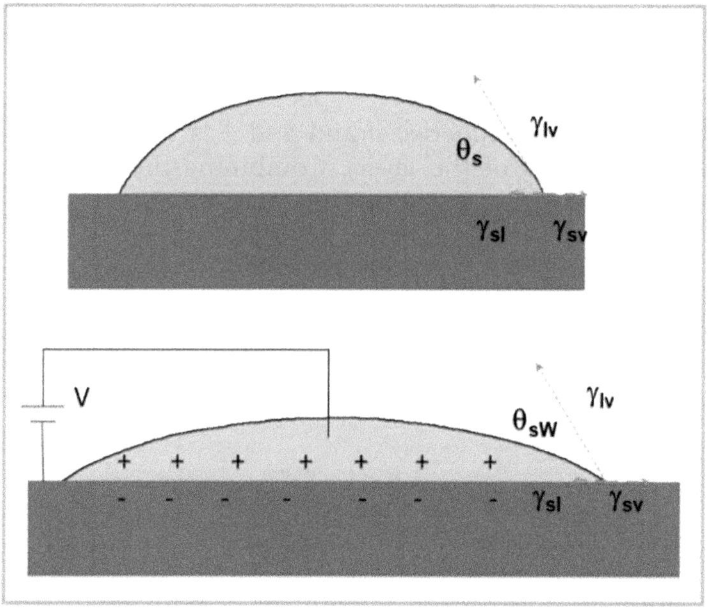

Figure 9.9 Top: the shape of a droplet in the absence of an electric field. Bottom: the shape of a droplet when a potential difference, V, is applied between the liquid and the substrate. θ_{sW} is reduced, showing an increase in the droplet wettability.

$$\cos\theta_{\mathrm{sW}} = \frac{CV^2}{2\gamma_{\mathrm{lv}}} + \cos\theta_{\mathrm{s}} \tag{9.17}$$

where θ_{s} is the Young contact angle.

9.4.2 Thermodynamics and EWOD

In this section, the theory of EWOD using a thermodynamic approach based on the formation of a double layer is discussed. In the absence of any applied voltage, the double-layer coating on the droplet presents a perfectly smooth surface on contact with the conductive liquid. The solid is a metal that is in direct contact with the liquid with a potential difference that is so small that there is no electric current flowing through the liquid. On application of an electric field, a potential difference builds up at the interface and an electric double layer forms in the liquid at the contact with the surface.

The droplet will present an electric double layer at a fixed distance, λ, from the surface and this double layer will also have a fixed specific capacitance, or capacitance per unit area. The specific capacitance of the electric double layer is given by

$$C = \frac{\varepsilon_r \varepsilon_0}{\lambda} \tag{9.18}$$

where ε_0 is the permittivity of free space or vacuum, ε_r is the relative permittivity of the conductive liquid and λ is the Debye screening length (or thickness of the layer). Combining the aforementioned Berge–Lippmann–Young equation (eqn (9.17)) with eqn (9.18) leads to

$$\cos\theta_{sW} = \frac{\varepsilon_r \varepsilon V^2}{2\lambda\sigma_{lv}} + \cos\theta_s \tag{9.19}$$

Sometimes, a spontaneous charge can form when a conducting surface is immersed in an electrolyte. Under this condition, the difference in potential in eqn (9.19) needs to account also for a secondary difference in potential:

$$\cos\theta_{sW} = \frac{\varepsilon_r \varepsilon (V - V_z)^2}{2\lambda\sigma_{lv}} + \cos\theta_s \tag{9.20}$$

where V is the potential difference under the electric field and V_z is the potential difference at no applied charge. Eqn (9.20) shows that the contact angle decreases with increase in the applied voltage, also depicted in Figure 9.9. However, direct application of the Berge–Lippmann–Young equation to a liquid contacting a metallic surface is of little use because of the limiting hydrolysis phenomena.

Essential values for water are $\lambda \approx 2$ nm, $\varepsilon_r \approx 80$ and $\gamma_{sl} \approx 0.040$ N m^{-1} and the maximum voltage differences are of the order of 0.1 V, so the relative change in the surface tension is $\Delta\gamma_{sl}/\gamma_{sl} \approx 2\%$. Assuming that γ_{lv} is equal to 0.072 N m^{-1} and using the listed values in eqn (9.20), it is possible to calculate that $\cos\theta_{sW} - \cos\theta_s$ is less than 0.01. Usually, the electric double layer has a relatively high capacitance and therefore large changes in contact angle can be achieved using relatively small applied voltages. However, these changes are limited as the high voltages cause electrolysis, therefore changing the volume and composition of the droplet.

To overcome the problem of electrolysis, a thin dielectric layer can be used to insulate the conductive liquid from the electrode. The main advantage of EWOD over conventional electrowetting is that higher potential differences and, therefore, greater contact angle changes can be employed without any detrimental electrochemical reactions on the electrode. In addition, the surface of the microfluidic chip can be coated with highly hydrophobic compounds to enhance the contact angle changes due to electrowetting. Figure 9.10a illustrates

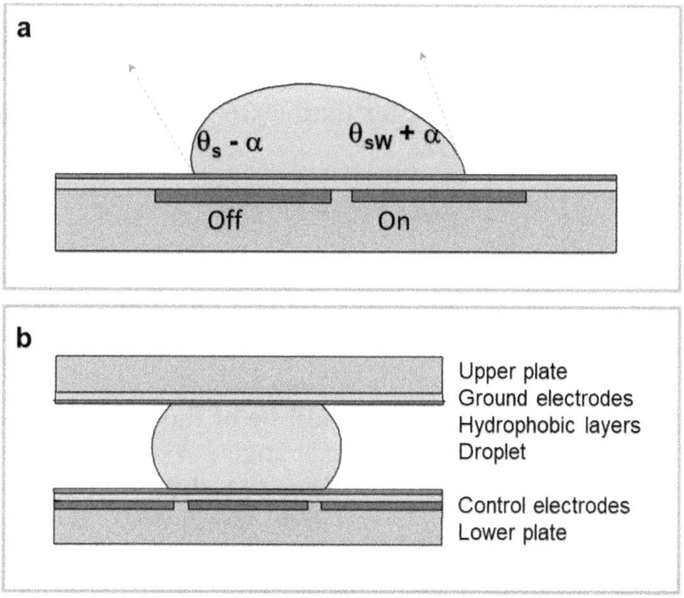

Figure 9.10 (a) Illustration of a droplet on top of an EWOD microfluidics chip. The contact angle of the droplet situated above the actuated electrode will be an advancing contact line, $\theta_{sW} + \alpha$, whereas the contact angle above the non-actuated electrode will be a receding contact line, $\theta_s - \alpha$. (b) Illustration of a covered EWOD microfluidic chip.

the change in the contact angle depending on which part of a droplet is under an electric field. The contact angle decreases because of the changes in the surface tension and double layer as discussed previously.

The inclusion of a second dielectric layer, as illustrated in Figure 9.10b, transforms the EWOD system to one composed of two capacitors in series with each other, namely the capacitance due to the electrical double layer at the solid–liquid interface, C_H, and the capacitance due to the dielectric layer, C_d, which is given by

$$C_d = \frac{\varepsilon\varepsilon_d}{d} \tag{9.21}$$

where ε_d, the relative permittivity of the dielectric layer, and d, the thickness of the dielectric layer, replace ε_0 and λ, respectively. Calculating the ratio C_H/C_d gives a direct function of the permittivity and of the ratio d/λ. On the other hand, the capacitors are in series and consequently

$$\frac{1}{C}=\frac{1}{C_H}+\frac{1}{C_d} \tag{9.22}$$

and seeing as C_H is much larger than C_d, $C \approx C_d$. For these EWODs, eqn (9.20) can be rewritten as

$$cos\theta_{sW}=\frac{\varepsilon_d \varepsilon V^2}{2d\sigma_{lv}}+cos\theta_s \tag{9.23}$$

where the first term is dimensionless and is called the electrowetting number, η.

Typical EWODs use 1 μm of Teflon or Parylene as the dielectric layer and a droplet of water, so that the ratio $\varepsilon_0\varepsilon_d/2d\sigma_{lv}$ is of the order of 10^{-4} V^{-2}. Applied electric potentials should be of the order of 30–80 V to obtain a substantial change in contact angle.

One of the disadvantages of EWODs is that the lower specific capacitance of the dielectric layer requires significantly larger applied potentials to achieve a given reduction in contact angle. One of the other disadvantages of using a dielectric layer is that biomolecules can sometimes bind non-specifically to the hydrophobic surface. Particularly after long-term use, the electrodes become coated.

9.4.3 Experimental Determination of Contact Angle Hysteresis

The electrowetting and thermocapillary processes exhibit dynamic hysteresis whenever the three-phase contact line is in motion because of the different contact angles at the advancing and receding menisci. Hysteresis is defined as the deviation of the contact angle from its mean value due to physical phenomena such as microscopic surface defects and roughness. During the dynamic motion of an interface, dynamic hysteresis refers to the difference between advancing and receding contact angles.

Direct calculation of the capacitance of the dielectric layer can be avoided by fitting the Berge–Lippmann–Young equation (eqn (9.17)) to experimental data for contact angle *versus* applied voltage. By plotting $\cos\theta_{sW} - \cos\theta_s$ against V^2, the resultant gradient of the fitted line provides C/γ_{lv} and therefore, with prior knowledge of the liquid–vapour surface tension, leads to an estimate of the capacitance. This technique is sometimes preferred to direct computation, owing to uncertainties in estimating the thickness and relative permittivity of the deposited layers.

9.4.4 Electrowetting Hysteresis and the Threshold Potential

The changes in the contact angle with respect to the equilibrium value are defined by the electrowetting hysteresis angle, which depends on the surrounding fluid. The hysteresis is essential for droplet manipulation in an EWOD-based system. Because of the extent of the contact angle hysteresis, a threshold potential is generated, V_{min}. A droplet standing between the two electrodes, as illustrated in Figure 9.10a, with one electrode switched off will not begin to move until the electrical potential exceeds the threshold. At the onset of droplet motion, the contact angle above the actuated electrode will be an advancing contact line, $\theta_{sW} + \alpha$, whereas the contact angle above the non-actuated electrode will be a receding contact line, $\theta_s - \alpha$. This asymmetry of the contact angle causes the droplet to move.

An estimation of the threshold potential for droplet motion can be determined by the net positive electrowetting force towards the activated electrode, where $\theta_{sW} + \alpha \leq \theta_s - \alpha$. This condition brings about the modification of the Berge–Lippmann–Young equation that permits the determination of the minimum actuation potential. The modified Berge–Lippmann–Young equation is

$$\frac{CV_{min}^{2}}{2\sigma_{lv}} = \tan\alpha\left(\sin\theta_{sW} + \sin\theta_{s}\right) \tag{9.24}$$

The theoretical descriptions of electrowetting discussed so far are applicable to the static conditions that occur in dc actuation. In the case of a very slowly varying ac voltage, the contact angle will follow the instantaneous value predicted by the Berge–Lippmann–Young equation. However, if the frequency of the applied ac voltage exceeds the hydrodynamic response frequency of the droplet, then the liquid will have insufficient time to adjust to the voltage variations and the response then depends on the time-averaged voltage on the control electrode. This is valid at low and moderate frequencies, but above a critical frequency the liquid will start to behave as a dielectric material and the Berge–Lippmann–Young equation will begin to break down.

Often ac actuation is preferred for transporting solvents such as acetone, chloroform, dimethylformamide and ethanol, which are difficult or impossible to transport with dc voltages. In addition, ac actuation has other advantages, including reduced dielectric hysteresis of the capacitance–voltage characteristics and increased reliability by avoiding the build-up of charges on the insulators.

9.5 EWOD-based Digital Microfluidics

EWOD-based microfluidic systems can be divided into two distinct categories depending on whether the system is open to the atmosphere or is covered.

A covered EWOD microfluidic chip is illustrated in Figure 9.10b. The lower plate of the chip contains the individual addressable electrodes and is usually fabricated from silicon or glass. The electrodes are then covered with a thin dielectric layer (Parylene, SiO_2 or Si_3N_4), typically with a thickness of about 1 μm, and finally the surface is coated with a highly hydrophobic layer. The upper confining plate is coated with the hydrophobic layer containing a continuous electrode. It is fabricated from conductive indium tin oxide (ITO) glass to provide the necessary electrical contact with the droplets. Silicon oil is used as an outer fluid.

An open EWOD microfluidic chip is illustrated in Figure 9.11.[13] The electrical contact with the droplets is achieved either using a network of conducting wire electrodes located a suitable distance above the surface of the chip or *via* the use of a coplanar design in which the lower plate contains both the buried activation electrodes and the contact electrodes.

One of the most important parameters in EWOD devices is the voltage required to manipulate the droplets. A typical control voltage is around 100 V, but control voltages can be substantially reduced by carefully selecting a dielectric material with high relative permittivity and by carefully controlling the thickness of the dielectric layer. Better dielectric materials allow for potentials below a few tens of volts. Also, the better selection of the hydrophobic layer can improve the working conditions.

9.6 Ohnesorge and Weber Numbers and Their Relevance to Droplet Manipulation in EWODs

As was mentioned earlier, in EWOD applications the gravitational forces are small in comparison with the surface tension forces; the Bond number is usually <1. Meanwhile, electrowetting, viscous and inertial forces play vital roles. The ratio between the viscous force and the inertial and surface tension forces in a droplet can be expressed by the non-dimensional Ohnesorge number:

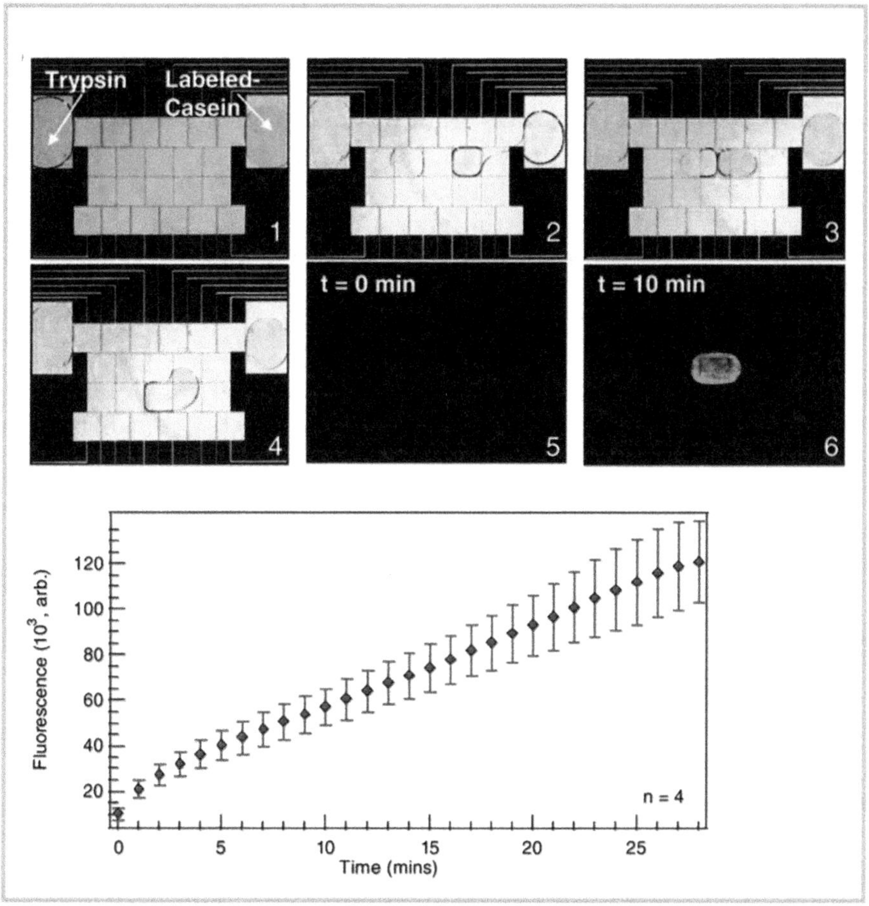

Figure 9.11 Top: an open EWOD microfluidic chip used for a digital microfluidic-driven proteolysis assay. The sequence of images depicts the fluorescent assay for tryptic digestion. (1, 2) Droplets containing BODIPY™-labelled casein (1 mg mL^{-1}) and trypsin (100 µg mL^{-1}) are dispensed from reservoirs, (3) merged and (4) actively mixed. (5, 6) Fluorescent images of a merged droplet before and after digestion. Bottom: the reported curve of the kinetics of the reaction measured from a fluorescence plate reader. Reproduced from ref. 13 with permission from American Chemical Society, Copyright 2008.

$$\mathrm{Oh} = \frac{\mu}{\sqrt{\rho a \sigma_{\mathrm{lv}}}} \tag{9.25}$$

where μ is the dynamic viscosity of the liquid and a is the characteristic dimension of the droplet. It has been demonstrated that above a critical Ohnesorge number ($\mathrm{Oh}_{\mathrm{crit}} \approx 0.03$), the viscosity is found to play an important role in droplet handling operations. However, a droplet of water in an EWOD application has an Oh that is much lower than the critical value of 0.03. Water is a very common solvent in EWOD systems so the effects of the viscosity of the liquid are generally fairly small. However, highly viscous ionic liquids have an Oh that may exceed the critical value, therefore significantly damping the droplet movement due to the viscous forces.

The ratio between the inertial and surface tension forces can be determined from the non-dimensional Weber number:

$$\mathrm{We} = \frac{\rho a v^2}{\sigma_{\mathrm{lv}}} \tag{9.26}$$

where v is the velocity of the droplet. Above a critical Weber number, $\mathrm{We}_{\mathrm{crit}}$, of approximately 1.1, the droplet will start to become distorted by the inertial forces and droplet breakup is a strong possibility. The We values in electrowetting applications are ~0.022, which is well below the critical value. Inertial effects are therefore relatively unimportant in most EWOD systems.

9.7 Motion, Splitting and Merging of Droplets in EWODs

For actuating the droplet motion in EWOD systems, the control electrodes are sequentially activated to set up suitable surface tension gradients on the surface of the microfluidic chip. This produces the necessary changes in the electrowetting forces to make the droplets move between neighbouring electrodes. Because of the small inertial force present, the droplets respond quickly to small changes in potential. However, the droplet motion can start only when the applied voltage exceeds the specific threshold voltage, required by the angle hysteresis phenomenon described earlier.

The droplet transport velocity is found to depend on the magnitude of the applied electrowetting force, in addition to many other factors including the viscosity of the liquid, the electrowetting hysteresis angle and the separation distance between the upper and lower plates of the chip.

9.7.1 Droplet Velocity

The maximum velocity of the droplet is reached when the electrowetting force, causing the motion, and the viscosity forces, causing the reduction in motion due to the shear stress along the surface of the chip, reach a balance. Velocity, v, is expressed as

$$v = \frac{h\sigma_{\mathrm{lv}}}{6\pi\mu r}(\cos\theta_{\mathrm{a}} - \cos\theta_{\mathrm{r}}) \tag{9.27}$$

where h is the separation between the upper and lower plates and r is the radius of the droplet in the horizontal plane. Under the hypothesis of the actuation voltage being lower than the saturation voltage and small contact angle hysteresis, the Berge–Lippmann–Young equation can be substituted into the velocity equation, leading to

$$v = \frac{h}{12\pi\mu r}CV^2 \tag{9.28}$$

Finally, the droplet velocity varies with the square of the applied voltage. A similar analysis can be performed using the modified Berge–Lippmann–Young equation accounting for the contact angle saturation. Under this condition,

$$v = \frac{h\sigma_{lv}}{6\pi\mu r}(\cos\theta_{\mathrm{s}} - \cos\theta_{\mathrm{eq}})L\left[\frac{CV^2}{2\sigma_{\mathrm{lv}}(\cos\theta_{\mathrm{s}} - \cos\theta_{\mathrm{eq}})}\right] \tag{9.29}$$

One of the most common operations used in EWODs is the splitting and merging of droplets, an example of which has already been shown in Figure 9.11. For splitting the droplet, the outermost electrodes are activated, which reduces the contact angle. This creates two hydrophilic zones that stretch the droplet in the longitudinal direction. At the same time, the non-actuated central electrode is a hydrophobic region, and exerts a 'pinching' force on the droplet which causes the neck of the droplet to split. Splitting is difficult to achieve in open EWOD systems since the droplet tends to move to only one of the hydrophilic electrodes. However, in closed EWODs, it is possible to force specific splitting locations by optimization of the distance between the top and bottom layers. By considering the pressure difference inside the droplet, caused by the different surface curvatures above the hydrophilic and hydrophobic electrodes:

$$\frac{h}{d} = -\cos\theta_{\mathrm{eq}} \tag{9.30}$$

where d is the characteristic dimension of the electrode, we can calculate, assuming that the hydrophobic contact angle is 110°, that the vertical separation of the plates should not exceed 340 μm.

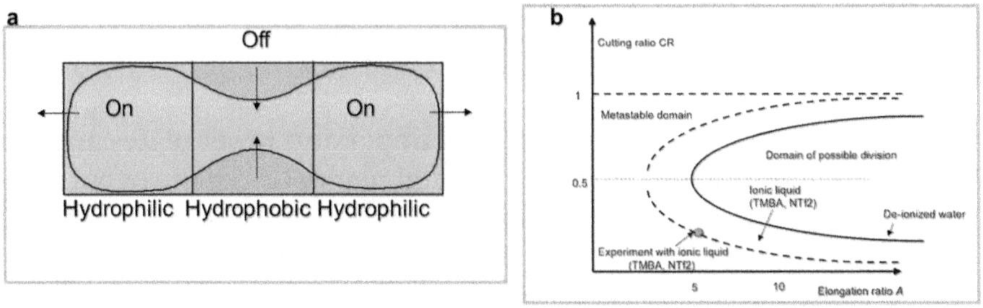

Figure 9.12 (a) The principle of droplet division. (b) A useful diagram for determining *A* and CR.

The principle of droplet division in a covered EWOD is shown in Figure 9.12a. The liquid is simultaneously stretched by two actuated electrodes at both ends of the droplet, creating two hydrophilic areas with electrocapillary forces that pull the droplet in opposite directions, forming an elongated droplet with a neck. The non-actuated hydrophobic electrode exerts a pinching force on the triple contact line of the newly formed neck. Depending on the force balance and the elasticity of the interface, the droplet can be cut in two. Ionic liquids may be split in an open EWOD system, because ionic liquids have very small contact angles with the actuated electrodes and inherently have an elasticity that is larger than that of aqueous liquids.

Two dimensionless parameters rule the division of a droplet in EWODs: the initial elongation ratio, A, between droplet length and droplet width and the cutting ratio, CR, between pinching length and initial droplet length. Figure 9.12b shows a useful diagram for understanding the relation between A and CR.

Open EWOD systems do not allow the division of many droplets, and for closed EWOD systems there exists a limiting value for the vertical gap above which splitting no longer occurs. The generic expression for determining the critical value of the gap is

$$\frac{\delta}{a} = -\cos\theta \tag{9.31}$$

where δ is the critical parameter and a is the characteristic dimension of the electrode. This expression is obtained by applying the Laplace law in the pinching region and correlating the pressures with the differences of the inverse of the curvature radii, while the minus sign considers the concavity of the drop surface. Eqn (9.31) also considers that $R_1 = \delta/2\cos\theta$ and that because the width of the pinching region goes to zero, $R_2 = a/2$. Considering all this, the modified Laplace equation is

$$\Delta P = 2\gamma_{\mathrm{lv}}\left(-\frac{\cos\theta}{\delta}-\frac{1}{a}\right) \tag{9.32}$$

which shows that the pressure inside the drop decreases when the vertical gap increases. The lowest possible pressure difference is zero, which gives $\delta/a = -\cos\theta$.

9.8 Geometry of the Electrodes

Throughout this entire chapter, the importance of the electrode has been mentioned many times. Clearly, the size and quality of the electrodes are essential factors for a high-quality EWOD. The gap between two contiguous microelectrodes is usually of the order of 10–30 μm due to the fabrication restrictions. The gap provides a hydrophobic region between two electrodes. If the droplet has a volume pinned in the boundaries of the electrode because of the presence of the hydrophobic gap, the motion of the droplet cannot occur until the neighbouring electrode is actuated. Jagged or crenellated electrodes have been designed.[14] The idea behind such a design is that the droplet contact line with the electrode plane extends over onto the next electrode, as shown in Figure 9.13a. When a dent is designed, parameters such as the size of the indentations, the width of the electrode, the separation gap between the electrodes, the total number of indentations along one side of the electrode, n, and the contact angles on the non-activated and activated electrodes must all be considered.

As soon as the next electrode is actuated, electrocapillary forces produce the motion of the droplet. Such jagged electrodes require more complicated microfabrication but are very efficient for droplet motion; Figure 9.13b is an example of such an EWOD.[15]

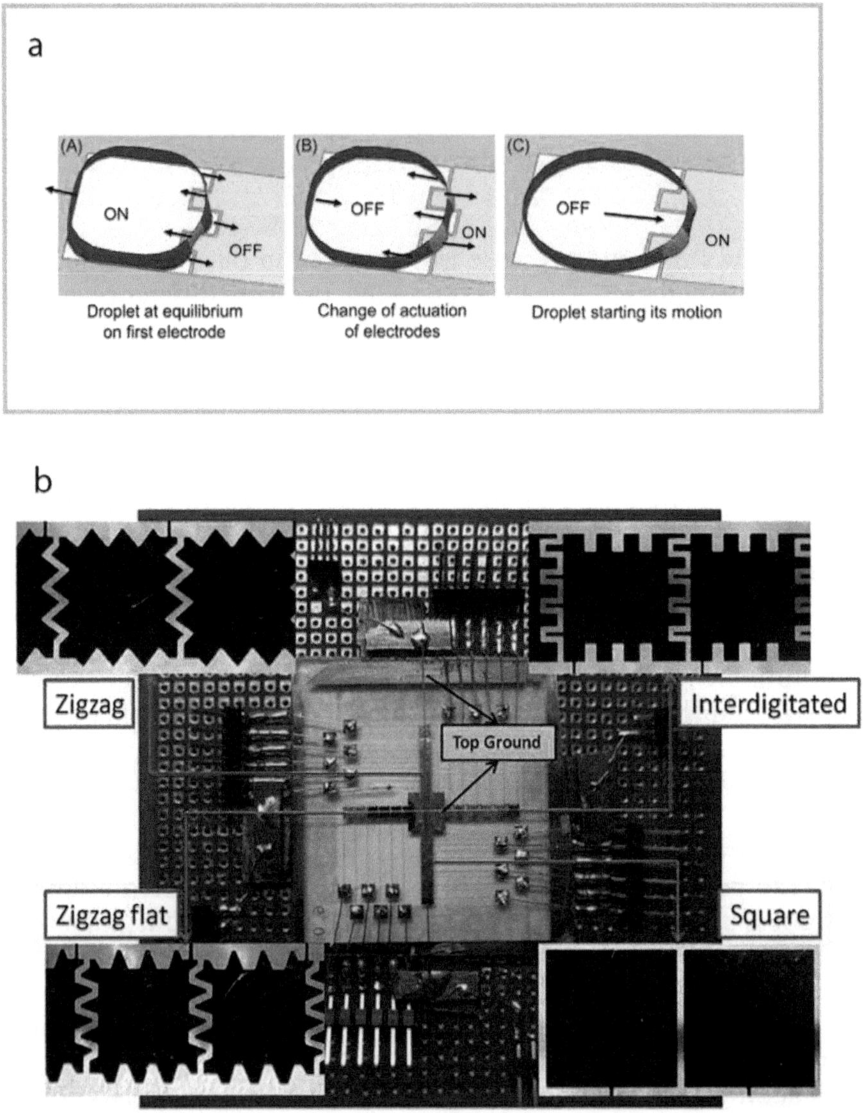

Figure 9.13 (a) Model of the droplet behaviour in the presence of a crenellated electrode. (b) An example of crenellated electrodes. Reproduced from ref. 15 with permission from Elsevier, Copyright 2017.

References

1. M. J. Fuerstman, P. Garstecki and G. M. Whitesides, *Science*, 2007, **315**(5813), 828.
2. Y. C. Tan, V. Cristini and A. P. Lee, *Sens. Actuators, B*, 2006, **114**(1), 350.
3. I. Shestopalov, J. D. Tice and R. F. Ismagilov, *Lab Chip*, 2004, **4**(4), 316.
4. W. C. Nelson and C. J. Kim, *J. Adhes. Sci. Technol.*, 2012, **26**(12–17), 1747.
5. B. J. Hindson, K. D. Ness, D. A. Masquelier, P. Belgrader, N. J. Heredia, A. J. Makarewicz, I. J. Bright, M. Y. Lucero, A. L. Hiddessen, T. C. Legler, T. K. Kitano, M. R. Hodel, J. F. Petersen, P. W. Wyatt, E. R. Steenblock, P. H. Shah, L. J. Bousse, C. B. Troup, J. C. Mellen, D. K. Wittmann, N. G. Erndt, T. H. Cauley, R. T. Koehler, A. P. So, S. Dube, K. A. Rose, L. Montesclaros, S. Wang, D. P. Stumbo, S. P. Hodges, S. Romine, F. P. Milanovich, H. E. White, J. F. Regan, G. A. Karlin-Neumann, C. M. Hindson, S. Saxonov and B. W. Colston, *Anal. Chem.*, 2011, **83**(22), 8604.
6. B. W. Colston, B. J. Hindson, K. D. Ness, D. A. Masquelier, F. P. Milanovich, D. N. Modlin, V. Riot, S. Burd, A. J. Makarewicz and P. Belgrader, *US pat.*, 9,156,010, Bio-Rad Laboratories Inc, Hercules, CA, US, 2015.
7. H. Moon, A. R. Wheeler, R. L. Garrell, J. A. Loo and C.-J. Kim, *Lab Chip*, 2006, **6**(9), 1213.
8. T. Nisisako, T. Torii, T. Takahashi and Y. Takizawa, *Adv. Mater.*, 2006, **18**(9), 1152.
9. R. Salomon, D. Kaczorowski, F. Valdes-Mora, R. E. Nordon, A. Neild, N. Farbehi, N. Bartonicek and D. Gallego-Ortega, *Lab Chip*, 2019, **19**, 1706.
10. J. Wang, J. Wang, L. Feng and T. Lin, *RSC Adv.*, 2015, **5**, 104138.
11. J. D. Tice, H. Song, A. D. Lyon and R. F. Ismagilov, *Langmuir*, 2003, **19**(22), 9127.
12. M. J. Jebrail, A. H. C. Ng, V. Rai, R. Hili, A. K. Yudin and A. R. Wheeler, *Angew. Chem.*, 2010, **49**, 8625.
13. V. N. Luk, G. C. H. Mo and A. R. Wheeler, *Langmuir*, 2008, **24**(12), 6382.
14. J. Berthier and K. A. Brakke, *The Physics of Microdroplets*, John Wiley & Sons, Hoboken, NJ, 2012, ch. 10.
15. V. Jain, V. Devarasetty and R. Patrikar, *J. Electrost.*, 2017, **87**, 11.

10 Extraction and Reactions

10.1 Extraction

In most cases, real samples contain contaminants, *e.g.* large organic or inorganic particles, that must be removed. Contaminants can cause many problems, including signal distortions and disruptions to fluid handling caused by fouling or blockage of the analytical system. Biological investigations often target one specific sample that could be contained within a cell. This must be removed by means of a cell membrane rupture technique. The sample must then be purified from the complex background matrix. Extracting and preconcentrating a sample before analysis can alleviate the sensitivity demands placed on the detector. In some cases, only trace amounts of sample are available, making the required extraction and preconcentration more sensitive, and thus more complicated.

10.1.1 Solid-phase Extraction

Solid-phase extraction refers to processes in which an analyte is retained by an appropriate solid stationary phase while the interfering sample matrix passes unhindered. Subsequently, the analyte is eluted in a purer and more concentrated form (see Figure 10.1).

Solid-phase extractions are split into two categories, termed non-selective (Figure 10.2a)[1] and selective (Figure 10.2b).[2] A non-selective solid phase is capable of extracting a wide range of compounds

Microfluidics and Lab-on-a-chip
By Andreas Manz, Pavel Neužil, Jonathan S. O'Connor and Giuseppina Simone

Published by the Royal Society of Chemistry, www.rsc.org

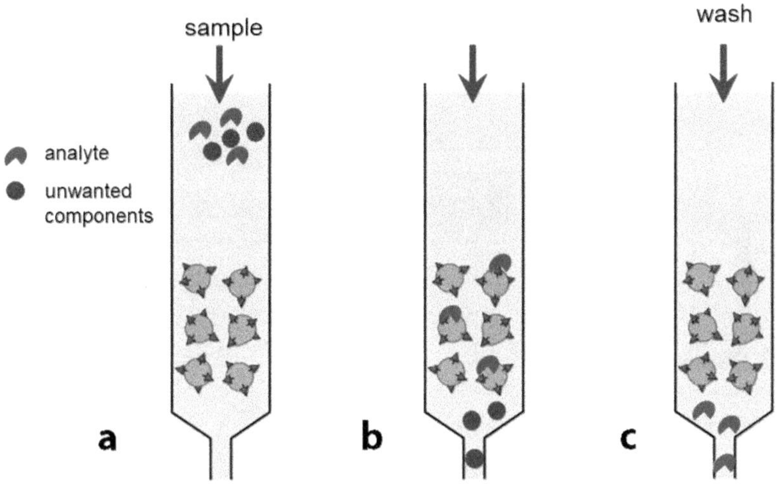

Figure 10.1 Schematic of a solid-phase extraction. a) Injection of sample. b) Extraction of analyte. c) Elution of analyte.

whereas selective solid phases are, as the name suggests, highly specific. Figure 10.2a illustrates an example of a non-selective solid phase, where the channel structure features two weirs at the intersection. These form a cavity that is loaded with 1.5–4 μm octadecylsilane-coated silica beads. A 1 nM solution of 4,4-difluoro-1,3,5,7,8-pentamethyl-4-bora-3a,4a-diaza-*s*-indacene (boron-dipyrromethene; BODIPY™ 493/503) is flushed across the cavity and binds to the beads. The BODIPY concentration on the beads then builds up as the solution flows, until the BODIPY is eluted with acetonitrile, leading to an increase in concentration of up to 500-fold. Such a process is an ideal example of extraction for preconcentration purposes.

Specific phases present functional groups or structures that immobilize only one specific target within the sample, which is essential for the targeting of bioactive species, as per the 'lock and key' model. Several forces can be responsible for the extraction; but for the scope of this chapter we will focus on the polymerase chain reaction (PCR) and the necessary DNA extractions. PCR is discussed in greater detail in Chapter 15, but for now the key principle to know is that PCR is a method to copy DNA and increase the concentration present. Silica is a very common solid phase for DNA extractions. Silica can be used in many forms, but beads and sol–gels are the

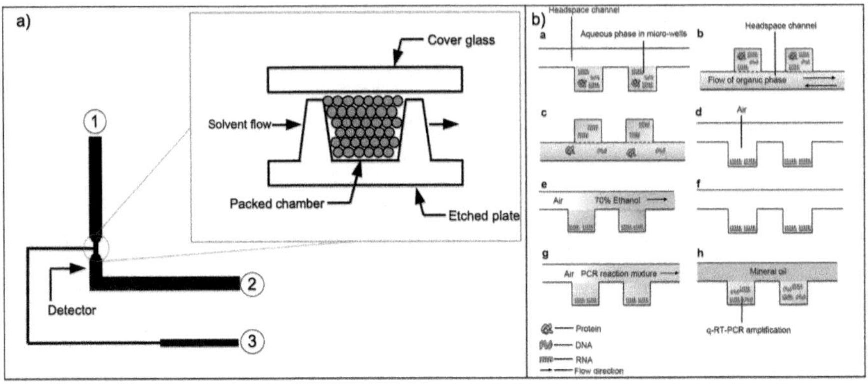

Figure 10.2 Microfluidics solid-phase extraction. (a) Top: schematic of microchip device used for microchannel packing and solid-phase extraction: 1, outlet buffer channel; 2, inlet sample/buffer channel; 3, bead introduction channel. Bottom: schematic and operation of a microfluidic chip that implements weir placement cavities and a hooked bead channel entrance. Etched channels are 580 μm wide in the wide channels and 30 μm wide for the bead introduction channel. The approximate chamber volume is 330 pL. Channels are 10 μm deep; the weirs are 9 μm high. (b) Schematic of the RNA purification process using a liquid-phase nucleic acid purification chip. Description of the steps from a to h. An aqueous phase containing DNA, RNA and protein as the model macro-biomolecules of bacterial lysate was isolated in the microwells. An organic phase of PCI (phenol–chloroform–isoamyl alcohol) equilibrated at pH 4.6 was then introduced into the headspace channel with continuous forward and reverse flow. In the case of on-chip DNA extraction, an organic phase with a pH of 8.0 was introduced into the headspace channel instead. The nucleic acid purification chip was inverted in this step as the organic phase has a higher density than the aqueous phase. Subsequently, the protein and DNA were transferred from the aqueous phase into the organic phase, while RNA was retained in the aqueous phase and the organic phase was expelled from the headspace channel and evaporated under vacuum, and purified RNA in the microwell was concurrently dried. Residual organic phase was further decontaminated by repeated washing and vacuum evaporation with 70% ethanol. Finally, a q-RT-PCR (quantitative reverse transcriptase PCR) reaction mixture was loaded into the microwells. Finally, the microwells were covered with mineral oil followed by on-chip q-RT-PCR amplification. Reproduced from ref. 1 and 2 with permission from American Chemical Society, Copyright 2000 and 2013.

most prevalent. DNA binds to silica because, in the correct buffer, both show a polarity whereas other contaminants remain non-polar. Therefore, this polar interaction is the specific extraction force. Extracted DNA can then be eluted by shifting the pH, hence changing the polarity presented by either the DNA or the solid phase and terminating the adsorption.

Microfluidic structures allow a very large volume-to-surface area ratio to be fabricated. The larger the ratio, the more sites are available to the extract, therefore increasing the likelihood of site binding and extraction. It is highly recommended that the reader peruses a paper by Wen *et al.*[3] as an introduction to the methods of silica-based DNA extractions on microfluidic chips.

The integration of solid-phase extractions into a microchip provides higher extraction efficiencies, reproducibility and sample stability. Another advantage is the simple compatibility with electrophoresis on microfabricated chips. Also, the efficiency of microchip-based solid-phase extraction is comparable to that of the standard methods based on the use of a microcentrifuge, with a drastic shortening of the experimental time.

10.1.2 Liquid-phase Extraction

Liquid-phase extraction, also referred to as solvent extraction, is the act of extracting at least one component from a liquid mixture by solubility preference into a different immiscible solvent. Sometimes an extractant is used. Extractants are chemical species that facilitate extraction by binding or forming molecular complexes with components of interest. This binding then allows transfer into the solvent to occur. There are many key factors in liquid–liquid extractions, including the contact time of different phases, pH and extractant concentration, and these will be discussed as necessary within the chapter.

A typical liquid–liquid extraction functions by extracting the desired compound or solute, B, out of the feed, A + B, into the solvent, C, forming the extract. The remaining A + B is called the raffinate, which is now depleted of B. The component A left behind in the raffinate is called the diluent. We can assess whether the chosen solvent is suitable by examining the partition ratio. The partition ratio, partition coefficient or distribution ratio, K_D, is defined as the ratio of the solute concentration in the extract phase to that in the raffinate phase after one equilibrium extraction. The partition ratio is dimensionless, but it is important to note the concentration units used in the calculation of K_D. For the extraction of the diluted solution, K_D is constant and it is convenient to use an extraction factor, E:

$$E = \frac{c_1 V_1}{c_2 V_2} = K_D \frac{V_1}{V_2} \tag{10.1}$$

where V_1 and V_2 are the volumes of two immiscible liquids and c_1 and c_2 are the respective concentrations of a given solute in either phase.

There are two types of liquid–liquid extractions, which are often defined using ternary diagrams: *Class I*, an immiscible pair of solvents with a distinct envelope, and *Class II*, two pairs of immiscible compounds – therefore two crossing envelopes exist. This adds complexity to the equilibrium relationship as some of each component is present in each phase. An extensive insight into the importance of the ternary diagram can be gained from the literature,[4–6] but for the purpose of this chapter we shall assume simplistic *Class I* extractions. The use of the ternary phase diagrams is important for the design of the liquid extractor.[7] Laboratory bench continuous-phase separations are often driven by the difference in the densities of the two fluid phases within a settler tank. A typical experiment is the extraction of caffeine from tea using dichloromethane (DCM). However, the gravitational forces are small compared with surface forces at the microscale, making gravitational density separations more challenging, unless alternative forces for driving phase separations are considered. Surface tension effects are particularly attractive as these dominate gravity and viscous forces at the microscale, as the Bond number (Bo) and the capillary number (Ca) are both about 1.

Extractions using multiple solvents focus on the extraction of multiple solutes, with the several solvents providing a method of highly controllable extractions. Extraction can be operated as either a single-stage process, where most of the solute is extracted in one cycle, or a multi-stage process, where the feed is cycled through multiple extractions to increase efficiency. Microfluidic counter-current and co-current extractions are capable of implementing sequential extractions, where the raffinate from a previous stage is fed into another extraction with fresh solvent. Counter-current extraction requires an arrangement where the feed and the extraction solvent have opposite flows. Surface treatments are required to establish stable counter-flowing phases. It is also possible for extractions to have efficiencies as high as 98.6%.[8] However, these advantages compete against the relatively low interfacial surface area-to-microchannel volume ratios and hence low yields.[9] Small interfacial areas have to be used for the extraction in order to maintain a sufficient capillary pressure to counterbalance the imposed driving pressure. However, this opposes the important principle that the larger the surface area of contact between the two phases, the shorter is the length of diffusion required for extraction. Additionally, the surface treatment can degrade over time, causing the flow stability to degrade. Modifying the wetting characteristics of the surfaces can stabilize interfaces. Surface modification can degrade over time and it is difficult to operate counter-flow extractors efficiently for more than one equilibrium stage. However, extractors exploiting coating with hydrophobic molecules have been successfully designed and used. More complicated extraction systems have also been developed, such as a three-phase flow extraction.[10]

Finally, membranes can be implemented to separate the liquid phases, while still allowing the selective transport of solute. The working principle of membrane-based liquid–liquid extractions can be demonstrated by considering Figure 10.3.[11] B, dimethylformamide (DMF), was selectively separated from a non-aqueous feed solution of A, dichloromethane (DCM), into C, water, into a receiving phase within a few seconds by employing a PTFE membrane formed in a microfluidic device. The PTFE membrane is wetted by the non-aqueous components, whereas the aqueous components do not wet the membrane. This wetting area allows the diffusion of B into the extract without water also crossing the membrane. A segmented flow, as shown in Figure 10.3a, can be used to increase the efficiency of membrane microextractors and facilitate convective mass transfer between the phases, because the surface area of contact increases, allowing more diffusion. To overcome this limiting factor, a liquid–liquid phase separation by using a thin, porous fluoropolymer membrane that selectively wets non-aqueous solvents has been designed and implemented, as shown in Figure 10.3b. In the integrated system shown in Figure 10.3c, the partially miscible fluids, A and B, are mixed, then brought into contact with the selectively immiscible liquid C, where component A partitions from B and C. Finally, the two phases A and B are separated by the membrane.

The working principle of the extractor described above is based on the equilibrium between the pressure and the capillary force. The pressure drop across the membrane is given by the Hagen–Poiseuille

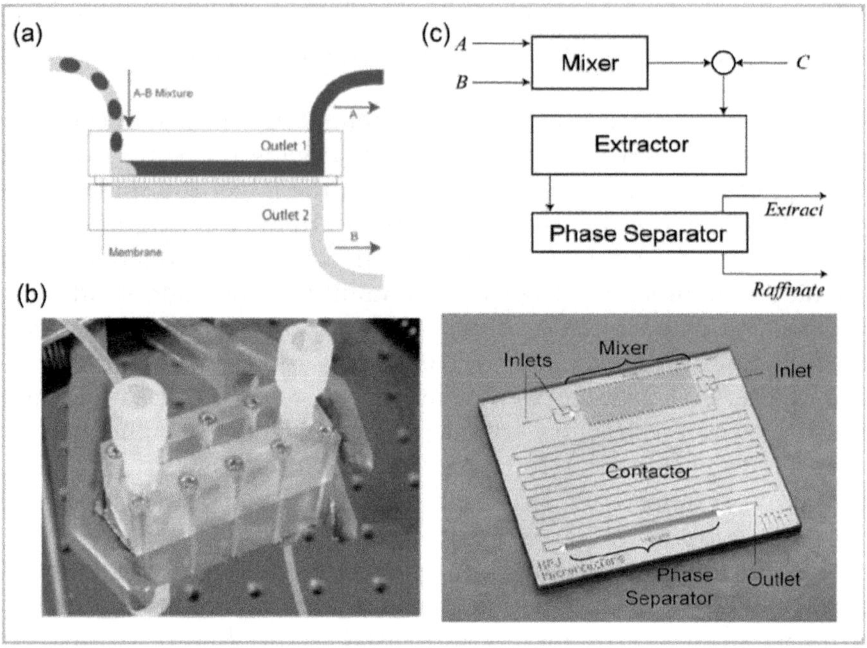

Figure 10.3 Two examples of a microfluidic extractor. (a) Schematic of the proof of principle for a PTFE membrane device with a segmented flow of a non-aqueous solution (A) dispersed in an aqueous phase (B). The organic phase wets the hydrophobic membrane and is driven through the membrane pores by the imposed pressure difference, leaving the aqueous solution behind in the bottom portion of the device. (b) The device fabricated in polycarbonate. (c) Top: schematic of the operation of a integrated silicon extractor device. Bottom: the fabricated device, 35 × 30 × 1.5 mm (W × L × H). Reproduced from ref. 11 with permission from the Royal Society of Chemistry.

equation, which is a function of the flow rates flowing throughout channel 1. Assuming that all of the pores are perfect cylinders, the capillary pressure of the pore (ΔP_c) is

$$\Delta P_c = \frac{2\gamma}{R}\cos\theta \tag{10.2}$$

where γ is the interfacial tension, R the pore size and θ the wetting angle. The actual pressure difference between the two sides of the membrane is $\Delta P_1 - \Delta P_2$. Flow across the membrane can only occur, if the pressure across the membrane is greater then the capillary pressure of the pore, $\Delta P_1 - \Delta P_2 > \Delta P_c$. Eqn (10.2) is used when design considerations are taken for membrane fabrication, as the pore size, materials and surface energies play such a large role in the cross-membrane flow.

There are two extreme cases of pressure drop based on the length of segments relative to the length of the membrane's pores: (i) small alternating aqueous and organic segments and (ii) large aqueous segments followed by large organic segments. In the case of small alternating aqueous and organic segments, the flow rate through outlet 1 will be Q_1, where all of the organic phase can go through the membrane and out from outlet 2. In the second case, the organic phase is longer than the separator region, so the pressure difference is not great enough, hence there is no flow through the membrane and outlet 1 will include organic segments with a flow rate Q_2. The actual operational Q falls between the two cases. Any flow throughout the membrane produces a pressure drop. However, membranes with a high pore density (n) aid the pressure drop, as the hydrodynamic resistance of the wetting phase is distributed across all of the flows across all pores. Increasing n leads to an increase in the flow of organic solutions across the membrane.

10.2 Reactions and Microreactors

10.2.1 General Introduction to Reactions and Reactors

When designing a microreactor, one must consider the thermal character of the reaction within. Many reactions require a constant temperature or specific temperatures depending on the part of the process. Extensive work has been performed in this area.[12] Reactions can also be classified as homogeneous, if the components exist in the same phase (gas, liquid or solid), or heterogeneous, if the components exist in different phases.

Three typologies of ideal reactors exist, the first of which is called the batch reactor, in which the reactants are initially loaded into a container, mixed and left to react for a certain period and then the products are recovered. This is an unsteady-state operation in which the composition changes with time, but at any instant the composition throughout the reactor is uniform. Although examples of these exist for specific reactions, *e.g.* a batch positron emission tomography (PET) tracers synthesis,[13] they are very uncommon.

The second ideal reactor is called a flow reactor, which is a steady-state operation. Flow reactors are variously known as plug flow, slug flow, piston flow, ideal tubular and unmixed flow reactors. The characteristics of a flow reactor are given by the flow of fluid through the reactor, which is assumed to be orderly with no element of fluid overtaking or mixing with any other element ahead or behind. In fact, it is possible that some lateral mixing of fluids in a plug flow reactor occurs; however, there must be no mixing or diffusion along the flow path. The condition for plug flow is for the residence time in the reactor to be the same for all elements of fluid.

The third reactor, which is also another ideal steady-state flow reactor, is called the mixed reactor or the ideal stirred tank reactor. This is a reactor in which the contents are well stirred and uniform throughout. Thus, the exit stream from this reactor has the same composition as the fluid within the reactor. Droplets have also been used as microreactors, as discussed in Chapter 9.

10.2.2 Microreactors

The three models of ideal reactors presented above are operable at the microscale, allowing many ways of processing a fluid: in a single batch or flow reactor, in a chain of reactors possibly with inter-stage feed injection or heating, in a reactor with recycling of the product stream using various feed ratios and conditions, and so on. Several factors have to be considered in order to design the best scheme for maximizing the yield of the reaction, such as the reaction type, planned scale of production, cost of equipment and operations, safety, stability and flexibility of operation, equipment life expectancy, length of time that the product is expected to be manufactured and ease of convertibility of the equipment to modified operating conditions or to new and different processes.

It is fair to assume, considering the factors already listed, that most of the microfluidic reactor research has been carried out in industrial settings, motivated by the hope of employing millimetre-scale reactors to allow a relatively smooth transition between discovery systems with low volumetric throughput and production systems with higher volumetric throughput. Analysis of the characteristic dimensionless numbers associated with reactors, such as Reynolds and Péclet numbers, shows that for a certain range of residence time (t = volume/flow rate), a reactor channel of 500 μm or 1 mm diameter operates within this microchannel-like zone of equivalency, acting much like a classic 100 μm microreactor.[14] It is also worth noting that mixing and thermal diffusion (best characterized by Fourier numbers) and convection or buoyancy effects (governed by Grashof or Rayleigh numbers) scale non-linearly with the characteristic dimension. Therefore, if these phenomena are to be controlled, then the correct scale of the reactor must be used. As stated in Chapter 1, lab-on-a-chip systems seek to perform complete processes, in this case a chemical synthesis, in one device. Therefore, functions of the device should include the controlled addition of reagents to a reaction mixture, mixing of the reagents, monitoring the reaction, transduction, *etc.*

Microreactors have found applications in environmental and toxicity analysis, catalyst screening and synthetic chemistry. Microfluidic reactors offer several crucial advantages over the conventional systems used in chemistry and biology, such as reduced requirements for reagents and solvents, which has the double advantage of requiring fewer reactants and producing less waste, hence the microreactors are environmentally friendly, as the process consumes less material and reduces the amount of hazardous chemicals. The large surface area featured by microchannel systems is advantageous for many chemical processes, such as extractions, catalytic reactions, highly exothermic reactions and reactions in which hazardous compounds are generated. Owing to the high surface-to-volume ratio, microfluidic reactors show better efficiency in heat and mass transfer and yield of the reaction.

Typical examples of microreactors are shown in Figure 10.4.[15–18] A strong trend in microfluidics is the development of commercially available microfluidic reactors, with robust designs with multiple syntheses stages, in what effectively becomes one module in a scale-out system. Another advantage of microreactors is the small operational volumes required, allowing for the rapid screening of very limited amounts of reagents/products, and finally, observing and manipulating several operational conditions simultaneously.

Green chemistry is a typical field where microreactors have been of keen interest. Green chemistry aims to decrease the energy requirements, increase the throughput per unit production area, decrease reagent consumption and use less hazardous chemicals, all of which are inherent to microfluidic reactor systems. *m*-Bromoanisole is converted into *m*-anisaldehyde in a highly exothermic reaction that is normally carried out in semi-continuous and batch reactors in industrial applications. The synthesis was scaled to microscale continuous reactors.[19] The durable stainless-steel microreactor was designed as a stemmed microfluidic reactor

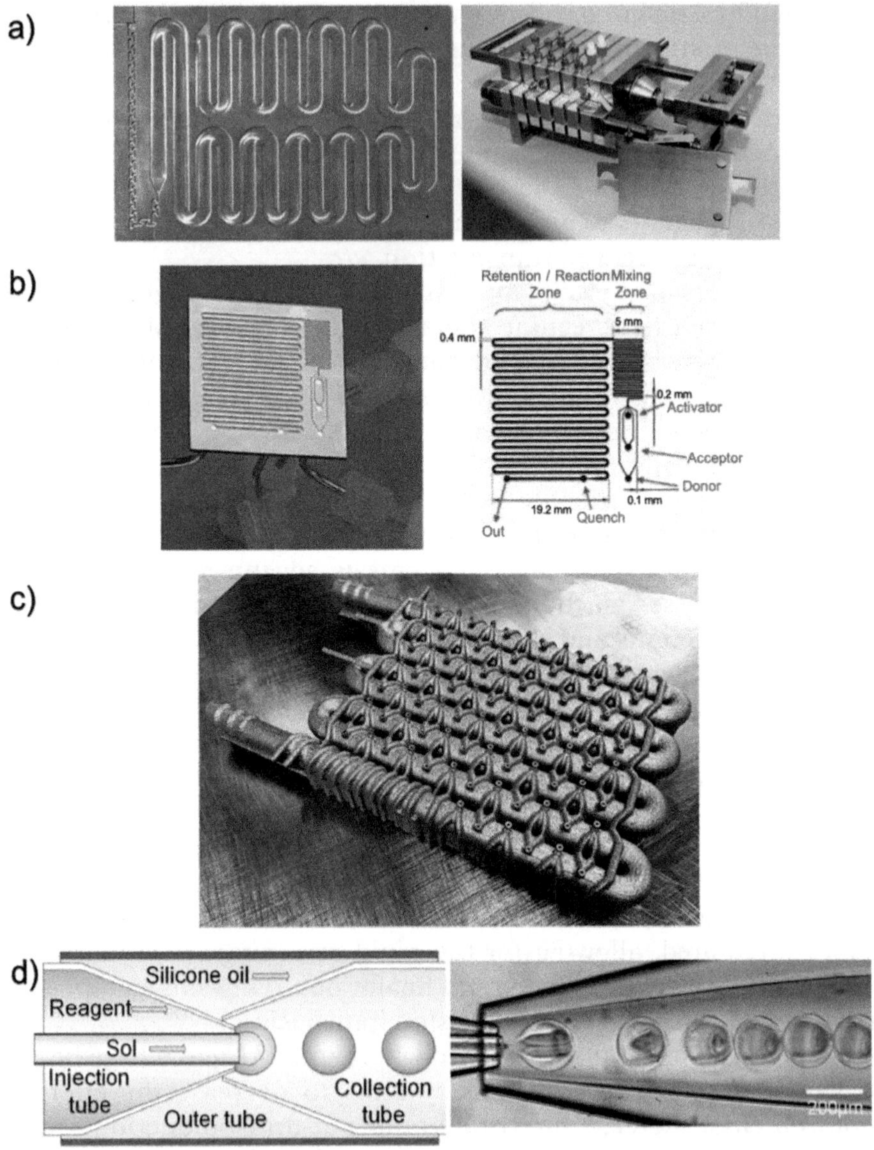

Figure 10.4 Several typologies of microfluidic reactors. (a) Lonza microreactor technology based on continuous plug flow reactors. Left: a single open-faced A6 plate. Right: multiple closed-faced A6 plates stacked between heat exchanger plates and assembled in a rack with a vice. (b) A photograph and schematic of a silicon and Pyrex microfluidic reactor used for glycosylation reactions. (c) Stainless-steel reactor for the difluoromethylation of

to improve the heat-exchange capabilities, saving in energy consumption for cooling, reduce the amounts of solvents consumed and increase the reaction yield.

Semi-batch microreactors also exist. An example is the production of size-controlled zeolitic imidazolate framework (ZIF-8) nanoparticles using a T-type micromixer.[20] It was found that the particle size could be finely tuned, as smaller particles are produced with an increase in flow rates while remaining laminar but a sufficiently higher mixing performance is accomplished at Re >2000, where the flow transitions between laminar flow and turbulent flow. At that Re point, the size and shape of the resulting nanoparticles are no longer dependent on the flow rates. Additionally, lowering the temperature increases the number of nuclei during the nucleation process, leading to smaller nanoparticles. However, the reagent ratios affect the subsequent particle growth process. An excess of 2-methylimidazole (2-MeIM) prevents particles from growing by covering the ZIF-8 particle surface. This process acts as a clear example of the benefits of microreactors to industrial production: continuous operation, reduced costs and highly controlled.

Finally, plate-stack reactors are also used, in which different volumes within the reactor are used to control the residence times, increase the speed of mixing, enhance heat exchange and minimize pressure drops.[21]

---

a lithiated nitrile with fluoroform as a major component. The reactor has an imprint of 164 × 93 × 16.4 mm (including the support structure). The reactor consists of a serpentine cooling core surrounded by the reaction channels with an inner diameter of 0.8 mm and a total length of 4 m, with four inlets, two defined reaction zones and one outlet. (d) A droplet microreactor to realize the internal sol–gel reaction of ZrO_2 inside the droplet. Left: schematic of the droplet microreactor in a microfluidic device. Right: image of droplet microreactors formed in the microfluidic device. (a) Reproduced from ref. 15 with permission from John Wiley and Sons, Copyright © 2019 Canadian Society for Chemical Engineering. (b) Reproduced from ref. 16 with permission from the Royal Society of Chemistry. (c) Reproduced from ref. 17 with permission from the Royal Society of Chemistry. (d) Reproduced from ref. 18 with permission from John Wiley and Sons, Copyright © 2016 The American Ceramic Society.

References

1. R. D. Oleschuk, L. L. Shultz-Lockyear, Y. Ning and D. J. Harrison, *Anal. Chem.*, 2000, **72**(3), 585.
2. R. Zhang, H. Q. Gong, X. Zeng, C. Lou and C. Sze, *Anal. Chem.*, 2013, **85**(3), 1484.
3. J. Wen, L. A. Legendre, J. M. Bienvenure and J. P. Landers, *Anal. Chem.*, 2008, **80**, 6472.
4. R. M. Price, *Liquid Extraction Lecture*, California Baptist University, 2003, http://facstaff.cbu.edu/rprice/lectures/extract.html, Date accessed: 8/2/2020.
5. U. Roessner and T. Rupasinghe, Liquid Extraction: Phase Diagram, in *Encyclopedia of Lipidomics*, ed. M. Wenk, Springer, Dordrecht, 2016, DOI: 10.1007/978-94-007-7864-1_87-1.
6. L. A. Robbins and R. W. Cusack, Section 15: Liquid-Liquid Extraction and Other Liquid-Liquid Operations and Equipment, in *Perry's Chemical Engineers' Handbook*, ed. D. W. Green, The McGraw-Hill Companies, New York, 7th edn, 1999.
7. W. L. McCabe, J. C. Smith and P. Harriott, *Unit Operations of Chemical Engineering*, McGraw-Hill, New York, 5th edn, 1993.
8. A. Aota, M. Nonaka, A. Hibara and T. Kitamori, *Angew. Chem., Int. Ed.*, 2007, **46**, 878.
9. https://www.zaiput.com/product/multistage-extraction/.
10. K. K. R. Tetala, J. W. Swarts, B. Chen, A. E. M. Janssen and T. A. van Beek, *Lab Chip*, 2009, **9**, 2085.
11. J. G. Kralj, H. R. Sahoo and K. F. Jensen, *Lab Chip*, 2007, **7**(2), 256.
12. V. Miralles, A. Huerre, F. Malloggi and M.-C. Jullien, *Diagnostics*, 2013, **3**, 33.
13. A. Elizarov, C. Meinhart, R. Miraghaie, R. M. van Dam, J. Huang, A. Daridon, J. R. Heath and H. C. Kolb, *Biomed. Microdevices*, 2011, **13**(1), 231.
14. K. S. Elvira, X. C. i. Solvas, R. C. Wootton and A. J. Demello, *Nat. Chem.*, 2013, **5**(11), 905.
15. A. Macchi, P. Plouffe, G. S. Patience and D. M. Roberge, *Can. J. Chem. Eng.*, 2019, **97**(10), 2578.
16. D. M. Ratner, E. R. Murphy, M. Jhunjhunwala, D. A. Snyder, K. F. Jensen and P. H. Seeberger, *Chem. Commun.*, 2005, **36**(5), 578.
17. B. Gutmann, M. Köckinger, G. Glotz, T. Ciaglia, E. Slama, M. Zadravec, S. Pfanner, M. C. Maier, H. Gruber-Wölfler and C. O. Kappe, *React. Chem. Eng.*, 2017, **2**, 919.
18. P. Wang, J. Li, J. Nunes, S. Hao, B. Liu and H. Chen, *J. Am. Ceram. Soc.*, 2017, **100**, 41.
19. D. Kralisch and G. Kreisel, *Chem. Eng. Sci.*, 2007, **62**(4), 1094.
20. D. Yamamoto, T. Maki, S. Watanabe, H. Tanaka, M. T. Miyahara and K. Mae, *Chem. Eng. J.*, 2013, **227**, 145.
21. S. J. Haswell and P. Watts, *Green Chem.*, 2003, **5**(2), 240.

11 Separations On-chip

11.1 Chromatography

The separation of different types of molecules can be based on the distribution equilibrium between a mobile and a stationary phase. The principle is the same as for extractions, discussed in depth in Chapter 10, but the process is now continuous and uses injected sample plugs that are much smaller in volume than the volume of the separation area. The fundamentals of chromatography can be found in many general textbooks on analytical chemistry.[1,2] The separation can be performed on the length scale, as in thin-layer chromatography, for example, whereby a fixed length is given for the separation to occur, or on the time scale, which is the more common case, as used in high-performance liquid chromatography (HPLC). A schematic diagram of an HPLC separation is shown in Figure 11.1. It consists of a separation column of fixed length, a sample injector at one end and a detector at the other end.

The sample injection volume and also the detection volume must be much smaller than the volume of the separation column, otherwise the full separation resolution cannot be achieved. Conventional HPLC columns are steel tubes or fused-silica capillaries packed with particles in the micrometre range. These particles carry the stationary phase on their surface. The mobile phase is pumped by a high-pressure pump through the injector, separation column and detector. Different transit times occur depending on the differences in the extraction equilibrium for each compound. Detection is mostly carried out using optical absorbance or mass spectrometry (MS).

Microfluidics and Lab-on-a-chip
By Andreas Manz, Pavel Neužil, Jonathan S. O'Connor and Giuseppina Simone

Published by the Royal Society of Chemistry, www.rsc.org

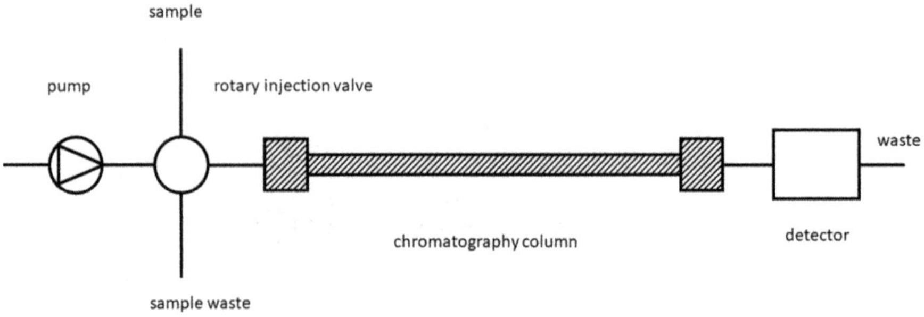

Figure 11.1 Schematic of the chromatographic process as used in high-performance liquid chromatography (HPLC).

Liquid chromatography has been implemented in microfluidic chips by three different methods: as a packed bed of microparticles, strictly in analogy with conventional HPLC, as a polymer monolith separation column similar to polymer monolith capillary columns and as microfabricated fluidic channel networks (Figure 11.2). The last approach exists only in microfluidic systems.

The separation efficiency, or resolution, depends strictly on the particle size, the capillary channel size and the pressure applied. The higher the pressure and the smaller the characteristic dimension (particle size and capillary channel size), the better is the separation. The substrate material and the interface to the pump need to be pressure tolerant. The most common substrate materials are silicon, glass and rigid polymers. Although these can function at high pressures, given the correct fabrication, they are not impervious to deformation and rupture. Microfluidic chip HPLC separations do not normally show superior resolution or shorter analysis times compared with the use of conventional HPLC columns. However, the small internal volumes are very attractive for small-scale separations. Many biological samples, *e.g.* protein extracts from cells, are small by definition. To obtain larger samples is very time consuming and therefore microfluidic HPLC is mainly commercialized for protein separations and MS analysis. For this purpose, sample preparation, injection, separation and electrospray are integrated in a single device.[3] The preferred separation column is the packed-bed type. Sample preparation consists in passing the material through a micromachined frit, which filters out particulate matter from the sample components, and an enrichment column for extracting the proteins from a larger sample volume onto

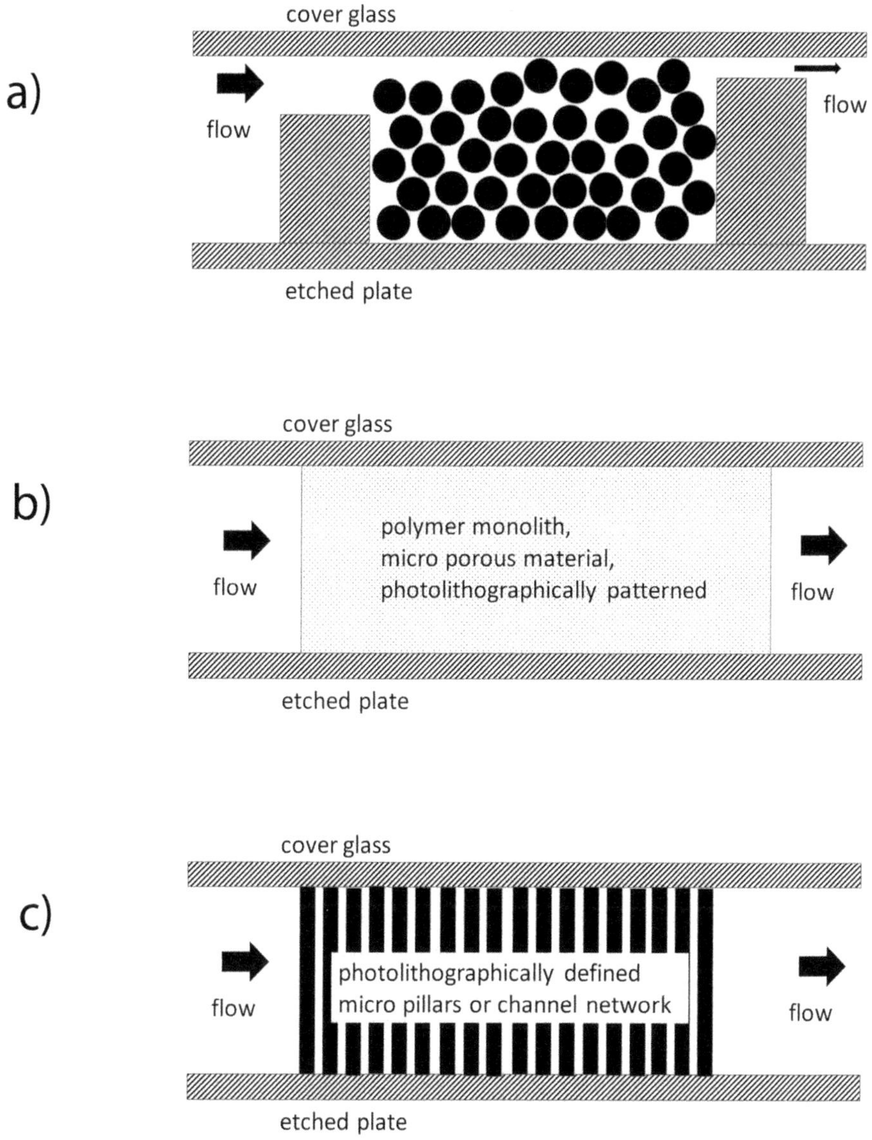

Figure 11.2 Side-by-side comparison of the different implementations of HPLC columns on-chip. (a) Packed bed; (b) polymer monolith; (c) channel network.

a smaller volume surface (see Figure 11.3). For sample loading and injection, an integrated rotary injection valve is used. The microfluidic device is made from multiple layers of polyimide and is pressure tight up to 200 atm.

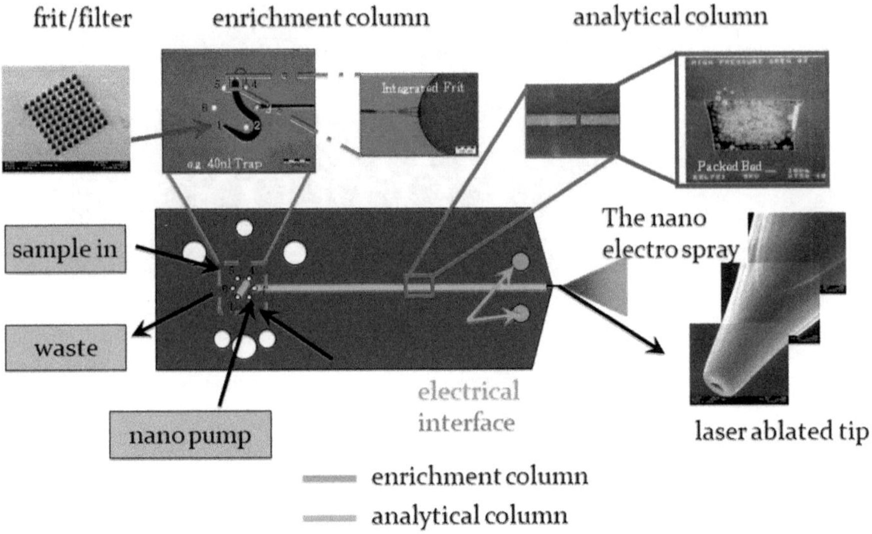

Figure 11.3 Schematic of a commercialized HPLC system for protein separations. The location of the rotary valve is on the left, with frit and enrichment column built in. The tip on the right-hand side serves as an electrospray interface to a conventional mass spectrometer. Image courtesy of Agilent Technologies, Inc., © Agilent Technologies, Inc.

11.2 Electrophoresis

Electrophoresis is commonly used for separations of charged biomolecules (ions). Briefly, the separation principle is defined by the applied electric voltage and depends on charge divided by the friction of the ion in the separation medium. This means that positive and negative ions move in opposite directions and neutral molecules cannot be separated. Ions with the same charge are separated by the differences in friction, whereas differently charged ions will have different velocities under the applied electric field. The fundamentals of electrophoresis can be found in many general textbooks on bioanalytical chemistry.[2,4] A common setup uses a slab gel for the separation with a few hundred volts applied. Such experiments are inexpensive but labour intensive. The resolution is limited by Joule heating, as the heat generated changes the diffusion and viscosity of the medium. The separation is usually performed for a fixed time and the distance travelled is measured for each compound.

To achieve much higher electric fields, fused-silica capillaries have been introduced for capillary electrophoresis.[5] Owing to the smaller electric currents in such capillaries, Joule heating occurs only at much higher voltages and therefore better separation efficiencies can be obtained. A typical capillary is 1 m long and a voltage of 20 kV is applied. One side of the capillary serves for sample introduction and a detector is used on the opposite side (Figure 11.4). Fluorescence detection is typically applied.

Capillary electrophoresis can be applied to separations of small ions, peptides/proteins and oligonucleotides. The main application is in DNA sequencing, where up to 1000 different fragments can be separated according to their lengths. Capillary electrophoresis was the key technology for the human genome project in the early days.

For special problems, capillary electrophoresis is used in the isoelectric focusing mode. The instrumentation is very similar to the setup shown in Figure 11.4. However, isoelectric focusing depends upon a pH gradient for the separation. Isoelectric focusing is mainly used for peptides and proteins. Proteins will have a charge that is a function of the surrounding pH. This means that, under an electric field, the charged protein will move along the pH gradient until it reaches a buffer zone pH where the sample presents a net charge of zero. That is called the isoelectric point, at which the molecules will not move any longer in the electric field. Each type of protein can be focused at a different location (pH). For detection, the pH gradient will have to be moved past the detection point.

In capillary electrophoresis, the properties of the wall can play a significant role in the movement of ions and molecules. Depending on the wall charge (zeta potential), a so-called electroosmotic flow (EOF) can occur. Glass surfaces, for example, are typically negatively

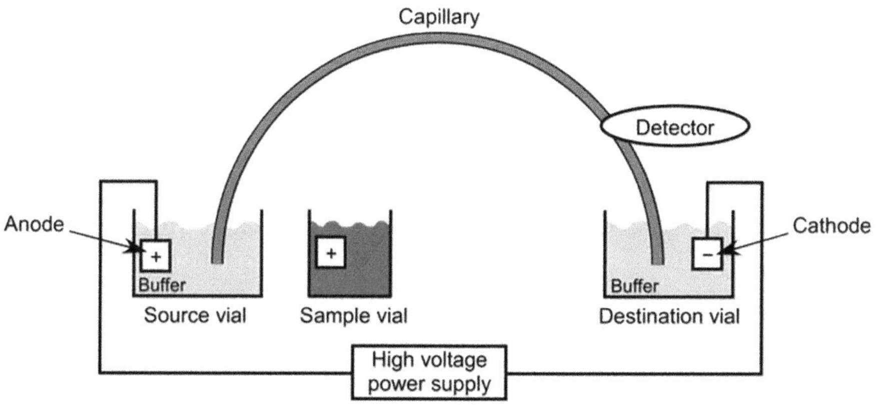

Figure 11.4 Schematic of a capillary electrophoresis setup. For injection, the capillary has to be moved from the buffer vial to the sample vial and back.

charged. This results in an EOF in the direction of positive ions (cations) towards the negatively charged electrode (cathode). A more in-depth description of EOF can be found in Section 6.3 in Chapter 6. Such EOFs are superimposed in electrophoretic separations and therefore, in the best case, anions and cations move into the same direction and can be easily injected and detected.

The separation efficiency in electrophoresis depends on the applied voltage and is limited by Joule heating. The speed of a separation depends on the separation length. The separation can be hidden if the sample injection volumes or detection volumes are too large compared with the volume of the separation capillary. In the setup shown in Figure 11.4, injection can be carried out using a voltage or a pressure. In the case of injecting by voltage, there will be a so-called electrophoretic bias. That means that the injected amount of faster moving ions is much larger than that of slower moving ions, which can be problematic. In the case of pressure injection, a parabolic flow profile forms during injection, which results in a larger volume of sample injected.

Electrophoresis on a microfluidic chip can solve these injection problems by using a simple channel manifold. As shown in Figure 11.5, the sample is brought to the injection point in a channel perpendicular to the separation channel, which is easily achieved by applying an electric voltage across the sample and waste wells.[6] Electrophoretic bias can be avoided by giving the sample sufficient time for the slowest moving components to arrive at the injection point. Then the voltages can be switched for the separation. The injection area is defined by the channel width and depth and can define volumes as small as picolitres. This allows the use of short separation lengths and high voltages. In contrast to chromatography on-chip, electrophoresis on-chip is much faster and high separation resolutions are possible compared with conventional capillary electrophoresis and slab gel electrophoresis. Separations in a few seconds are common.

The commercial version of an electrophoresis chip has a more complicated layout, for multiple samples (see Figure 11.6). The applied voltages are in the low kV range and the separation distance is 12 mm. The devices are used mainly for RNA and DNA fragment sizing.[7]

More recent commercial products use polymer film and blister technology, to provide prefilled straight electrophoresis channels, without any sophisticated channel manifold for injection. The channels are larger in cross-section and the voltages applied are lower.[8]

A more in-depth review of recent developments and applications can be found in the literature.[9]

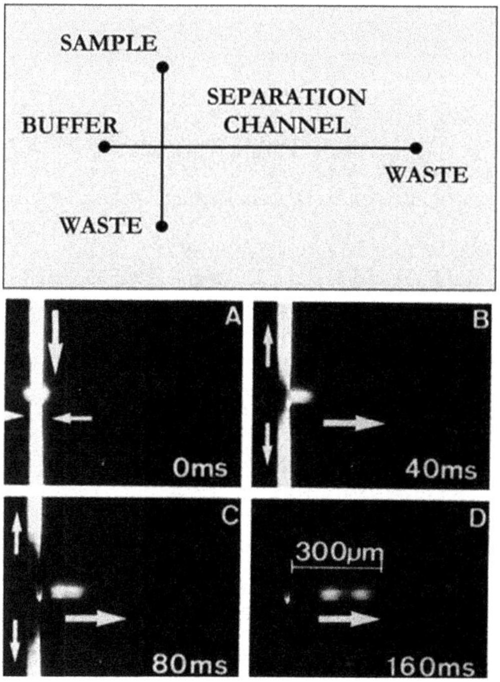

Figure 11.5 Top: schematic of a simple design for sample injection and electrophoresis on-chip. The black spots are the access holes (microwells) used to apply a voltage. Bottom: the micrographs show a time sequence of the injection process. Arrows indicate the direction and intensity of flow. Reproduced from ref. 6 with permission from American Chemical Society, Copyright 1996.

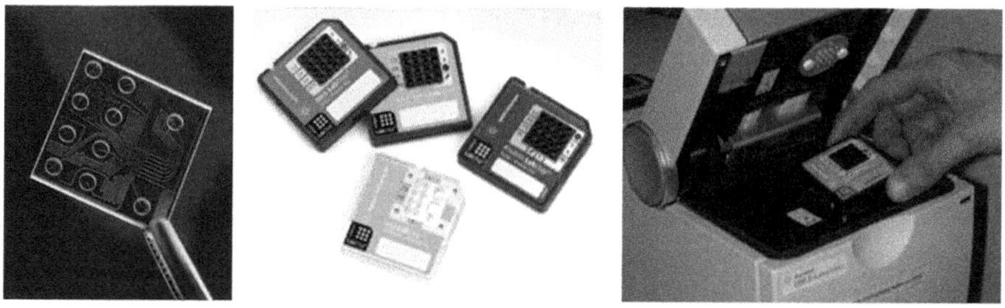

Figure 11.6 Commercial capillary electrophoresis devices are manufactured using glass, and a plastic holder and a benchtop instrument for voltage programming and fluorescence detection are shown. Images courtesy of Agilent Technologies, Inc., © Agilent Technologies, Inc.

References

1. R. Kellner, J. M. Mermet, M. Otto, M. Valcarcel and H. M. Widmer, *Analytical Chemistry*, Wiley, 2nd edn, 2004.
2. A. Manz, P. S. Dittrich, N. Pamme and D. Iossifidis, *Bioanalytical Chemistry*, Imperial College Press, 2nd edn, 2015.
3. H. Yin, K. Killeen, R. Brennen, D. Sobek, M. Wierlich and T. van de Goor, *Anal. Chem.*, 2005, **77**, 527.
4. F. Lottspeich and J. W. Engels, *Bioanalytics*, Wiley, 2018.
5. J. W. Jorgenson and K. D. Lukacs, *Anal. Chem.*, 1981, **53**, 1298–1302.
6. F. von Heeren, E. Verpoorte, A. Manz and W. Thormann, *Anal. Chem.*, 1996, **68**, 2044–2053.
7. N. J. Panaro, P. K. Yuen, T. Sakazume, P. Fortina, L. J. Kricka and P. Wilding, *Clin. Chem.*, 2000, **46**, 1851–1853.
8. A. Padmanaban, A. Inche, M. Gassmann and R. Salowsky, *J. Biomol. Tech.*, 2013, **24**(Suppl), S41.
9. E. R. Castro and A. Manz, *J. Chromatogr. A*, 2015, **1382**, 66.

12 Optical Detection

12.1 Fluorescence

A beam of excitation light with higher energy (shorter wavelength) interacts with fluorescent molecules. Following the Jablonski diagram depicted in Figure 12.1A, the photons are absorbed, exciting electrons from a ground state (energy E_1) to an excited state (energy E_2). After a short time, the excited electrons relax to E_1 while emitting photons of lower energy and longer wavelengths. The lifetime of fluorescence is generally dependent on the fluorophore type, ranging from 10^{-8} to 10^{-11} s. The process of energy transfer can be described *via* the following function:

$$E_1 + hv \rightarrow E_2 \rightarrow E_1 + hv' \quad (12.1)$$

where h is Planck's constant, v is the exciting light frequency and v' is the emitted light frequency.

The energy of the emitted light is always lower than that of the exciting light owing to the loss of heat. As a result, the wavelength of the emitted light is longer than that of the exciting light, with the wavelength difference being the Stokes shift (Figure 12.1B).

Typical light sources for fluorescence were traditionally water-cooled lasers or mercury burner lamps. Lasers have excellent coherence and low divergence, making them easy to focus on samples with a volume measured in microlitres or smaller. However, lasers

Microfluidics and Lab-on-a-chip
By Andreas Manz, Pavel Neužil, Jonathan S. O'Connor and Giuseppina Simone

Published by the Royal Society of Chemistry, www.rsc.org

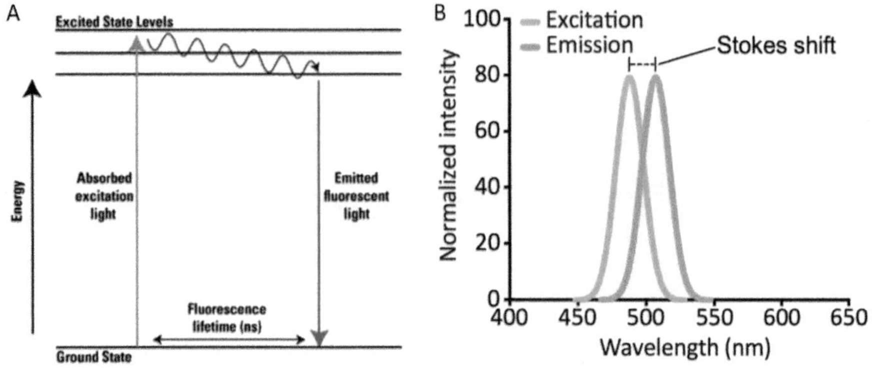

Figure 12.1 (A) Jablonski diagram. (B) Wavelength change between excitation light and emission light.

are expensive and their maintenance is costly. Modern and cheaper laser diodes (LDs) operate at blue light wavelengths, such as a nominal wavelength of 471 nm, using second harmonics from the first at 942 nm. Even a light source at 471 nm is monochromatic, hence the first harmonic has to be eliminated by a low-pass filter. Also, the laser light has to be typically coupled into an optical fibre and delivered to the location of interest *via* a fibre-optic system. This family of detector systems is typically called laser-induced fluorescence (LIF).

Mercury burner lamps produce a number of different wavelengths such as 365 nm (i-line), 405 nm, 436 nm (g-line) and others, thus offering a good selection of different wavelengths for a wide variety of fluorophores. The wavelength of interest is selected by an optical bandpass filter. However, mercury burner lamps are bulky, with a typical operational lifetime of 2000 h.

Recently, developments in light-emitting diode (LED) technology have enabled these devices to dominate as light sources for fluorescence as they have a longer operational lifetime (over 50 000 h), are cheaper and can easily be electrically modulated. There is also an option to connect LEDs in sequence using dichroic mirrors to merge different lights with different wavelengths into a single optical path and, in a similar fashion, to decouple the excited light and detect each of the wavelengths separately. LEDs are also small, making them ideal for integration with portable microfluidic systems.

The amplitude of the emitted fluorescence is typically captured by a photodiode (PD), a photomultiplier tube (PMT) or a charge-coupled device (CCD). A PD is a passive optical detector, converting light into an electrical current, which is generated when photons are absorbed in the PD. However, the quantum efficiency of a PD may be not sufficient to detect the low power of emitted fluorescence. In contrast, a photoemissive device, such as a PMT, is designed to detect weak signals. A PMT works by using the photoelectric effect, absorbing photons and emitting multiplied electrons by a photocathode exposed to a photon flux, to amplify the observed signal. In addition, a CCD is a device that can be used for imaging, not only detection, as it is composed of an array of photodiodes transmitting the captured signal amplitude in the form of a change *via* a cascade of potential wells. As their fabrication is challenging, CCDs cannot be made in a standard semiconductor foundry, and have recently been replaced with complementary metal–oxide–semiconductor (CMOS) imagers.

Fluorescence detection is the most popular detection strategy in biosensing owing to its sensitivity and selectivity. Biomolecules are labelled with different kinds of fluorophores such as fluorescent probes, fluorescent proteins, fluorescent dyes and quantum dots. The specificity of these fluorophores improves the detection sensitivity.

Miniaturized optical elements are of keen interest for integration and application in microfluidic chips. For instance, a microfluidic chip was developed to detect label-free circulating tumour cells (CTCs) based on fluorescence detection and live CTCs could be quantified by lactic acid production.[1] The chip was designed and fabricated to generate uniform emulsion droplets, which were placed under a fluorescence microscope equipped with a mercury lamp, a specific optical filter and a PMT (Figure 12.2). The captured fluorescence could then be digitized by a data acquisition card connected to a personal computer. The results showed that the intensity of the fluorescence signal was proportional to the number of cancer cells. Cancer cells with a known number were also utilized to verify the method, revealing no significant difference between the detected number and the known number. The proposed device provided a new method for label-free CTC detection. However, the optical detection system is off-chip, so it is difficult to integrate these optical parts in a chip, especially with a lamp as a light source.

LDs offer an option for miniaturized microfluidic systems and LIF is considered a highly sensitive detection method, capable of detecting a single molecule. A hand-held LIF detector has been proposed for multiple applications such as flow cytometry, capillary electrophoresis and scanning detection.[2] The detector's design, based on an LD, includes a principal wavelength of 450 nm associated with bandpass filters, a dichroic mirror, a collimating lens, an aperture with diameter of 1 mm, a PMT and an electronic module composed of recording,

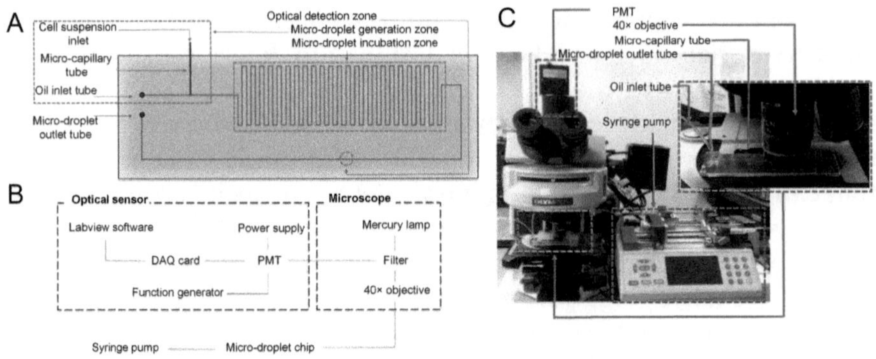

Figure 12.2 (A) Schematic of the microfluidic device. (B) Schematic of the platform. (C) Photograph of the platform. Reproduced from ref. 1, https://doi.org/10.3390/s150306789, under the terms of a CC BY 4.0 license, https://creativecommons.org/licenses/by/4.0/.

processing and displaying units. The whole configuration of the detector had an approximate size of 9.1 × 6.2 × 4.1 cm (Figure 12.3). The performance of the detector was verified using sodium fluorescein solution, achieving a limit of detection (LOD) of ~0.42 nM. The applications of the detector were also demonstrated in capillary electrophoresis, separating fluorescein isothiocyanate (FITC)-labelled amino acids, and in flow cytometry for tumour cell detection. The proposed detector could become the portable universal detector applied in biosensing systems. However, the cost and maintenance of LDs increase the cost of the portable device and the high power of an LD makes it dangerous if not operated properly. As a result, LEDs are the most popular light source applied in optofluidics.

A low-cost miniaturized fluorescence detection system was developed for lab-on-a-chip (LOC) applications.[3] A built-in lock-in amplifier was included in the system, allowing measurements under ambient light with a sensitivity in the low nanomolar range. The lock-in amplifier was improved upon and subsequently applied in a hand-held polymerase chain reaction (PCR) system with an approximate size of 100 × 60 × 33 mm to detect synthetic complementary deoxyribonucleic acid (cDNA) of the H7N9 avian influenza virus (Figure 12.4).[4] It was capable of detecting a single DNA copy. The proposed device may have broad, technologically relevant applications, extending to the rapid detection of infectious diseases in small clinics.

A real-time fluorescence nucleic acid testing device was developed for malaria detection.[5] It was composed of a disposable microfluidic disc with four reaction chambers and an analyser. The sample preparation element and the unit for subsequent real-time loop-mediated

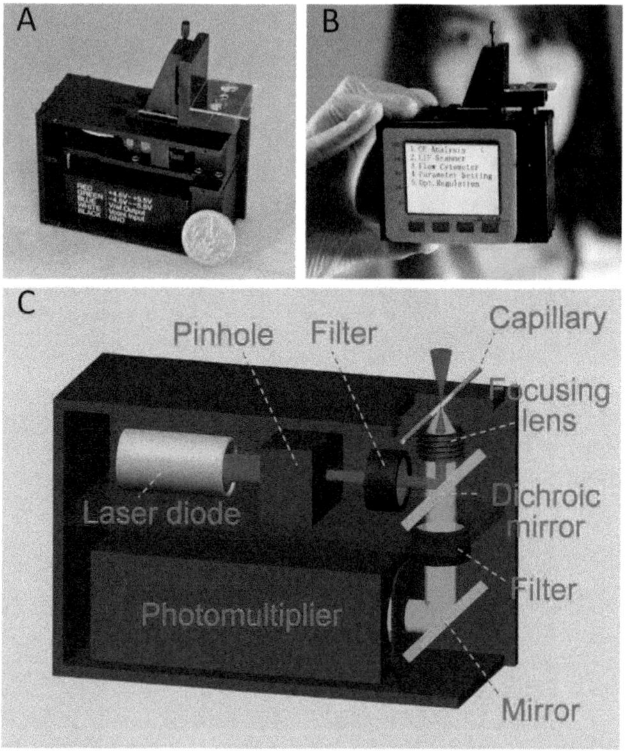

Figure 12.3 (A) Top view of the LIF detector. (B) Side view of the LIF detector. (C) Schematic of the LIF detector. Reproduced from ref. 2 with permission from Elsevier, Copyright 2016.

isothermal amplification (LAMP) were integrated into a microfluidic compact disc, driven by magnetic field interactions. Each reaction chamber was coupled with an LED with a principal wavelength of 488 nm through a polymer optical fibre to track the amplification process in real time. Each LED was also connected to a potentiometer to confirm the uniformity of the incident light intensity and it was perpendicular to the optical sensors to minimize excitation interference (Figure 12.5). Four parallel testing units could be simultaneously performed in 50 min, achieving an LOD of ~0.5 parasites μL^{-1} for whole blood, which was sufficient for detecting asymptomatic parasites. The proposed device could be highly useful for malaria screening owing to its specificity, sensitivity and scalable sample preparation.

The development of organic optoelectronic devices could be an alternative to integrating optical sensors on a chip. For instance, a microfluidic system integrated with organic LEDs (OLEDs) and organic photodetectors (OPDs) was designed and fabricated to detect

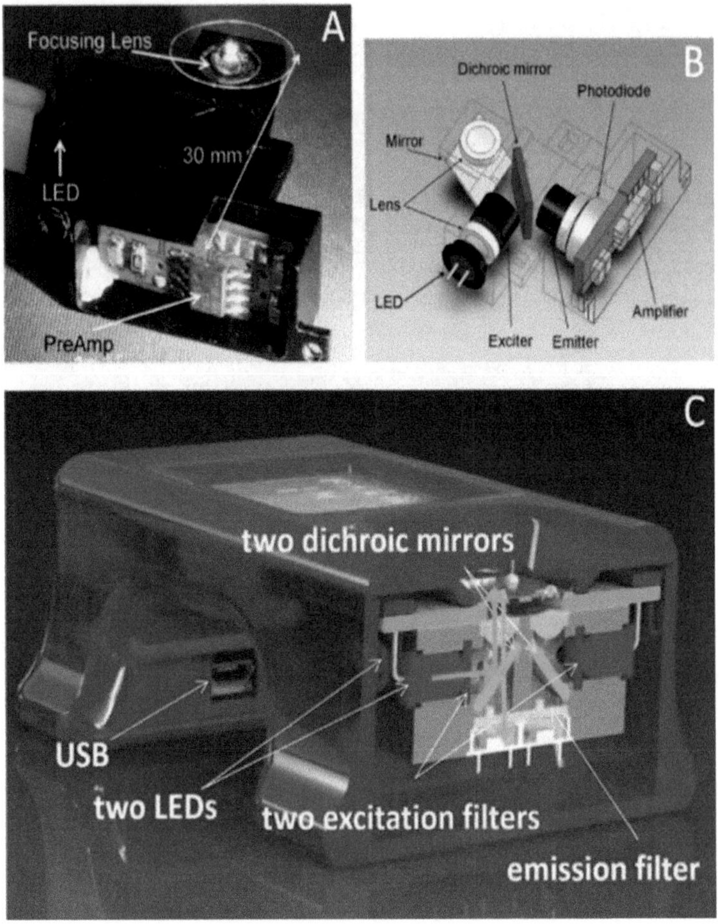

Figure 12.4 (A) Photograph of the integrated detection system. (B) Schematic of the detection system. (C) Schematic of the hand-held PCR system. Reproduced from ref. 3 and 4 with permission from the Royal Society of Chemistry.

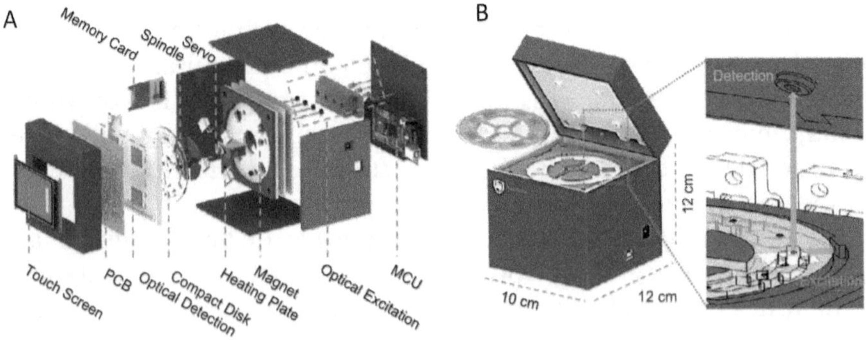

Figure 12.5 (A) Section view of the device. (B) Diagram of the assembled device. Reproduced from ref. 5 with permission from Elsevier, Copyright 2018.

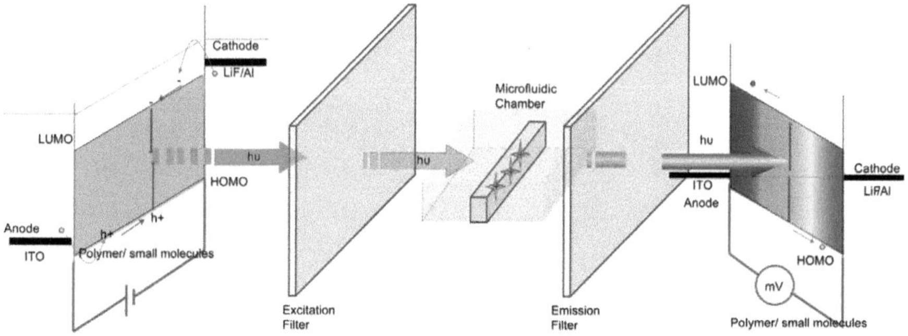

Figure 12.6 Schematic of a microfluidic system integrated with an OLED and an OPD. Reproduced from ref. 6 with permission from Elsevier, Copyright 2015.

phytoplankton fluorescence.[6] The system consisted of an OLED as a light source, a microfluidic chip with reaction chambers, optical filters and an OPD as a fluorescence detector. The optical excitation filter and emission filter were positioned on either side of the chamber, allowing the OPD to detect pure fluorescence emitted by fluorophores (Figure 12.6). Different concentrations of algae were utilized to test the system.

A compact and miniaturized microfluidic system integrated with an optical system could be a trend for point-of-care (POC) systems. More convenient detection systems such as paper-based microfluidics can further reduce the size of such systems, making them especially useful in undeveloped regions of the world. In this respect, a paper-based device was used to perform rapid and precise diagnostics of malaria in rural Uganda at low cost.[7] It combined vertical flow sample-processing steps, LAMP and lateral flow detection. Anti-FITC antibodies and immobilized streptavidin were in the test and control lines in the strip, respectively (Figure 12.7). Experiments were performed with individual diagnoses in <50 min; the detection took <1 min and the results were analysed by eye, achieving a detecting sensitivity of ~98%. The device removes the benchmark instruments and compact optical elements, simplifying the system design and operation, but the results were ambiguous at low levels of infection.

12.2 Absorption

Light absorption is a process whereby light is absorbed by an irradiated object and converted into other forms of energy. Commonly, light cannot be absorbed completely by matter and some is replaced by refraction or reflection. Absorptivity is the quantification of how

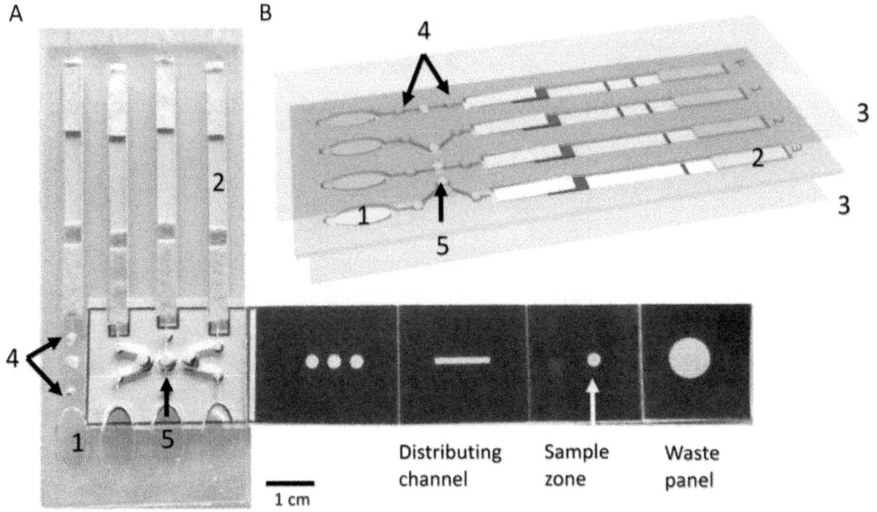

Figure 12.7 (A) Photograph of the paper-based device: 1, buffer chambers; 2, lateral flow DNA detection strip; 4, filter-paper-based valves; 5, filter-paper for LAMP. (B) Schematic of the paper-based device: 3, acetate films. Reproduced from ref. 7, https://doi.org/10.1073/pnas.1812296116, under the terms of a CC BY 4.0 license, https://creativecommons.org/licenses/by/4.0/.

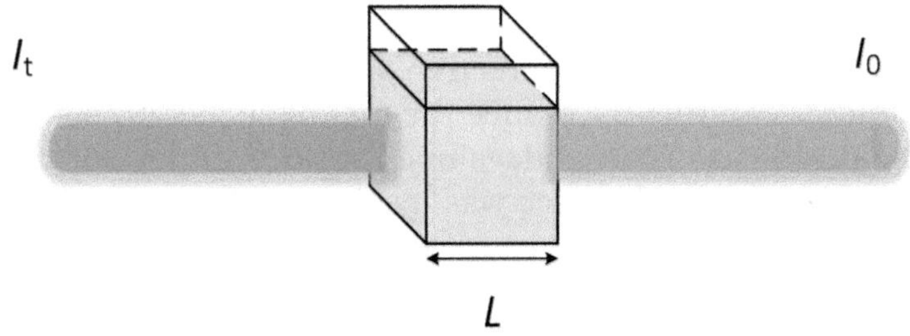

Figure 12.8 Principle of optical absorption and the Beer–Lambert law.

much light an object absorbs and it depends on both the properties of light and the object (Figure 12.8). Absorptivity can be calculated using the Beer–Lambert law:

$$A = -\log_{10}\left(\frac{I_t}{I_0}\right) = \varepsilon Lc \tag{12.2}$$

where I_t is the intensity of transmitted light, I_0 is the intensity of incident light, ε is the molar absorption coefficient, L is the optical path-length and c is the concentration of the absorbing matter.

The Beer–Lambert law demonstrates the optical attenuation while the light is travelling through the material. The law is commonly applied to chemical analysis measurements and used in understanding attenuation in physical optics, for photons or neutrons.

The wide application of UV–visible absorption in microfluidic devices is hindered by the low sensitivity due to the short optical pathlength. However, researchers are beginning to focus more on this technique, exploring methods to improve sensitivity. A microfluidic system for thiourea detection was developed using optical fibre-based UV absorption to study the separation of peptides.[8] The separation channel in the chip was partly filled with a sol–gel stationary phase using C_4-modified silica particles of 5 μm size. UV absorption detection was performed at the end of the channel without any stationary phase. The chip was sandwiched between two optical fibres. The top fibre was connected to a deuterium–tungsten source and the light was collected by the bottom fibre and transmitted to the CCD (Figure 12.9). An LOD of 167 μM was achieved, which could be further improved by using a noise filter.

Further, an optical LOC device was developed to detect ammonia based on absorption. The core part of the device was a thin film made of a ninhydrin–polydimethylsiloxane (PDMS) composite.[9] The microfluidic channels in the PDMS substrate were fabricated to guide the ammonium gas to pass across the film. Detection was based on the reaction of ninhydrin with ammonia (ammonium ion), producing a purple-coloured stable compound. It also resulted in a change in optical absorption. An LED and a photoresistor were placed perpendicular to the film, detecting the change in absorption and permitting the quantification of ammonia

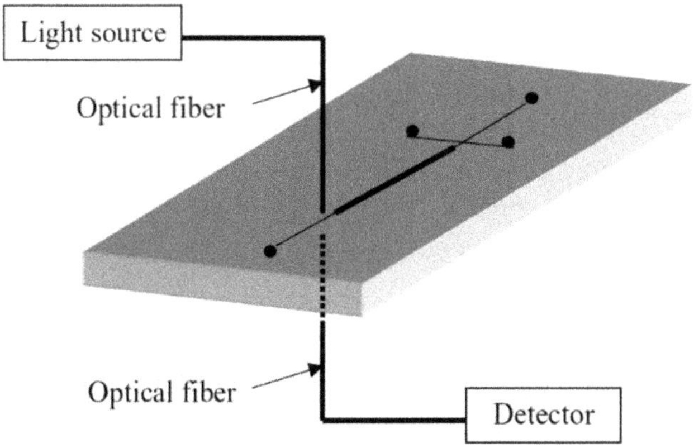

Figure 12.9 Schematic of the setup based on UV absorption. Reproduced from ref. 8 with permission from Elsevier, Copyright 2004.

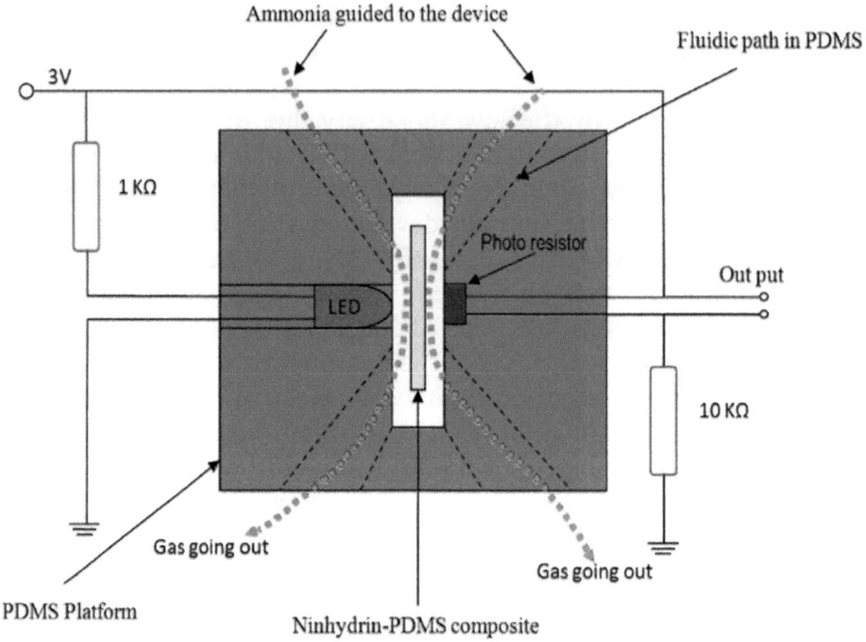

Figure 12.10 Schematic of the device based on ninhydrin-PDMS. Reproduced from ref. 9, https://doi.org/10.1149/2.0121808jes, under the terms of a CC BY 4.0 license, https://creativecommons.org/licenses/by/4.0/.

(Figure 12.10). The response time of the device was linear for a wide range of ammonia concentrations and the LOD was as low as 2 ppm.

Rapid detection of salivary potassium was performed with a hybrid integrated optical microfluidic biosensor based on colorimetric optical absorption.[10] The biosensor was composed of two inlets and a micromixer. Two different kinds of fluids were pumped into the inlets and mixed by the piezo-actuated micromixer. The mixed solution was subsequently transported to the optical detection element. The input waveguide and output waveguide made of SU-8 were integrated near the microfluidic channel, coaxially with each other and connected to the light source and a spectrometer, respectively (Figure 12.11). As a result, the incident light could be directly coupled with the output waveguide. The sample mixed with chromogenic substrates resulted in a colour change associated with the analyte concentration, which could be detected by light absorbance. The detection results showed that the error was within 5%.

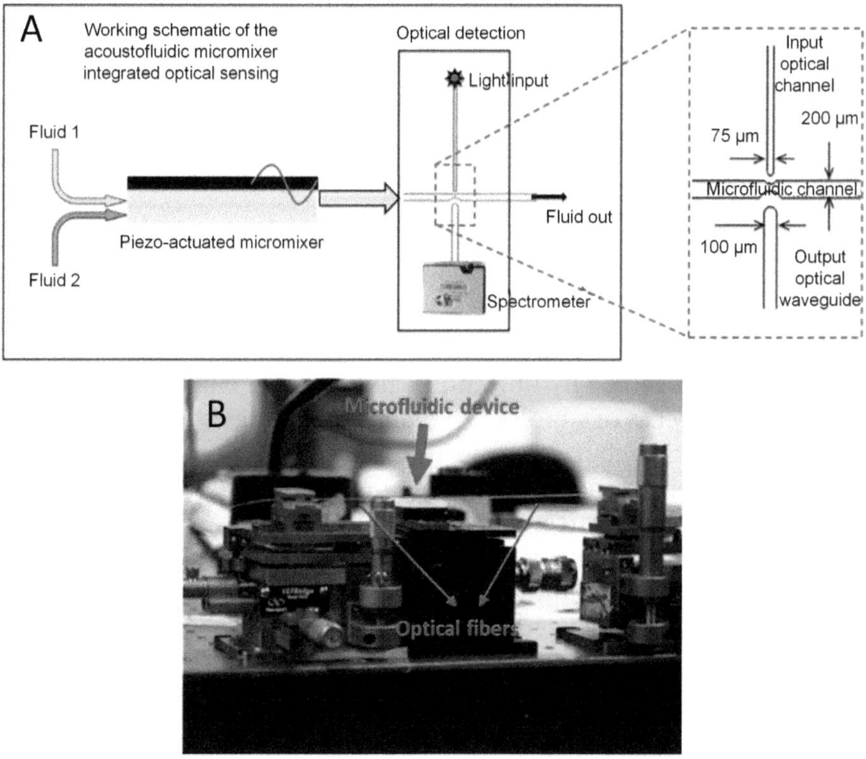

Figure 12.11 (A) Principle of the setup. (B) Photograph of the proposed platform. Reproduced from ref. 10, https://doi.org/10.3390/bios9020073, under the terms of a CC BY 4.0 license, https://creativecommons.org/licenses/by/4.0/.

12.3 Surface Plasmon Resonance

Surface plasmon resonance (SPR) is a technique to detect molecular interactions between surfaces and proteins or other interacting molecules. The protein immobilized on a suitable surface changes the nature of the plasmons generated from that surface (Figure 12.12).

When the light is incident at a critical angle on the interface of two media with different refractive indexes and the frequencies of the incident light and free electrons (plasma) on the metal surface are consistent, the plasma on the surface of the metal absorbs the light energy and resonates. This results in a significant decrease in reflected light intensity at a certain angle, known as the SPR angle.

The SPR angle changes based on the refractive index of the metal film. As a result, we can acquire the specific signal of the interaction

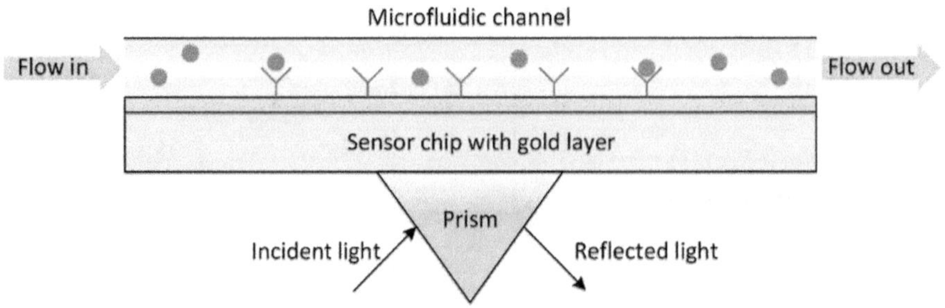

Figure 12.12 Principle of SPR.

between biomolecules by obtaining the dynamic change of the SPR angle in the process of a biological reaction. SPR has been widely used in highly sensitive chemical and biomolecular detection as there is no need to label the molecules, thus eliminating possible changes to their molecular properties.

A real-time protein detection device based on SPR was designed and fabricated with the advantages of portability and ease of use.[11] The SPR sensor in this work used a capillary with an inner wall coated with self-assembled gold nanoparticles (GNPs) to form a sensing layer for high-throughput screening of biomolecular interactions. It was functionalized using ligands bound to GNPs.

In the proposed device, two connectors were connected to the plastic cladding multimode optical fibres and the SPR sensors. Both ends of the fibres were polished to ensure that they were smooth. The end of the optical fibre was placed close to an LED with a principal wavelength of 595 nm. Two CMOS image sensors were placed at the side and at the end of the capillary sensors to measure the transmission light from the output fibre and scattered light from the capillary sidewall, respectively. The signal intensity captured from the two CMOS sensors was then transmitted to a computer and used to calculate the SPR sensor response in real time. The device shown in Figure 12.13 was used to detect transferrin and immunoglobulin G (IgG). Transferrin transfers iron from absorption sites to tissues and IgG is the most common type of antibody in the blood circulation,[12] making the measurement of IgG an important diagnostic tool for the detection of autoimmune hepatitis. The results demonstrated that the proposed device achieved qualitative identification and quantitative measurement of transferrin and IgG.

Current biomarkers are not efficient enough for the early diagnosis of acute kidney injury (AKI), hence effective treatment may be delayed.

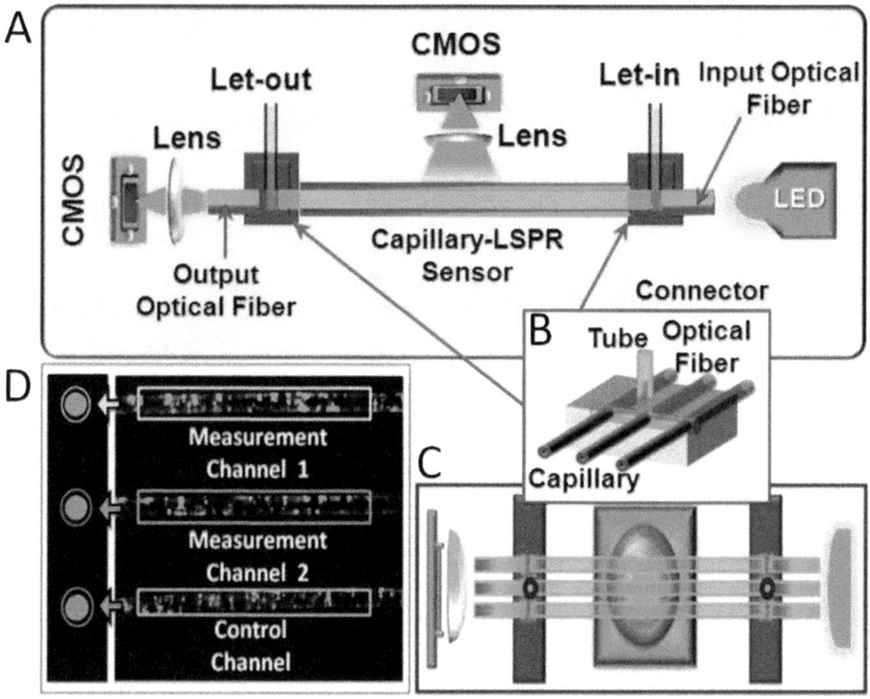

Figure 12.13 (A) Schematic of the sensing device. (B) Structure of connectors. (C) Top view of the sensing device. (D) Image captured by CMOS. Reproduced from ref. 11, https://doi.org/10.3390/nano9071019, under the terms of a CC BY 4.0 license, https://creativecommons.org/licenses/by/4.0/.

A microfluidic chip based on label-free nanostructure transmission SPR was developed for the early detection of urinary miRNA-16-5p, known as a biomarker for kidney injury.[13] In this work, a highly sensitive capped gold nanoslit (CG nanoslit) film was applied as the sensing platform. The patterned microfluidic chip was fabricated based on a poly(methyl methacrylate) (PMMA) substrate. The CG nanoslit film was sandwiched between the PMMA layer and the glass layer using optically clear double-sided tape. A hybridization method isolated the target molecules, enhancing the SPR signals using magnetic nanoparticles (Figure 12.14). A urine sample of ~1 mL was taken from a patient and applied to the device and the results were compared with data obtained by reverse transcriptase PCR (RT-PCR), confirming the validity of the method in clinical application. This work could be improved by developing a portable system that could be deployed in a clinical environment.

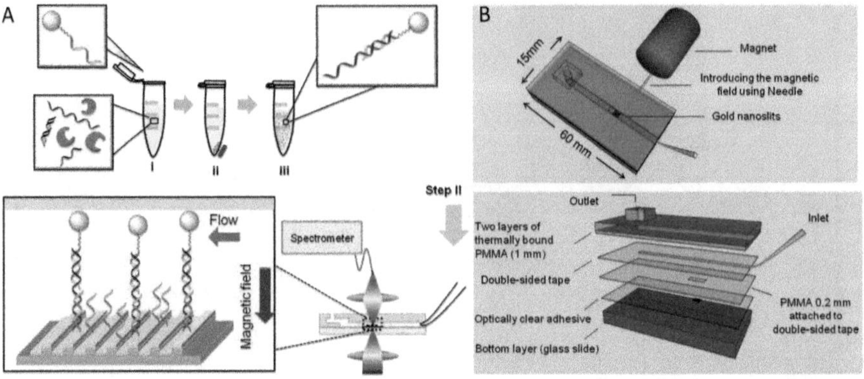

Figure 12.14 (A) Principle of target molecule isolation and capture. (B) Schematic of the device. Reproduced from ref. 13 with permission from the Royal Society of Chemistry.

12.4 Reflection

Reflection is the change in the direction of light at an interface between two different media. The angle of incidence is equal to the angle of reflection when the interface is flat and smooth; otherwise, the light bounces off in many directions, known as diffuse reflection.

A microfluidic disc with an optical pickup was proposed to perform enzyme-linked immunosorbent assays (ELISAs).[14] The disc was composed of three PMMA plates and a poly(ethylene terephthalate) (PET) film. A polystyrene solution was spin-coated on the top of the bottom PMMA plate to form a hydrophobic surface, immobilizing the antibodies. The top PMMA plate included eight holes as inlets and one hole in the centre for a washing buffer. Eight channels were laser machined in the PET film and connected to the corresponding inlets. There was also a PMMA plate held with a water-absorbing sheet between the top PMMA plate and the PET film (Figure 12.15). It was used for absorption of washing buffer. All the plates were assembled using double-sided tape. A solution of the sample to be tested was pumped into the channels and proteins were captured by the antibodies immobilized on the bottom PMMA surface. An ELISA was then performed between proteins and antibodies. An LD with a principal wavelength of 532 nm was focused onto the interface between the top PMMA plate and solution in the channel *via* a 60× objective lens. The intensity of the reflected light beam was captured by a PMT and recorded by a data acquisition board. Human C-reactive protein (CRP) associated with cardiovascular disease was

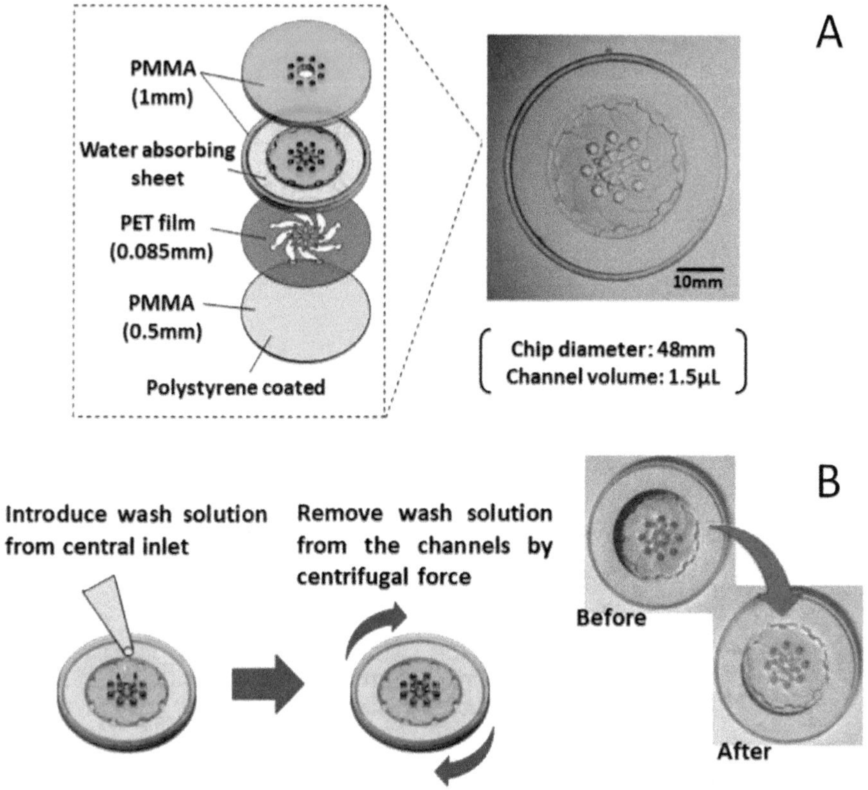

Figure 12.15 (A) Schematic of the ELISA chip. (B) Principle of channel washing. Reproduced from ref. 14 with permission from the Royal Society of Chemistry.

selected to test the device performance, obtaining an LOD of ~2 ng mL^{-1} in ~20 min.

An optical biosensing platform was built *in situ* to detect chemical pollutants in oceanic waters.[15] This platform consisted of a fluidic system, biosensing system and optical readout system. The fluidic chip, under the biosensing chip, was designed to allow the tested solution to flow through the biosensing units. The biosensing chip was composed of 12 biophotonic sensing cells (BICELLs) on a quartz substrate and each BICELL was an array of nanopillars used as resonant nanopillars (R-NPs) (Figure 12.16). The optical readout system was built with a broadband LED, illumination fibres, fibre splitter, collection fibres and spectrometer. The LED was coupled to 12 illumination fibres using a fibre splitter and 12 collection fibres were coupled to a 12-channel integral field spectrometer (IFS). Each fibre was connected to a BICELL to perform 12-channel detection simultaneously. The incident light from

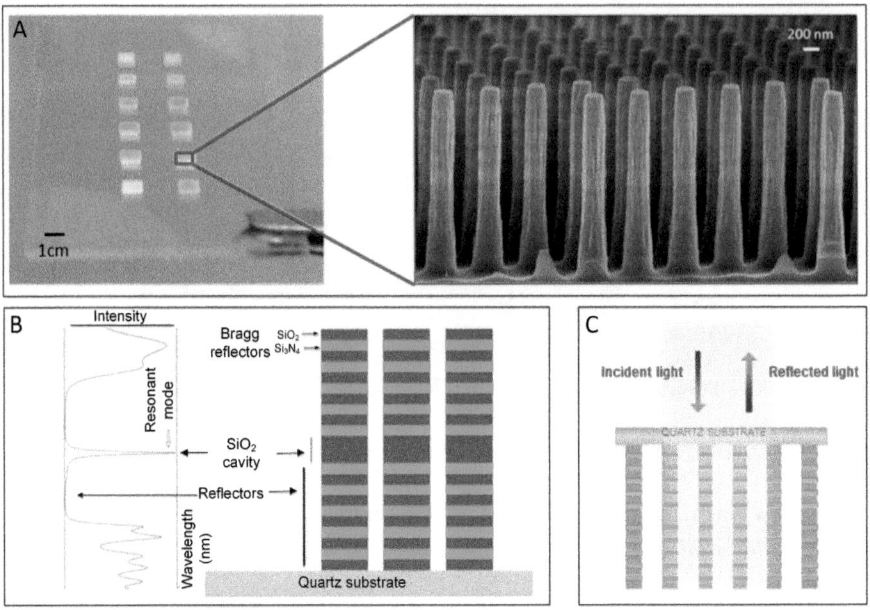

Figure 12.16 (A) Photograph of the sensing chip and scanning electron microscope (SEM) image of the R-NPs. (B) Principle of detection. (C) Schematic of the back side of the chip. Reproduced from ref. 15 https://doi.org/10.3390/s19040878, under the terms of a CC BY 4.0 license, https://creativecommons.org/licenses/by/4.0/.

the LED was reflected on the surface of the BICELLs. The reflected light was captured by the collection fibres and transmitted to the IFS for analysis. The IFS captured the signal, producing 12 spectra in parallel with a spectral range from 525 to 635 nm and the range overlapped the resonance bandwidth of the R-NPs. The time response of the R-NPs under optical reflectance was then measured and the sensitivity was calculated. As a proof-of-concept, NaCl solutions with different concentrations were flowed through the biosensing area, obtaining a sensitivity of 285.9 ± 16.4 nm per refractive index unit (RIU) and an LOD of ~2.95×10^{-6} RIU. These results indicate the high potential of the developed platform to be applied for *in situ* multiplex optical biosensing.

LOC devices have advantages in multiplexed analysis owing to the low sample volume requirements and optical detection methods that provide high sensitivity. However, LOC devices with multiplexed optical detection are limited by the chip complexity as it is difficult to integrate multiple light sources and detectors in a single device. A microfluidic-controlled optical router (μCOR) integrated with micro-optic and micro-optofluidic elements was designed and fabricated to perform

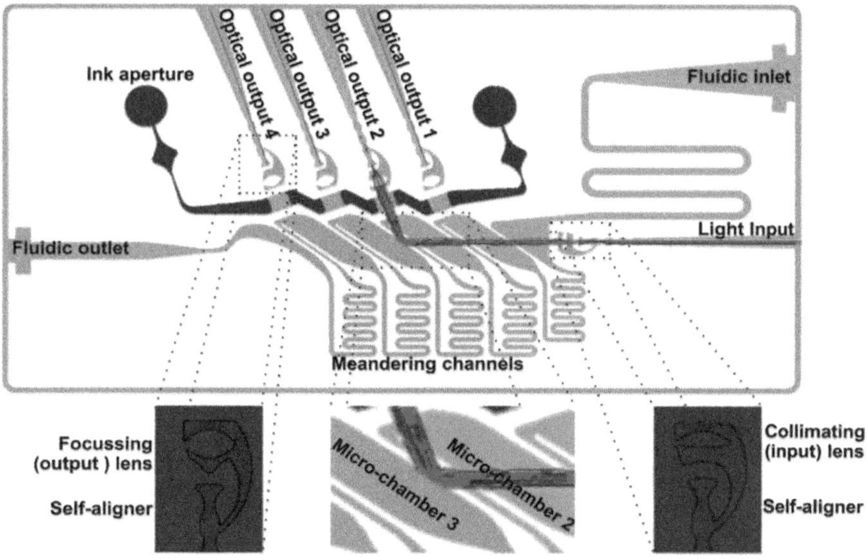

Figure 12.17 Schematic of μCOR. Reproduced from ref. 16 with permission from the Royal Society of Chemistry.

multiplexing measurements using a single light source and detector in a PDMS substrate.[16] Microfluidic parts of the μCOR were composed of a fluidic inlet and outlet, microchambers and phase guides, which were used for fluid management in chambers. All the micro-optical and microfluidic elements were applied for self-alignment, in-coupling, guidance and out-coupling of light to four independent optical channels (Figure 12.17). Air gap mirrors were generated by filling/emptying the microchambers with water using an external pump. In the device, incident light was in-coupled by an optical fibre, then collimated by a polymeric microlens and guided through four sequentially connected microchambers. If the chamber was empty, total internal reflection occurred at the interface of PDMS and air, guiding the light to the output optical fibre. Once one chamber was filled with water, light was transmitted to the next empty chamber. Fluorescein was selected to test the device performance, resulting in cross-talk of <2% and a high switching frequency up to 0.343 ± 0.006 Hz.

12.5 Interference

Optical interference is the interaction of two or more light waves based on the superposition of waves. If the frequency and phase of the light waves are the same, vibrating at the same rate, the net

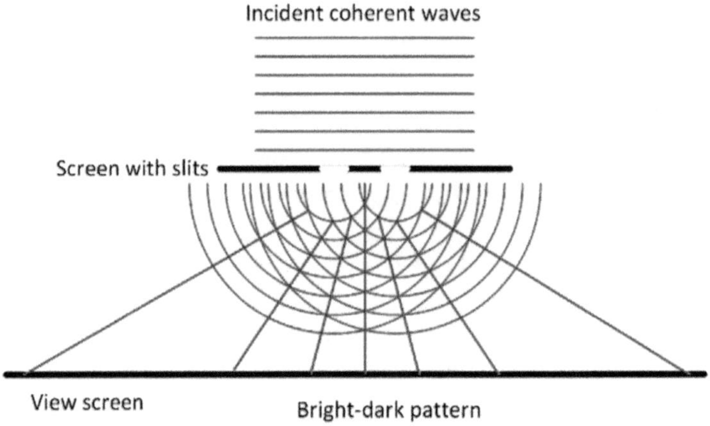

Figure 12.18 Principle of optical interference.

effect of the wave amplitudes is reinforced, producing constructive interference and showing bright patterns. When the phases of these two light waves are shifted by half a period, the net effect of the interference is destructive interference, displaying dark patterns (Figure 12.18).

A low-cost microfluidic platform with a nanophotonic sensor was developed for POC biomarker analysis.[17] A chip was made based on a silicon substrate composed of a Mach–Zehnder interferometer (MZI) and an on-chip optical spectral analyser with an arrayed-waveguide grating (Figure 12.19). In the sensing area, the waveguide was split into two parts: one was coupled with the Mach–Zehnder interferometer, called the sensing arm, and the other was left uncoupled, referred to as the reference arm. The effective refractive index of the sensing arm changed when a bioreaction occurring in the microfluidic channel was exposed to the sensing arm, resulting in a phase shift of the sensing arm and a subsequent spectral shift of the MZI. This shift was captured and recorded by the detector at the end of the sensor.

The nanophotonic sensor chip was first tested before the subsequent integration, resulting in an LOD of 6×10^{-6} RIU with a standard deviation of 2×10^{-6} RIU, indicating high reproducibility. The result was of the same order of magnitude as that reported using state-of-the-art evanescent wave sensors. The chip was then biofunctionalized to test for *Mycobacterium tuberculosis* lipoarabinomannan in clinical urine samples, subsequently integrated into a polymer microfluidic cartridge. Finally, a compact POC platform was built, including the cartridge, light source, optical detection and read-out systems, to analyse patient samples in real time. The clinical urine samples and a protein biomarker CRP were utilized to verify the platform.

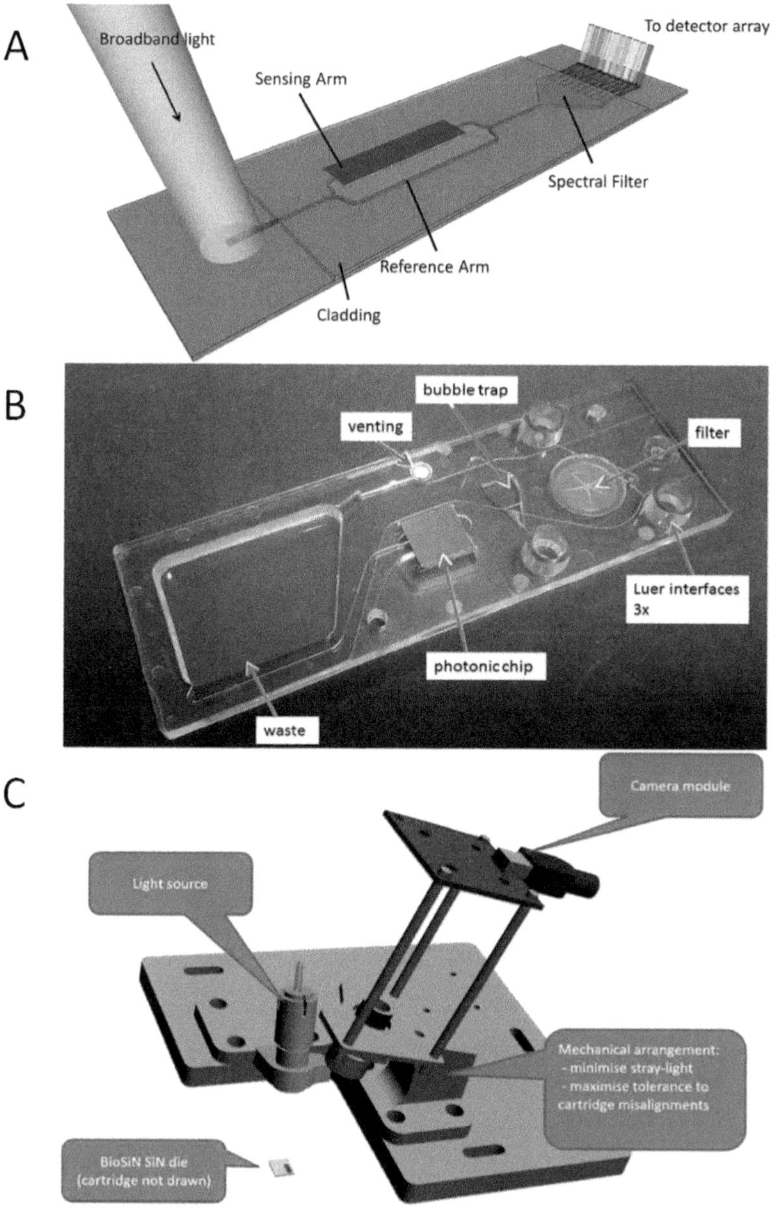

Figure 12.19 (A) Schematic of the sensing chip. (B) Schematic of the microfluidic cartridge integrated with the sensing chip. (C) Schematic of the readout module. Reproduced from ref. 17 with permission from the Royal Society of Chemistry.

References

1. T. K. Chiu, K. F. Lei, C. H. Hsieh, H. B. Hsiao, H. M. Wang and M. H. Wu, *Sensors*, 2015, **15**(3), 6789.
2. X. X. Fang, H. Y. Li, P. Fang, J. Z. Pan and Q. Fang, *Talanta*, 2016, **150**, 135.
3. L. Novak, P. Neuzil, J. Pipper, Y. Zhang and S. Lee, *Lab Chip*, 2007, 7(1), 27.
4. C. D. Ahrberg, B. R. Ilic, A. Manz and P. Neuzil, *Lab Chip*, 2016, **16**(3), 586.
5. G. Choi, T. Prince, J. Miao, L. Cui and W. Guan, *Biosens. Bioelectron.*, 2018, **115**, 83.
6. F. Lefevre, P. Juneau and R. Izquierdo, *Sens. Actuators, B*, 2015, **221**, 1314.
7. J. Reboud, G. Xu, A. Garrett, M. Adriko, Z. Yang, E. M. Tukahebwa, C. Rowell and J. M. Cooper, *Proc. Natl. Acad. Sci. U. S. A.*, 2019, **116**(11), 4834.
8. R. Jindal and S. M. Cramer, *J. Chromatogr. A*, 2004, **1044**(1–2), 277.
9. J. Ozhikandathil, S. Badilescu and M. Packirisamy, *J. Electrochem. Soc.*, 2018, **165**(8), B3078.
10. V. Surendran, T. Chiulli, S. Manoharan, S. Knisley, M. Packirisamy and A. Chandrasekaran, *Biosensors*, 2019, **9**(2), 73.
11. Y. Liu, N. Zhang, P. Li, L. Yu, S. Chen, Y. Zhang, Z. Jing and W. Peng, *Nanomaterials*, 2019, **9**(7), 1019.
12. G. Vidarsson, G. Dekkers and T. Rispens, *Front. Immunol.*, 2014, **5**(520), 1.
13. M. Z. Mousavi, H. Y. Chen, K. L. Lee, H. Lin, H. H. Chen, Y. F. Lin, C. S. Wong, H. F. Li, P. K. Wei and J. Y. Cheng, *Analyst*, 2015, **140**(12), 4097.
14. H. Yoshikawa, M. Yoshinaga and E. Tamiya, *RSC Adv.*, 2018, **8**(26), 14510.
15. A. L. Hernandez, F. Dortu, T. Veenstra, P. Ciaurriz, R. Casquel, I. Cornago, H. V. Horsten, E. Tellechea, M. V. Maigler, F. Fernandez and M. Holgado, *Sensors*, 2019, **19**(4), 878.
16. J. Dietvorst, J. Goyvaerts, T. Nils Ackermann, E. Alvarez, X. Munoz-Berbel and A. Llobera, *Lab Chip*, 2019, **19**(12), 2081.
17. D. Martens, P. Ramirez-Priego, M. S. Murib, A. A. Elamin, A. B. Gonzalez-Guerrero, M. Stehr, F. Jonas, B. Anton, N. Hlawatsch, P. Soetaert, R. Vos, A. Stassen, S. Severi, W. Van Roy, R. Bockstaele, H. Becker, M. Singh, L. M. Lechuga and P. Bienstman, *Anal. Methods*, 2018, **10**(25), 3066.

13 Electrochemistry

13.1 Introduction

Electrochemistry (EC) is a popular detection technique used in analytical chemistry[1] and is widely used in microfluidics. EC, as compared with laser-induced fluorescence (LIF), originally had the advantages of being more portable, inexpensive, with the possibility of miniaturization and excluding the optical path. Currently, with the advent of light-emitting diodes (LEDs) operating with blue light wavelengths between 450 and 490 nm and with ultraviolet wavelengths, the original advantages of EC over fluorescence are no longer as obvious. Nevertheless, EC methods still have their merits as the devices can be fabricated using automated production lines and are compatible with microfabrication systems and microfluidics.

EC is a physical and analytical chemistry method that combines chemistry and electrical engineering. EC deals with chemical reactions induced by electron transfers (*e.g.* electrodeposition and sensing) or *vice versa* (*e.g.* batteries) causing the oxidation or reduction of a substance in the reaction. It can also be stated that EC is a science that links the chemical reactions occurring at the interface of an electron conductor and an ionic solution involving electron transfer across the interface.

EC techniques used today were developed by many scientists and probably originated in the sixteenth century. Michael Faraday introduced the laws of electrochemistry and the key equation governing electrochemical processes, which is now known as Faraday's law of electrolysis:

Microfluidics and Lab-on-a-chip
By Andreas Manz, Pavel Neužil, Jonathan S. O'Connor and Giuseppina Simone

Published by the Royal Society of Chemistry, www.rsc.org

$$m = \frac{QM}{Fz} = \frac{ItM}{Fz} \quad (13.1)$$

where m is the mass of the oxidized/reduced substance due to charge transfer with an amplitude Q, I is the amplitude of the electric current passing through the system for a time t, M is the molar mass of the substance being oxidized or reduced, F is the Faraday constant with a value of ~96 485 C mol^{-1} and z is the number of electrons in the reaction causing either the oxidation or reduction of the substance.

A basic electrochemical system requires two electrodes: a cathode and an anode. The negative cathode attracts cations and subsequently reduces the cations (lowering their oxidation state), whereas the positive anode attracts anions and leads to the phenomenon known as oxidation. There are two fundamental types of electrochemical systems. The first is galvanic (voltaic) devices such as batteries. These devices have two electrodes made of different materials, one of which is electrochemically dissolved (oxidized), causing an electric current to flow between the electrodes until either the electrolyte is depleted or the electrode made from the less noble metal is completely dissolved.

The second fundamental system in EC is called an electrolytic cell, which requires an external power supply to work. The externally applied electric current causes the oxidation and reduction reactions that are frequently used in EC sensing. In principle, two electrodes are needed; however, this is only in theory. The desired electrochemical reaction is conducted on a working electrode (WE), which is always grounded. Voltage is applied to obtain a correctly polarized layer at the WE by biasing a second electrode immersed in the same solution and either a two- or three-electrode system is used.

The two-electrode system has an unknown potential drop at the second electrode, introducing error into the measurement, whereas the three-electrode system introduces an electric current from a third electrode, which is called either the auxiliary (AUX) or counter electrode. This current is sent into the WE *via* the solution to create the desired potential at the second electrode, which is called the reference (REF), using a feedback loop system created by a piece of equipment called a potentiostat. The REF has a precisely defined potential and is either an original hydrogen electrode with potential defined by hydrogen pressure or an electrode made from more convenient materials, such as $Hg/Hg_2Cl_2/KCl$ or $Ag/AgCl/KCl$. The latter combination is more common owing to the absence of toxic Hg. It consists of an Ag wire coated with AgCl in a solution of KCl and has an electrochemical reaction at the silver surface:

$$Ag^{+} \xleftrightarrow{e^{-}} Ag(s) \tag{13.2}$$

The potential (E) is given by the Nernst equation:

$$E = E_0 - \frac{RT}{nF}\ln\left(\frac{[C][D]}{[A][B]}\right) \tag{13.3}$$

where E_0 is the standard electrode potential for the reaction, R is the universal gas constant with a value of 8.314 J K^{-1} mol^{-1}, T is the thermodynamic temperature, n is the number of electrons in the reaction and [A], [B], [C] and [D] are the activities of compounds A–D, equal to concentrations for dilute solutions. In reality, the potential of an Ag/AgCl/KCl electrode is a function of the KCl concentration. The most commonly used type is a saturated Ag/AgCl/KCl electrode containing a saturated solution of KCl as the concentration and therefore the electrode potential are not affected by evaporation of water from the KCl solution.

Finally, the compartment containing the Ag/AgCl/KCl has to be electrically connected with the solution containing the WE. This occurs *via* the porous material forming the so-called liquid junction. There is a potential drop at this junction, which is a function of the electric current flowing through it and the difference between the diffusion coefficients of K^+ and Cl^-. The KCl electrolyte is selected as the similarity of the diffusion coefficients of K^+ and Cl^- minimizes the liquid junction voltage drop.

There have been numerous attempts to miniaturize Ag/AgCl/KCl electrodes for microfluidics, most of which have had very limited success or had limited applications. The major problem is the availability of the KCl solution (or similar) as the reservoir has a limited volume, resulting in short-term stability. Once the KCl concentration starts to change, the REF potential fluctuates. The same is true for micromachined Ag electrodes covered with a gel saturated with KCl. Nevertheless, there are still a few options for miniaturizing the REF without losing stability. First, KCl can be added to the tested solution. Then, a simple Ag wire coated with AgCl can work as a second kind of stable REF.

Can we use a two-electrode system by eliminating the AUX? This would greatly simplify the electronics as there would be no need to use a closed feedback loop system. It has been done before and the trick is to keep the electric current density at the REF's surface as low as possible, thus minimizing the unwanted voltage drop by making the REF with a much larger surface area than that of the WE.

13.2 Voltammetric Methods

13.2.1 Scanning Voltammetry

Cyclic voltammetry (CV) is the most commonly used voltammetric technique. CV is performed by a potentiostat while gradually increasing V_{REF} from V_{REF_MIN} to V_{REF_MAX} and back with a set value of the scan rate ν while monitoring the amplitude of i_{WE}. An array of Au nanostructured electrodes in a reversible system, such as the Fe^{2+}/Fe^{3+} system $K_4Fe(CN)_6/K_3Fe(CN)_6$, in KCl electrolyte were used to adjust the electrical conductivity of the solution (Figure 13.1A). The amplitude of i_{WE} and the potential value at the i_{WE} maximum [$i_{WE(PEAK)}$] determine the analyte concentration and its composition, respectively.

CV is typically the first test of a newly set up electrochemical system and is conducted using either an Fe^{2+}/Fe^{3+} system for most electrodes or an Ru^{2+}/Ru^{3+} system for a more negative range of potentials. Put simply, if the CV method in a conventional Fe^{2+}/Fe^{3+} or Ru^{2+}/Ru^{3+} system does not have reversible behaviour, there is something wrong with the system, such as the electrode integrity, setup, *etc.*, and it has to be fixed before performing further experiments.

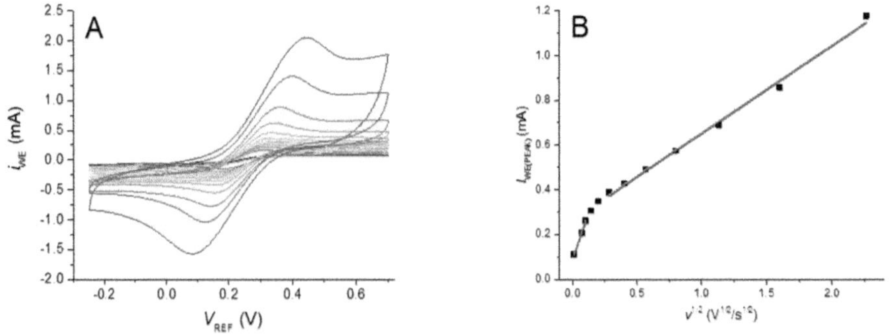

Figure 13.1 (A) Scanning voltammetry of an array of nanostructured Au electrodes with a saturated Ag/AgCl/KCl electrode as the REF with different values of v using an Fe^{2+}/Fe^{3+} solution in KCl electrolyte. (B) Extracted values of current peaks from CV as a function of v. There are two different slopes, showing that the system has two well-distinguished behaviours: microelectrode behaviour at low values of v and conventional macroelectrode behaviour at higher values of v. Reproduced from ref. 9 with permission from Elsevier, Copyright 2019.

A CV method with different values of ν can also be used to calculate the electrochemically active area of the WE and further estimate the behaviour of the WE.[2] We performed a set of CV tests with ν as a parameter (Figure 13.1A) and using an array of nanostructured Au electrodes.[3] $i_{WE(PEAK)}$ was then plotted as a function of $\nu^{½}$ (Figure 13.1B). The electrochemically active area of the WE is a function of the fitted line slope in Figure 13.1B according to the Randles–Sevcik equation:

$$i_{WE(PEAK)} = 0.4463nFAC\sqrt{\frac{nF\nu D}{RT}} \tag{13.4}$$

where $i_{WE(PEAK)}$ is the current maximum, n is the number of electrons transferred, F is the Faraday constant, A is the electrode area, ν is the scan rate, R is the gas constant, T is the absolute temperature, D is the diffusion coefficient of the Fe^{2+} ions and C is the concentration (analyte activity) with the assumption that an Fe^{2+}/Fe^{3+} system is used. There are two distinguished regions at the fitted line in Figure 13.1B, showing that the array has different electrochemically active areas with a ν_0 value below and above ~20 mV s^{-1}. The array of WEs behaves as an array of microelectrodes and as a conventional macroelectrode with a ν value below and above ν_0, respectively.

The CV technique is often used in an H_2SO_4 environment prior to the actual testing to clean the electrode surfaces. Between 30 and 60 cycles of CV are performed before the electrochemical cell is washed with deionized water and replaced with analyte to perform the analysis.

CV is not a particularly sensitive method for biosensing applications as the measured electric current has three components: reversible (i_R), charging (i_C) and kinetic (i_K). Each current depends on the value of ν in a different way, which led to a mathematical technique called elimination voltammetry being introduced[4,5] to express individual currents separately. Assuming that we performed the CV tests with a scan rate with principal values of ν, $\nu/2$ and 2ν, we can derive three equations:

$$F(i_R) = a_1 + \frac{1}{\sqrt{2}}a_{0.5} + 2a_2\sqrt{2} \tag{13.5a}$$

$$F(i_C) = a_1 + 0.5a_{0.5} + 2a_2 \tag{13.5b}$$

$$F(i_K) = a_1 + 0.5a_{0.5} + a_2 \tag{13.5c}$$

where a_1, $a_{0.5}$ and a_2 are coefficients related to nominal, half and double values of v, respectively.

We can extract a particular current such as i_R, i_C or i_K by eliminating two of the others by solving the set of eqn (13.5a,b,c), such as

$$F(i_R) = 1 \quad F(i_C) = 0 \quad F(i_K) = 0 \tag{13.6}$$

leading to

$$i_R = -\left(6+4\sqrt{2}\right)i_{0.5v} + \left(9+6\sqrt{2}\right)i_v - \left(3+2\sqrt{2}\right)i_{2v} \tag{13.7}$$

or

$$F(i_R) = 0 \quad F(i_C) = 1 \quad F(i_K) = 0 \tag{13.8}$$

leading to

$$i_C = -\left(2+2\sqrt{2}\right)i_{0.5v} - \left(4+3\sqrt{2}\right)i_v - \left(2+\sqrt{2}\right)i_{2v} \tag{13.9}$$

or

$$F(i_R) = 0 \quad F(i_C) = 0 \quad F(i_K) = 1 \tag{13.10}$$

leading to

$$i_K = -\left(4+2\sqrt{2}\right)i_{0.5v} + \left(4+3\sqrt{2}\right)i_v + \left(1+\sqrt{2}\right)i_{2v} \tag{13.11}$$

where i_v, $i_{0.5v}$ and i_{2v} are i_{WE} measured at nominal, half and double values of v, respectively. CV tests using a planar electrode system with v values set from 6.25 to 200 mV s^{-1} (Figure 13.2A) were performed. Then, eqn (13.7), (13.9) and (13.11) were used to calculate the values of i_R, i_C and i_K. These are plotted in Figure 13.2B–D, respectively. The peaks of i_R are significantly sharper than the peaks of i_{WE}, shown in the original CV measurement.

13.2.2 Stripping Voltammetry

Compared with other methods, stripping voltammetry (SV) is a powerful tool for determining trace amounts of ions in environmental, clinical and industrial samples. SV consists of two steps: preconcentration

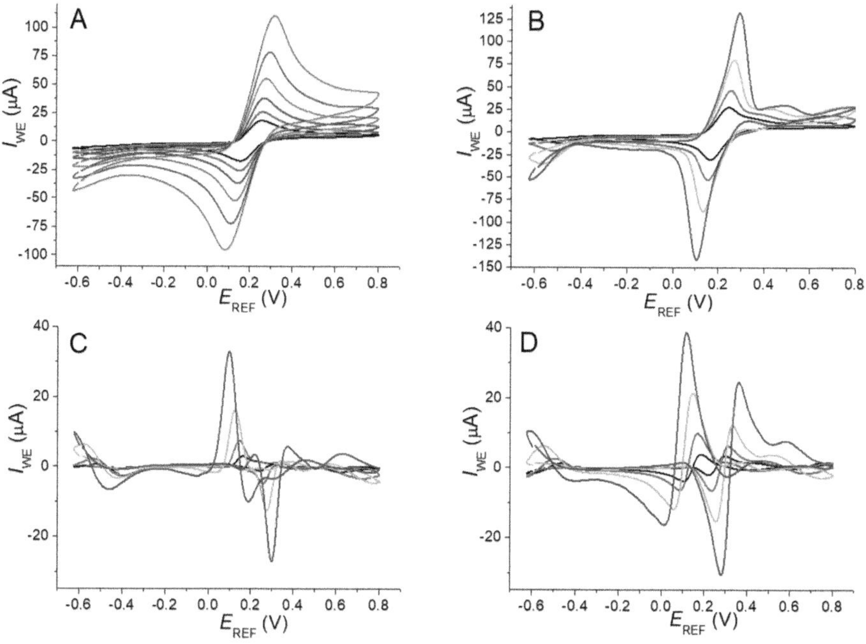

Figure 13.2 (A) CV performed in an Fe^{2+}/Fe^{3+} solution using planar electrodes made of Au with v values set from 6.25 to 200 mV s^{-1}. (B–D) are individual current components i_R, i_C and i_K, respectively.

(PR), where all the ions of interest are extracted from the solution, and stripping (ST), where the electric current required to remove the absorbed ions from the surface is measured.

During the PR step, the analytes are immobilized and concentrated on the surface of the WE by applying a limiting current potential. Analyte concentrations on the WE are affected by the electrode surface area, stirring speed and time. During the ST step, the polarity of the current is reversed to strip off the analytes and generate a peak current related to the concentration of the original analytes. SV consists of anodic stripping voltammetry (ASV) and cathodic stripping voltammetry (CSV), both of which follow the same principle of SV. The difference is that the WE is a cathode during ASV and an anode during CSV. ASV is used for metal detection and CSV is used for anion detection. The basic process for ASV can be described by the following equations:

$$M^{n+} + ne^- \rightarrow M^0 \tag{13.12}$$

$$M^0 \rightarrow M^{n+} + ne^- \tag{13.13}$$

This method is relatively simple and has a greater potential to be integrated into microfluidic devices compared with classical spectrometric methods, thus providing sensitive, qualitative and quantitative detection of rare analytes.

A nanostructured gold microelectrode array was fabricated for the ultrasensitive detection of heavy metal contamination using ASV. A nanostructured Au microelectrode array with a large surface area was made to improve the signal-to-noise ratio, enhance sensitivity and lower the limits of detection (LODs). The array was first patterned using a lithographic method and then selectively electrodeposited in circles. Subsequently, the array was dip-coated with a layer of gelatin and cross-linked using glutaraldehyde solution. As a result, the surface area increased by a factor of ~1440 before modification. The gelatin layer cross-linking was performed to characterize the Au surface properties (Figure 13.3). ASV was then performed to detect the [As^{3+}] ultrasensitively, with an LOD of ~0.0212 parts per billion (ppb) (signal-to-noise ratio = 3.3), 470 times lower than the acceptable limit in potable water recommended by the World Health Organization. These nanostructured arrays could also be used for the detection of other metals, such as Cr, Hg, Cu and Sb.

A paper-based microfluidic device combined with SV detection was developed to detect Pb^{2+},[6] for comparison with the previous chip.

Figure 13.3 Illustration of the microelectrode array, a schematic diagram of the ASV process and the detection results. Reproduced from ref. 3 with permission from American Chemical Society, Copyright 2018.

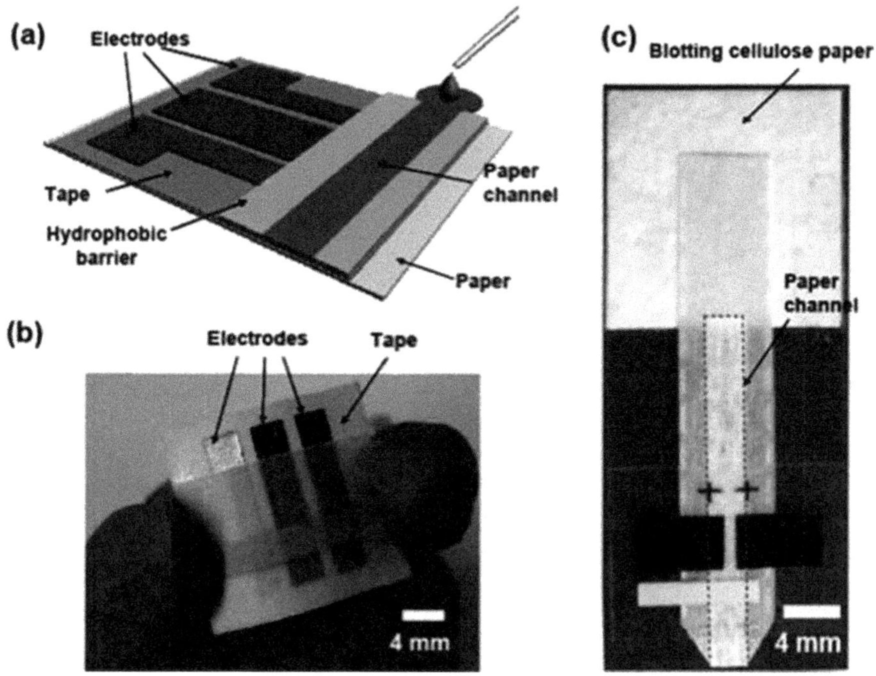

Figure 13.4 (a) Schematic of the paper-based device; (b) photograph of the hydrodynamic paper-based electrochemical sensing device used for glucose analysis; (c) photograph of the device used for heavy metal ion detection. Reproduced from ref. 6 with permission from the Royal Society of Chemistry.

The microfluidic channels were patterned by wax printing on polyester–cellulose blend paper and the electrodes were screen-printed by the Ag/AgCl ink on the paper (Figure 13.4). The PR efficiency was dramatically improved during the SV test as the sample solution was continuously wicked in the paper, enhancing the detection sensitivity and reliability. The device achieved an LOD of 1.0 ppb. Although this LOD is much higher than that with the nanostructured gold microelectrode array, the low-cost, disposable and portable device can be conveniently used in areas that lack bench-top instruments.

13.3 Impedance Measurement

Another technique frequently used is electrochemical impedance spectroscopy (EIS), which measures the response of an electrochemical system to an applied potential. The frequency dependence of this impedance can reveal underlying chemical processes.

In EIS, an alternating current (ac) perturbation with a small amplitude of ~10 mV with frequency as a parameter and a direct current (dc) potential are applied between the WE and REF. The impedance and the phase as a function of the frequencies are recorded to calculate the real component of impedance (Z_{re}) and the imaginary component of the impedance (Z_{im}).

The impedance (Z) can be expressed as

$$Z = Z_{re} + jZ_{im} \tag{13.14}$$

As the frequency decreases, the real impedance is equal to the charge-transfer resistance, $Z_{re} = R_{ct}$, representing the redox reaction rate at the surface of the electrode. The data are commonly displayed using a Nyquist plot of Z_{re} and Z_{im} with frequency as a parameter (Figure 13.5). EIS is a highly sensitive method and is frequently used in biosensing detection as it can directly monitor the binding events that occur on the surfaces of electrodes without interference.

A microfluidic device based on non-Faradaic impedance spectroscopy was developed to detect label-free deoxyribonucleic acid (DNA) in real time.[7] The microfluidic channel patterned in a chip was fabricated on a polydimethylsiloxane (PDMS) layer and sandwiched by two poly(methyl methacrylate) (PMMA) slides. Four interdigitated Ti/Pt microelectrodes were made based on the lift-off technique and integrated in the PDMS layer. The amino-terminated probe DNA was immobilized between the electrodes (Figure 13.6). The label-free DNA was detected by the change in the complex impedance of the electrodes and the concentration of target DNA could be detected at levels

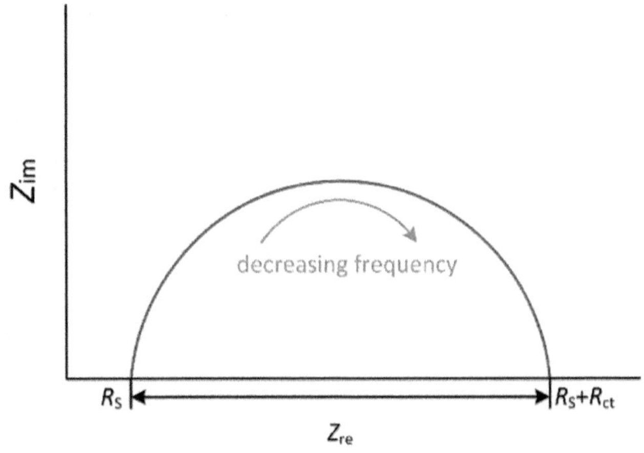

Figure 13.5 The principle of EIS.

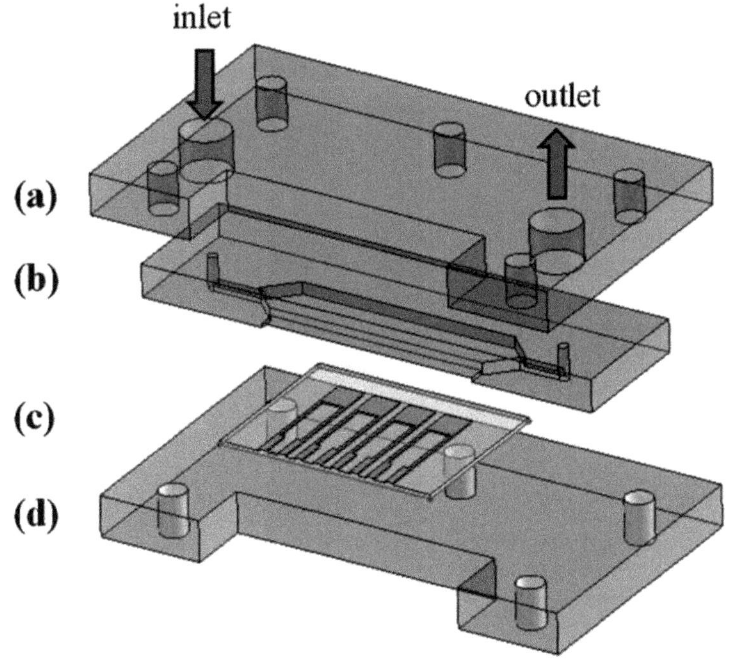

Figure 13.6 Schematic of the chip for biosensing. Top: (a) PMMA slide; (b) PDMS layer with microfluidic channels. Bottom: (c) microelectrodes layer; (d) PMMA slide. Reproduced from ref. 7 with permission from the Royal Society of Chemistry.

down to ~1 nM. The system was then used to detect the pathogen DNA from *Salmonella choleraesuis* (now *Salmonella enterica*) in dairy products.

Another microfluidic chip incorporating EIS was fabricated to characterize and identify suspended cells.[8] The device consisted of two glass layers; the microfluidic channels and two electrodes for impedance measurement were in the bottom layer, and the inlet and outlet were in the top layer. The chip was fabricated based on glass owing to the hydrophilic surface for the channels (Figure 13.7). As a result, the fluidic sample could be delivered by capillary force without an external pump. The microfluidic channels and the inlet and outlet were made based on the wet etching method. A spray coating-based lithographic technique was then used to pattern the Ti/Pt electrodes on top of the bottom glass layer. Finally, the two glass layers were bonded together by a very thin adhesive layer (SU-8). Subsequently, chips with different geometries of electrodes were tested using deionized water, phosphate-buffered saline (PBS) and dead and living cells suspended in PBS.

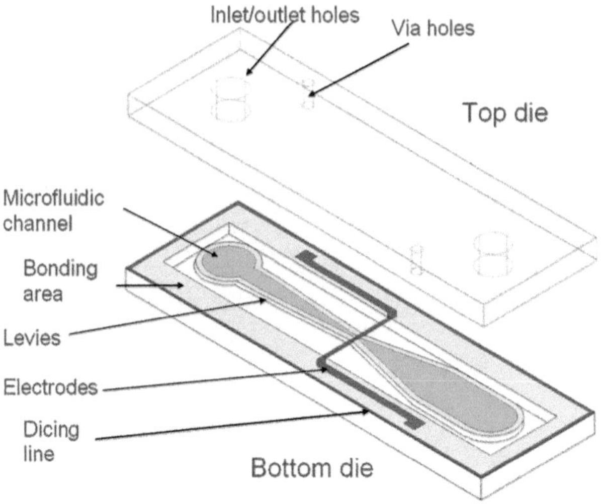

Figure 13.7 Schematic of the microfluidic chip used for the analysis of cell suspensions. Reproduced from ref. 8 with permission from Elsevier, Copyright 2007.

13.4 Electrochemiluminescence

Electrochemiluminescence (ECL) is a technique that produces light emissions due to chemical reactions. In ECL, the molecules are excited by an electrochemical redox reaction and emit light on returning to the ground state. Compared with fluorescence techniques, discussed in Chapter 12, ECL has the advantage of no background noise in the absence of analyte, resulting in a dynamic range wider than six orders of magnitude.

The ECL method was first used to detect the luminophore tris(2,2′-bipyridine)ruthenium(II) dichloride, $[Ru(bpy)_3{}^{2+}]Cl_2$, in aprotic electrolytes. ECL can be used in water-based electrolytes when a co-reactant pathway is introduced, making it a popular technique for DNA and protein detection. The ECL luminophores could be directly utilized as labels with analytes considered as co-reactants or indirectly with analytes quenching the ECL. It could also be used as part of the energy transfer system. Many factors affect the ECL amplitude, such as the composition and pH of the solution, the surfactant in the electrolyte, the co-reactant types, the luminophore immobilization on the surface of electrodes, the addition of ethanol to the solution and the electrode material. Here, we introduce a few examples of the ECL-based technique applied in microfluidics.

Microelectrodes and microarrays are popular in microfluidics owing to the small sample volume requirement and high sensitivity.

A nanostructured gold amalgam (Au_xHg_y) microelectrode array was fabricated using a single-step co-deposition method.[9] The nanostructured surface was fabricated using a lithographic mask with a 4 μm diameter opening area for selective Au_xHg_y deposition. This resulted in a surface area of ~43.5 mm^2, three orders of magnitude larger than the ~5.3×10^{-2} mm^2 defined by lithography (Figure 13.8). The reactions and kinetic details of ECL were studied by monitoring the ECL process from a single microelectrode in an electrochemical cell. The electrode array could be used for ultrasensitive biosensing detection as hydrogen evolution is suppressed and the electrodes can be operated at more negative potentials compared with Au or Pt. The properties of the surface material were also modified by Hg removal using a dealloying method so the array could operate in the positive potential region, similarly to conventional Au electrodes.

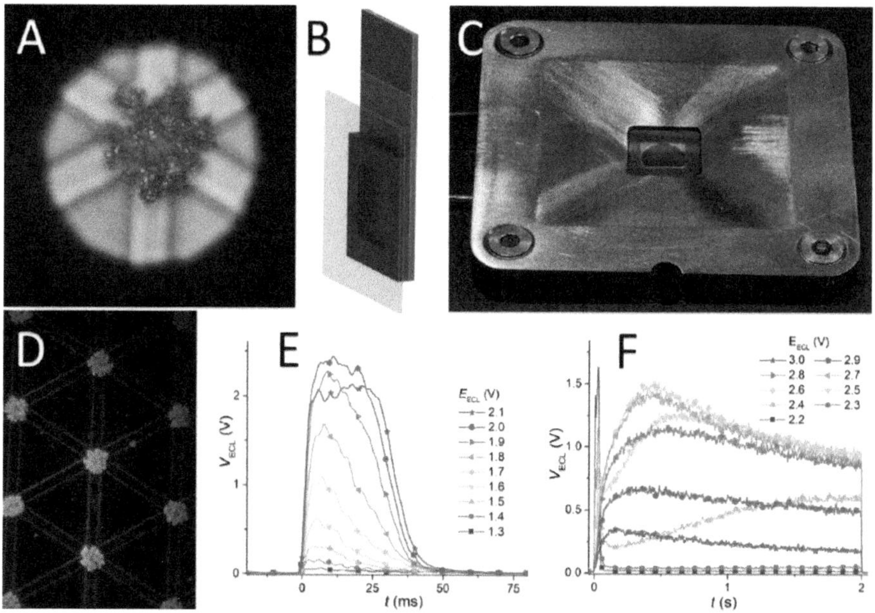

Figure 13.8 (A) Image of a single electrode for ECL; (B) schematic of the cell setup; (C) photograph of the cell setup for a single measurement; (D) microscope image of the array in the EC cell in a dark field; (E) V_{ECL} as a function of time in the E_{ECL} series in a range from 1.3 to 2.1 V; (F) V_{ECL} value measured by a photomultiplier tube using a 50x lens with a numerical aperture of 0.5 of an optical microscope from a single spot with nanostructured Au with E_{ECL} applied at an ITO electrode as a parameter and as a function of time. Reproduced from ref. 9 with permission from Elsevier, Copyright 2019.

Another example is a paper-based microfluidic device combining ECL for multiplex immunoassay.[10] $Ru(bpy)_3^{2+}$ and carbon nanodots (CNDs) were used as the ECL labels to obtain a high throughput. Two carbon WEs were screen-printed on the patterned paper and each was used to detect two analytes. ECL was triggered by a battery, aiding portability and disposability (Figure 13.9). A voltage controller was designed and fabricated for precise control of the output voltage. Simultaneous multiplex detection was performed by controlling the operational constant potential of $Ru(bpy)_3^{2+}$ at +1.2 V and CNDs at −1.2 V. Four tumour markers in human serum were simultaneously tested with the device and agreement was obtained with the benchmark parallel single-analyte tests. This device provides a new method for multiplex immunoassay with high throughput.

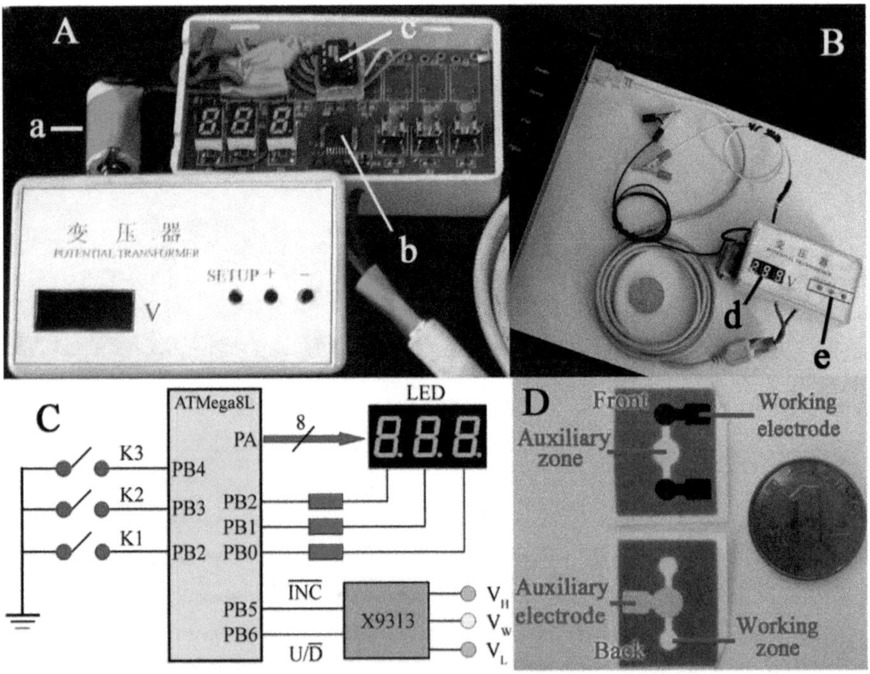

Figure 13.9 (A) Setup of the battery-triggered device; (B) photograph of the voltage-tuneable power device; (C) electronic circuit schematic of the voltage-tuneable power device; (D) photograph of the paper-based microfluidic device combining the ECL for multiplex immunoassay. Reproduced from ref. 10 with permission from the Royal Society of Chemistry.

13.5 Electrochemistry and Microfluidics

Great efforts have been made to develop lab-on-a-chip (LOC) and to integrate the bench-top instruments commonly used in microfluidics. For analytical chemistry applications, one challenge is the integration of detection components into a LOC system. EC detectors have potential as the diversity of the electrodes' geometry and characteristics lead to easy integration into a LOC. Additionally, EC detectors are reliable, strongly selective and highly sensitive.

13.5.1 Electrodes Modified with Nanomaterials

Electrodes modified with nanomaterials represent a promising approach for improving the detection performance in microfluidic systems for immunoassays.[11]

A microfluidic system for detecting the degradation of the pesticide atrazine (Atz) was fabricated based on PDMS and integrating a set of electrodes using a screen printer.[12] The set of electrodes was made of graphite (AUX), Ag/AgCl (REF) and boron-doped diamond electrodeposited with Pt nanoparticles (WE). The WE was used as a highly conductive catalytic transducer to perform detection with high sensitivity (Figure 13.10). The degradation was based on a competitive magneto-controlled enzyme reaction and was detected by measurement of the voltage while keeping the electric current constant (chronoamperometry), obtaining one of the lowest LODs of 3.5 pM for atrazine. The LOC platform may trigger interest in detecting diverse pesticides and other contaminants.

A microfluidic device was developed for the determination of glutathione in pharmaceutical products.[13] The microfluidic structure was fabricated on PDMS using a soft lithographic technique. The Ag/AgCl RE and graphene-based WE and AUX were then screen-printed on a glass substrate. Finally, the substrate was bonded to the PDMS using an oxygen plasma treatment to form the microfluidic device (Figure 13.11). Amperometry was performed to detect glutathione and a wide dynamic range of ~10–500 μM and a low LOD of ~3 μM were obtained. Also, the oxidation current of glutathione was about two times higher than that with carbon-based electrodes. The device was compared with a high-performance liquid chromatographic method and reasonable agreement between them was obtained. The proposed device has

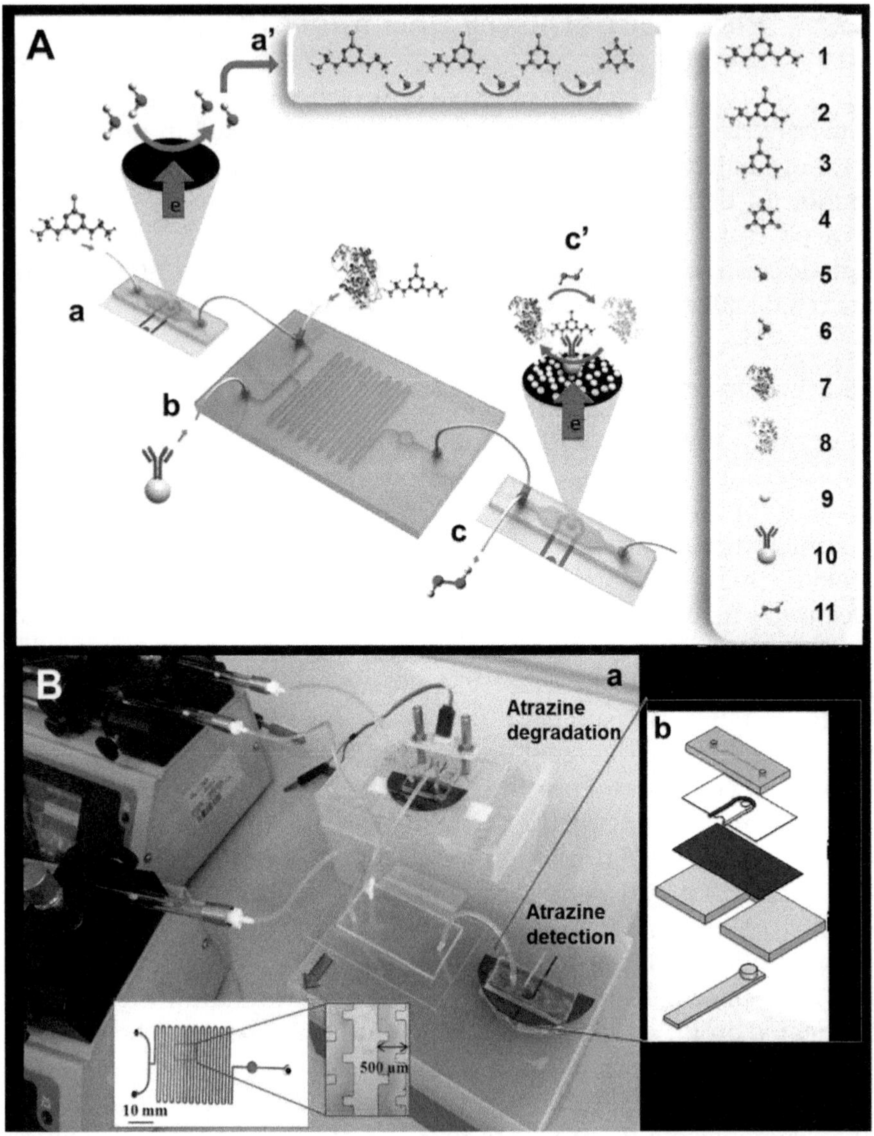

Figure 13.10 (A) Schematic of the microfluidic chip for detection of Atz degradation; (B) photograph of the microfluidic platform. Reproduced from ref. 12 with permission from Elsevier, Copyright 2016.

great potential for mass fabrication and wide application owing to its low cost, repeatability, high throughput and high sensitivity.

Another microfluidic device was described as an immunosensor with electrochemical detection for the quantitative determination of immunoreactive trypsinogen (IRT), also known as a marker for cystic

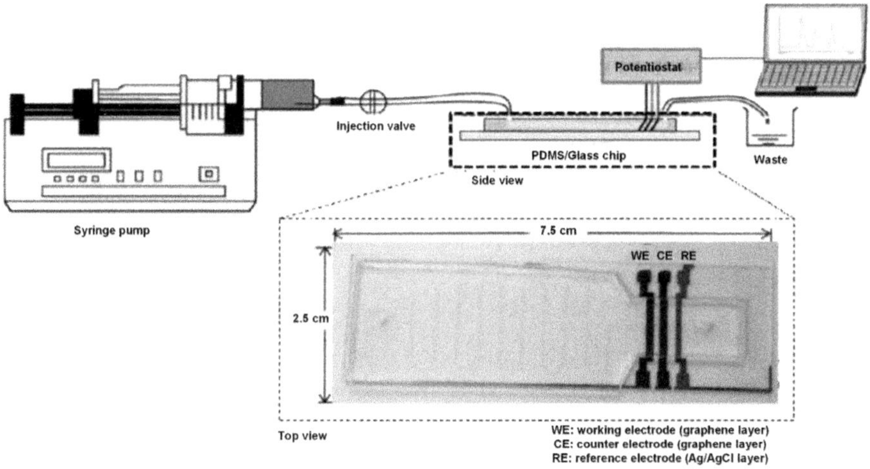

Figure 13.11 Schematic of the proposed setup and a photograph of the microfluidic chip with an integrated screen-printed graphene-based electrochemical electrode. Reproduced from ref. 13 with permission from the Royal Society of Chemistry.

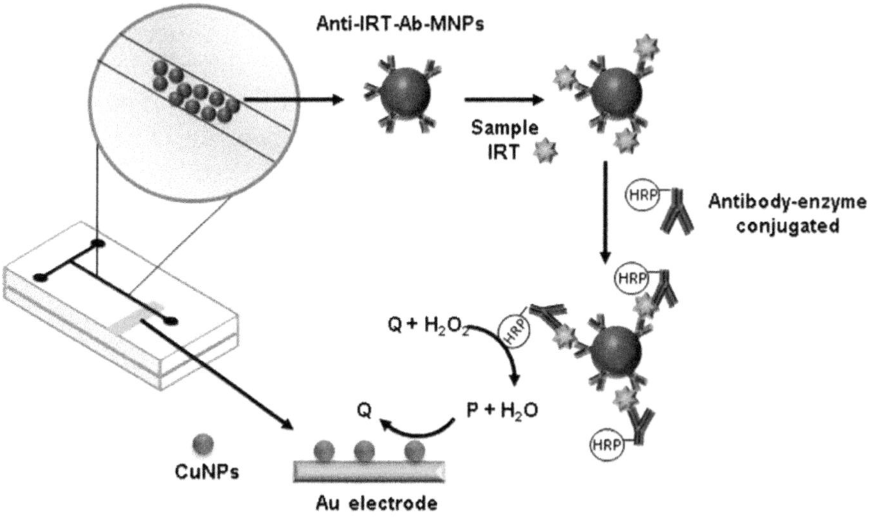

Figure 13.12 IRT detection procedure using a microfluidic device integrated with electrodes modified with nanoparticles. Reproduced from ref. 14 with permission from Springer Nature, Copyright 2015.

fibrosis.[14] The device was based on a sandwiched structure of glass/PDMS/glass integrated with electrodes. Copper nanoparticles were electrodeposited '*in situ*' on a gold WE, improving the detection sensitivity by increasing the active area (Figure 13.12). The IRT extracted from blood samples was injected into the device and then captured by anti-IRT-loaded nanoparticles and quantified electrochemically using CV, showing coefficients of variation of <5% for within-day assays and <6.4% for between-day assays. The immunosensor was also compared with a commercial enzyme-linked immunosorbent assay (ELISA) for IRT detection and the correlation coefficient was ~1. The proposed microfluidic-based sensor represents a portable alternative for screening cystic fibrosis.

13.5.2 Electrodes in Flexible Fabrication

The development of flexible electrode fabrication techniques, such as screen printing and stamp transferring, greatly improved the development of wearable biosensors. Different electrode geometries have been designed for use in physiological, environmental and security monitoring.[15] Flexible thick-film electrochemical sensors were characterized for future use in wearable and display applications.[16] Graphite-based WEs were screen-printed onto three different flexible substrates: Mylar, Kapton and poly(ethylene naphthalate) (Figure 13.13). Mechanical rolling, bending and crimping were then studied to test the electrical and electrochemical properties of the biosensors. The results showed that the sensors did not lose function even if they were bent to an extremely small radius of curvature of ~8 mm. They were also resistant to repeated bending. The electrochemical response was maintained at sub-millimetre bending radii and the device was still available at a 180° pinch. This illustrates that screen-printed electrodes on flexible substrates could be used for sensing with non-planer geometries integrated with various on-body wearable devices.

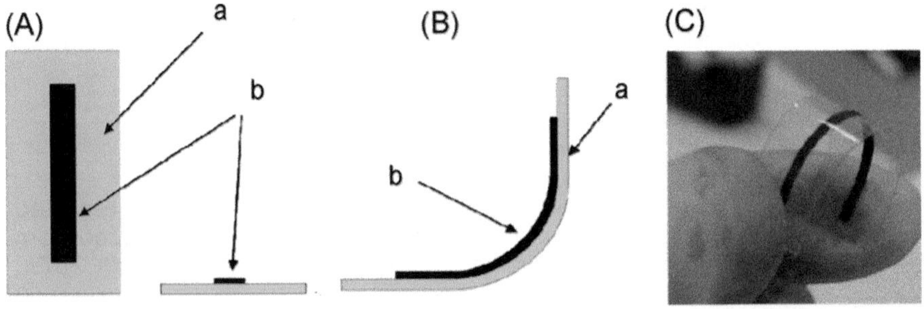

Figure 13.13 (A) Schematic of a screen-printed electrode (SPE) on a flexible substrate (a) with a thick-film carbon WE (b); (B) schematic of the 90° inward bend of the flexible SPE; (C) photograph of the 180° inward bend of an SPE on a Mylar substrate. Reproduced from ref. 16 with permission from Elsevier, Copyright 2009.

Flexible screen-printing-based sensors are not easy to attach to the body, owing to the incompatible elasticity between the substrate and skin blocking direct epidermal integration. A new methodology was proposed to allow printable electrodes that are compatible with the non-planarity and irregularities of the human anatomy to sense directly on the skin.[17] A printed temporary tattoo-based electrochemical sensor was developed to monitor the physiology and security of the chemical constituents, leading to a demonstration of electronic skin (Figure 13.14).

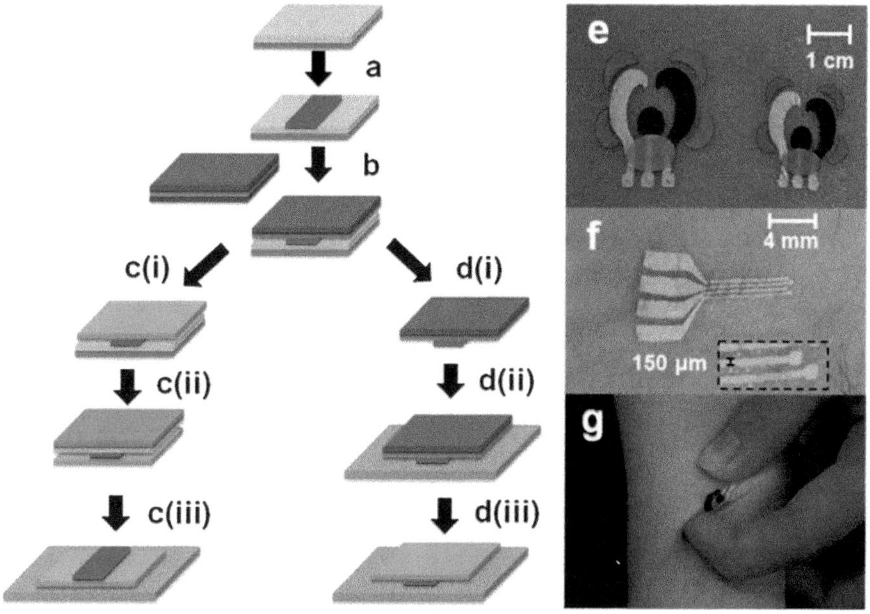

Figure 13.14 Left: schematic of the sensor fabrication. Step (a): the electrode design (in red) was screen-printed on the release agent-coated (olive) base paper (orange). Step (b): the adhesive sheet (blue) with a protective coating (maroon) was utilized for the printed electrochemical sensor. Step (c): the protective sheet was removed (i), the tattoo was flipped and applied to the skin (green) and tapped with water (ii) and the release agent-coated base paper was removed, exposing the adhered sensor pattern to the wearer's external environment for remote sensing (iii). If physiological monitoring is desired, the routine illustrated in (d) is followed: the tattoo paper is removed (i) and the tattoo pattern is applied to the skin (ii); the protective coating is then removed (iii). (e) Photograph of various tattoo sensor designs on human skin. (f) Photograph of the dimensions of the sensors. (g) Photograph of the sensors being twisted. Reproduced from ref. 17 with permission from the Royal Society of Chemistry, Copyright 2012.

Two designs were proposed for monitoring the environment and physiology. The tattoo biosensor was validated using 2.5 mM uric acid (UA) and 2.5 mM ascorbic acid (AA). In addition, the performance of the biosensor with a carbon fibre (CF)-reinforced electrode and with an unreinforced electrode was measured on porcine skin for 18 h (Figure 13.15). The results indicated that the CF-reinforced electrochemical sensor achieved high-fidelity electroanalytical operation under the severe demands imparted by epidermal wear. The reported novel device is compliant with skin, resulting in non-invasive chemical monitoring. This device has the potential for wide application in areas where true bionic integration is a core requirement.

Many other examples of electrochemistry utilized in microfluidic systems can be found in several reviews.[18–21]

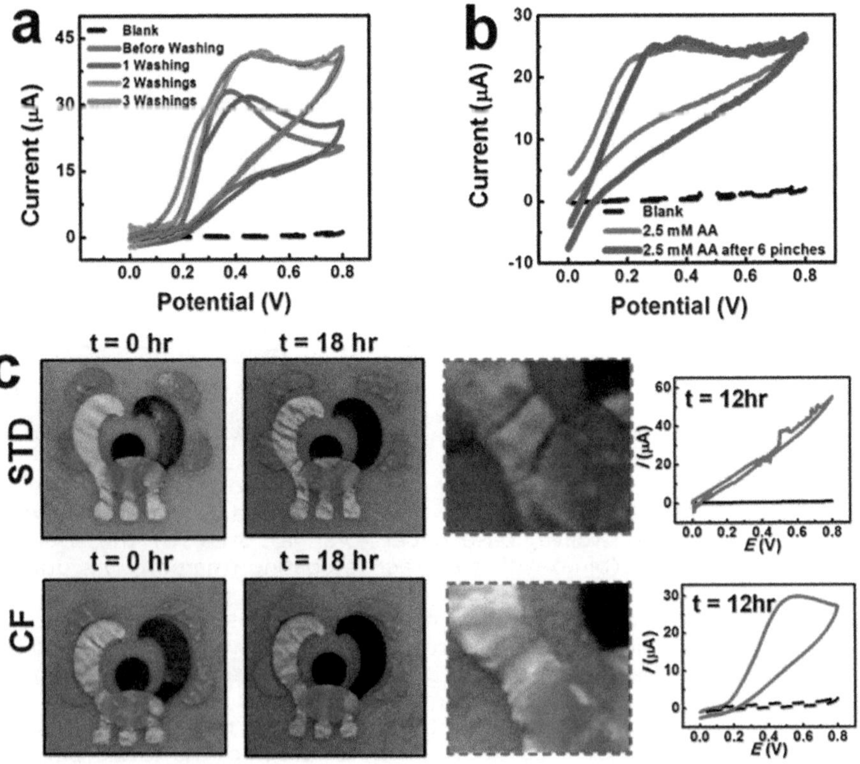

Figure 13.15 CV of UA was achieved using a CF-reinforced tattoo sensor on porcine skin with (a) washing and (b) pinching. (c) Photograph and electrochemical evaluation of an unreinforced and CF-reinforced T3 sensor on human skin. Reproduced from ref. 17 with permission from the Royal Society of Chemistry.

References

1. A. J. Bard and L. R. Faulkner, *Electrochemical Methods: Fundamentals and Applications*, Wiley India Limited, 2004.
2. M. Pumera, *et al.*, *Sens. Actuators, B*, 2007, **123**(2), 1195–1205.
3. P. Podesva, I. Gablech and P. Neuzil, *Anal. Chem.*, 2018, **90**(2), 1161–1167.
4. L. Trnková and O. Dračka, *J. Electroanal. Chem.*, 1996, **413**(1), 123–129.
5. O. Dračka, *J. Electroanal. Chem.*, 1996, **402**(1), 19.
6. Z. Nie, *et al.*, *Lab Chip*, 2010, **10**(4), 477–483.
7. D. Berdat, *et al.*, *Lab Chip*, 2008, **8**(2), 302–308.
8. C. Iliescu, *et al.*, *Sens. Actuators, B*, 2007, **123**(1), 168–176.
9. P. Podesva, X. Liu and P. Neuzil, *Sens. Actuators, B*, 2019, **286**, 282–288.
10. S. Wang, *et al.*, *Lab Chip*, 2012, **12**(21), 4489–4498.
11. T. Sierra, A. G. Crevillen and A. Escarpa, *Electrophoresis*, 2019, **40**(1), 113–123.
12. M. Medina-Sanchez, *et al.*, *Biosens. Bioelectron.*, 2016, **75**, 365–374.
13. C. Karuwan, *et al.*, *Anal. Methods*, 2017, **9**(24), 3689–3695.
14. M. L. Scala Benuzzi, *et al.*, *Microchim. Acta*, 2016, **183**(1), 397–405.
15. J. R. Windmiller and J. Wang, *Electroanalysis*, 2013, **25**(1), 29–46.
16. J. Cai, *et al.*, *Sens. Actuators, B*, 2009, **137**(1), 379–385.
17. J. R. Windmiller, *et al.*, *Chem. Commun.*, 2012, **48**(54), 6794–6796.
18. W. B. Zimmerman, *Chem. Eng. Sci.*, 2011, **66**(7), 1412–1425.
19. J. Adkins, K. Boehle and C. Henry, *Electrophoresis*, 2015, **36**(16), 1811–1824.
20. E. Kjeang, N. Djilali and D. Sinton, *J. Power Sources*, 2009, **186**(2), 353–369.
21. J. Wang, *Talanta*, 2002, **56**(2), 223–231.

14 Cells in Lab-on-a-chip

14.1 Introduction

During the 1970s and 1980s, progress in micro- and nanofabrication technologies was developing, allowing quite complex and integrated microsystems to be fabricated. This attracted the interest of biologists for their use in biochemical assays and cell culture.

Genomics, more specifically DNA sequencing, was of keen interest for the differentiation of different living organisms. This was the trend until the 1990s, when the scientific interest of biologists moved towards understanding cells as the whole organism, to correlate concepts such as phenotype–genotype and structure-to-function.[1,2]

Microfluidic technologies underwent significant progress during this period, as seen in previous chapters, so their application to studying cells was a natural transition in the field of biology. Mainly the combination of chemistry in microfluidic systems with cell biology and tissue engineering has seen a rapid growth in interest.[3,4]

Nowadays, the main biological challenge is to move away from the analysis of populations of live or dead fixed cells towards single cells, specifically the high-throughput and multiplex analysis of single cells to further the understanding of how cellular functions are connected.

Microfluidics and Lab-on-a-chip
By Andreas Manz, Pavel Neužil, Jonathan S. O'Connor and Giuseppina Simone

Published by the Royal Society of Chemistry, www.rsc.org

14.2 Cell Trapping and Separation

One of the first steps in handling and analysing cells is termed trapping. Cell trapping is required to slow or completely stop the cell before the analysis is carried out, but some form of trapping is also required for cell sorting. Usually, cell trapping is followed by a separation step. The two operations can be developed independently, as individual steps, or simultaneously, integrated into a single device.

Cell trapping and separation can be achieved using fluid dynamics or microfabricated mechanical traps or it can be based on external forces; cell separation operations aim to isolate a subpopulation of cells before the analysis. It is not always simple to separate the two operations, so for ease of understanding these operations will be discussed together in this section.

14.2.1 Hydrodynamic Traps

Hydrodynamic traps are based on the fluid dynamics inside microfluidic channels and flow focusing. As discussed in Chapter 8, two-dimensional focusing is an essential consequence of laminar flow and independent streams (diffusive mixing). Figure 14.1 perfectly exemplifies hydrodynamic traps. Using the same principles as in flow focusing devices, the sample stream undergoes focusing, and the width and position of the output focused flow are a function of the focusing flow rate relative to that of the sample. In two-dimensional focusing, some cells can flow near the floor or the roof of the microfluidic channel and confuse the detector. Two-dimensional focusing has been employed for patterning cells in particular areas and/or next to each other, for additional experimentation such as exposing the cells to highly controlled stimuli (*e.g.* concentration of nutrients, dyes, lasers). An example is presented in Figure 14.1.[5]

For better detectability of cells in a hydrodynamic trap, three-dimensional (3D) focusing is required, creating a single-cell-wide, constant-velocity stream in the centre of the channel. Such 3D focusing can be achieved by introducing a focused stream at the inlet; this approach has been widely explored using different methods of fabrication, but can be achieved by exploiting the laws of fluid dynamics, aiming to create a zero-speed point for studying the cells. In a curved

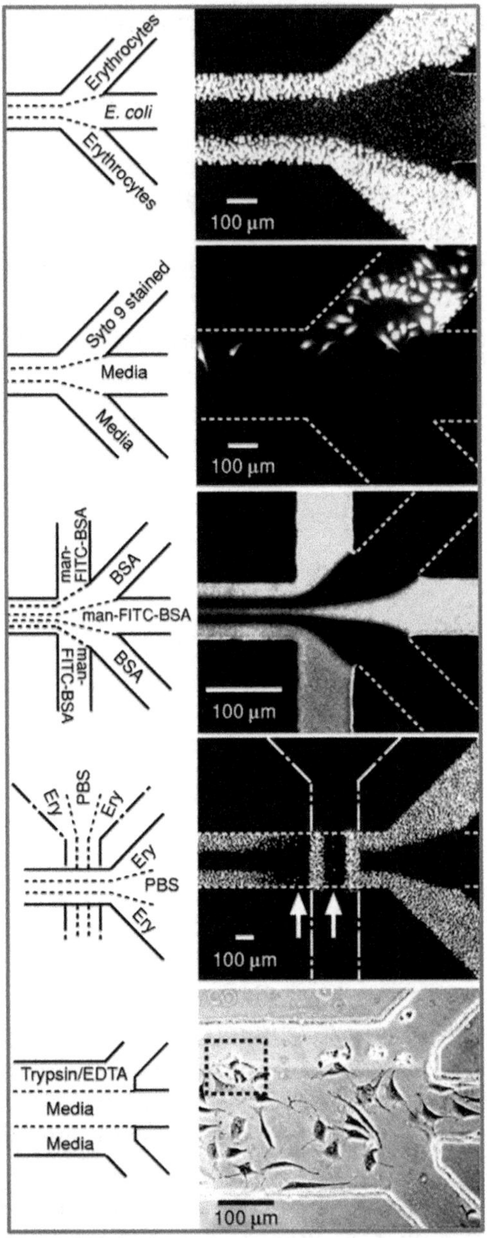

Figure 14.1 Two-dimensional focusing, with examples of using three- and five-port inlets. Reproduced from ref. 5 with permission from National Academy of Sciences USA, Copyright 1999.

channel, inertial forces are generated and applied to the fluid resulting in a secondary flow, also referred to as Dean flow.[6] Trapping by Dean flow has been applied to the 3D focusing of particles, as shown in Figure 14.2.[7] The fluid elements are forced centrifugally from the inside towards the outside, so that the centrifugal force density is greatest in the centre, where the primary flow is fastest. A drifting effect is generated when a stream of cells flows in a curved microfluidic channel at high velocity, of the order of metres per second, which focuses the particles or the cells in 3D, although a high shear stress acts on the cells, which might induce unwanted modifications of the cells. To overcome this problem, while maintaining the advantages of Dean flow, a drifting effect can be generated by a series of contraction–expansion structures in the microfluidic channel (Figure 14.2b).[8] Every time the fluid enters a widening of the microfluidic channel, it is forced to follow rotational lines to conserve momentum, thus concentrating the

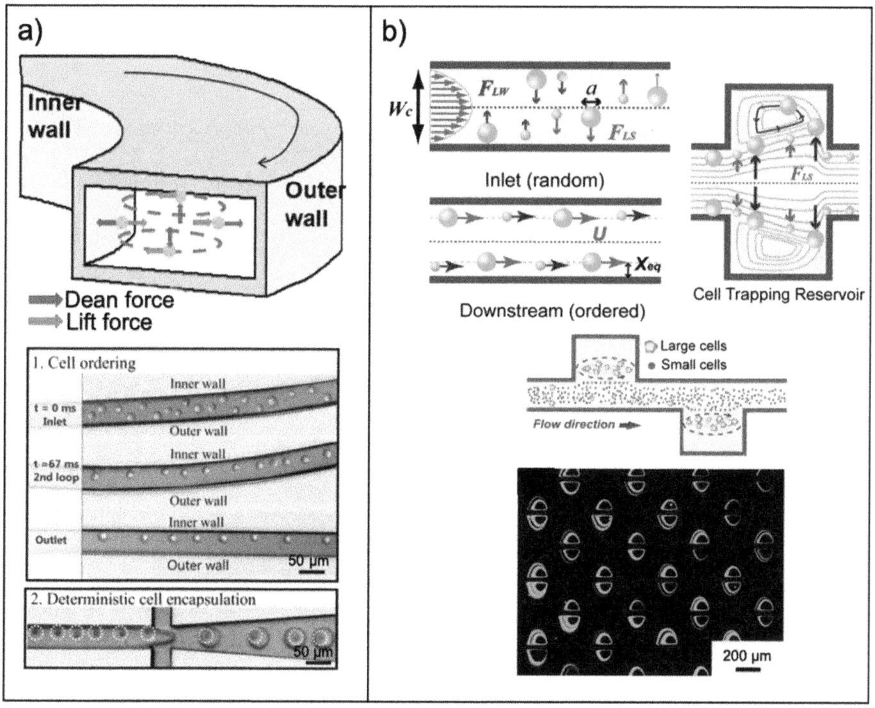

Figure 14.2 Three-dimensional focusing. (a) Dean flow in a curved channel. Drag and lift forces pull the cells towards the wall. (b) The drifting effect generated by a series of contraction–expansion structures in the microfluidic channel. (a) Reproduced from ref. 7 with permission from the Royal Society of Chemistry, Copyright 2012 (b) Reproduced from ref. 8 with permission from AIP Publishing, Copyright 2011.

sample flow in the 3D centre of the channel while also reducing the stress applied to the cells.

A combination of hydrodynamics and microfabrication can also be used for the trapping of a subpopulation of cells or a single cell. This approach is based on the fact that cell manipulation takes place in an aqueous solution that at the microscale has an apparent viscosity that is not negligible. Figure 14.3 shows how a cell flows into the cage. The flow cannot pass over the cage once a cell is trapped, causing the flow to travel only at the sides of the cage, contributing to cell confinement and forcing other cells towards other available cages.[9]

14.2.2 Micromagnetic Trap

Magnetic trapping techniques utilize magnetic fields and magnetic particles of different kinds and sizes. The magnetic force on a particle can be expressed as a function of the particle volume (V), the difference in magnetic susceptibilities (χ) between the media and the particle and the strength and gradient of the applied magnetic field, according to

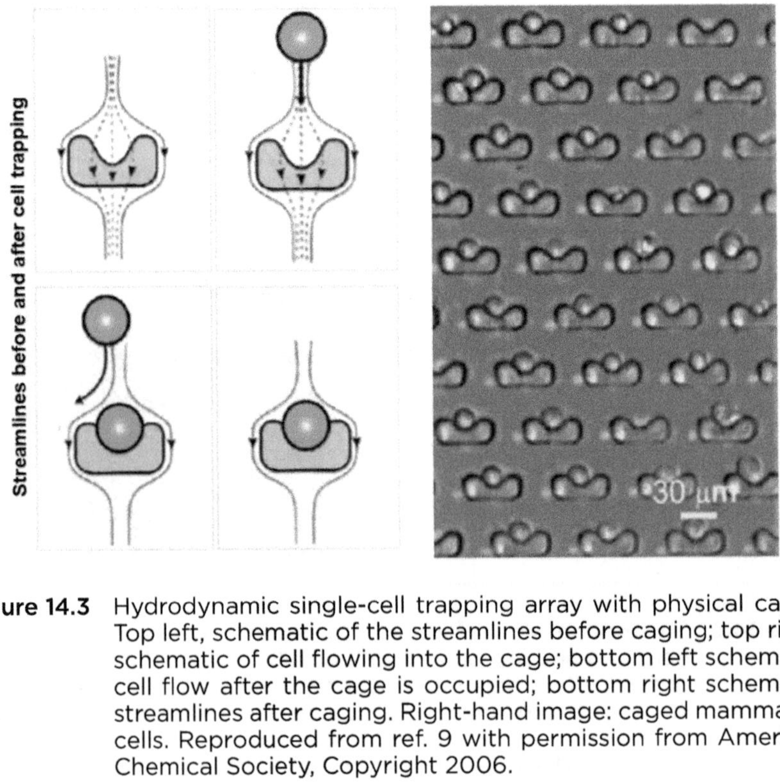

Figure 14.3 Hydrodynamic single-cell trapping array with physical cages. Top left, schematic of the streamlines before caging; top right, schematic of cell flowing into the cage; bottom left schematic, cell flow after the cage is occupied; bottom right schematic, streamlines after caging. Right-hand image: caged mammalian cells. Reproduced from ref. 9 with permission from American Chemical Society, Copyright 2006.

$$F_m = \frac{V(\chi_p - \chi_m)}{\mu_0}(B \cdot \nabla)B \tag{14.1}$$

where χ_p and χ_m are the magnetic susceptibilities of the particle and medium, respectively, μ_0 is the magnetic permeability of free space and $\boldsymbol{B}$ is the magnetic field vector. The volumetric magnetic susceptibilities and the volume of the particle can be adjusted by changing the medium or the material (*i.e.* permeability) of the object to be trapped; for a given set of materials, the adjustable parameter is the term $(\boldsymbol{B}\cdot\nabla)$, which is determined by the strength and configuration of the magnets.

In a homogeneous magnetic field, the magnetic gradient is zero and the force applied on the cells is zero. Hence the magnetic manipulation requires inhomogeneous magnetic fields. Typically, the magnetic force applied on the cells varies between a few and tens of piconewtons, and this is generated by permanent magnets rather than electromagnets because of their higher achievable efficiency, as shown in Figure 14.4, where a three-dimensional magnetic trap for diamagnetic objects in an aqueous solution of paramagnetic ions is

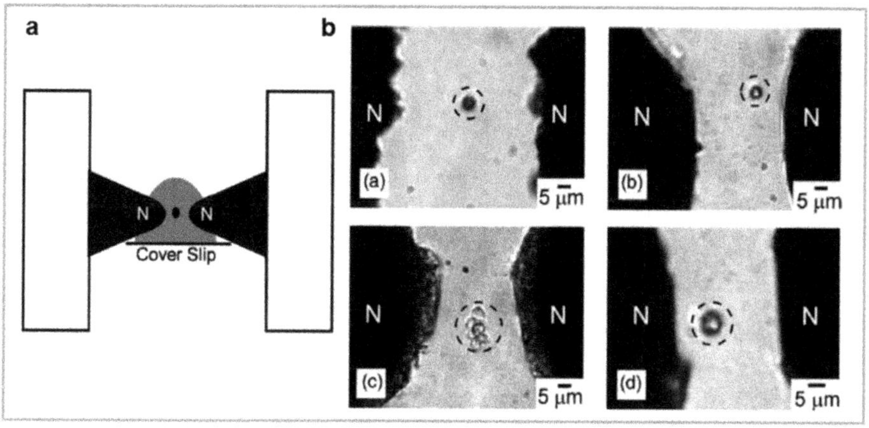

Figure 14.4 (a) Schematic of the experimental apparatus (radius of curvature 76 μm). N is the magnetic North pole. (b) The trapped particle in each image is indicated by a dashed circle. From top left to bottom right: trap of a single 6.5 mm bead in a 50 mM $GdCl_3$ solution; a single 3T3 fibroblast cell with 40 mM Gd·DTPA solution in 0.5 × DMEM culture media; multiple yeast cells with 40 mM Gd·DTPA × DMEM solution (pH 7.2) in 0.5 × YPD broth (1% yeast extract, 2% peptone, 2% dextrose); a single *Chlamydomonas* cell trapped with 40 mM Gd·DTPA solution (pH 7.2) in sporulation media. Reproduced from ref. 10 with permission from AIP Publishing, Copyright 2004.

used for trapping a cell.[10] However, a system with electromagnets is more flexible and there is also the possibility of using several magnetic poles and to use these to translate objects within the permanent magnetic trap.

14.2.3 Dielectrophoretic Cages

Dielectrophoresis (DEP) is the movement of a dielectric object due to forces generated by a non-uniform electric field. This is different from electrophoresis, which relates to the movement of a charged particle in an electric field. DEP refers to the force acting on induced polarization or dipoles in a non-uniform electric field, produced by embedded electrodes, of low voltage.

Mammalian cells are dielectric objects and DEP is a powerful tool for trapping them. In fact, mammalian cells can readily polarize when exposed to electric fields, whereas the cellular functions are not affected by electric fields. The induced polarization depends on a multitude of factors related to cellular physiological conditions, such as the bilipid membrane characteristics, internal structure or size of the nucleus. Depending on the frequency of the alternative electric field to which the cells are exposed, the cells may polarize in the direction of the field vector (positive DEP) or in the opposite direction (negative DEP). The different cell response to electric fields differentiates between cell types and different activation states of similar cells. There are other advantages related to the dielectrophoretic cell trap, the most important being that at the microscale the heat generated is negligible. The dielectrophoretic force is expressed as follows:

$$F_{\text{DEP}} = 2\pi\varepsilon_{\text{m}} r^3 \text{Re}(K(\omega))\nabla E^2 \tag{14.2}$$

where $\text{Re}(K\omega)$ is the real part of the Clausius–Mossotti factor, with

$$K(\omega) = \frac{\varepsilon_{\text{p}}^* - \varepsilon_{\text{m}}^*}{\varepsilon_{\text{p}}^* + 2\varepsilon_{\text{m}}^*}; \quad \varepsilon^* = \varepsilon - \frac{j\sigma}{\omega}$$

where ε_{m} is the dielectrophoretic constant of the medium, ε_{p} is the dielectrophoretic constant of the particle or cell, E is the electric field, ω is the frequency and σ is the electrical conductivity.

Usually, in highly conductive aqueous media the cells are less polarizable than the media, at all frequencies, so that the relationship $\varepsilon_{\text{p}} < \varepsilon_{\text{m}}$ exists between the dielectrophoretic constant of the cell and the medium, and the dielectrophoretic force is negative so that the cells move towards the field minima (Figure 14.5a).[12] Owing to the high permittivity of water (7×10^{-10} C^2 N^{-1} m^{-2}), living cells can also show negative DEP at high fields and high frequencies. At the same time, when operating at high frequencies, there are several advantages related to the ac field, such as the absence of electrophoresis and electrochemical reactions that produce bubbles and electrode corrosion.

The most interesting aspect is that different cell types respond differently to the dielectrophoretic force depending on the frequency of the applied alternative electric fields. Thus, by changing the field frequency or amplitude in dielectrophoretic devices, cell subtypes can be separated. Because of this display of favourable factors, a large number of microscale devices have been proposed for different applications involving the manipulation of cells in suspensions. Depending on the size of the sample, the strategies recommended for dielectrophoretic separation are different. For small samples containing few cells, the cells of interest may be separated by trapping them within predefined locations. For larger samples, cells can be diverted from the mixture

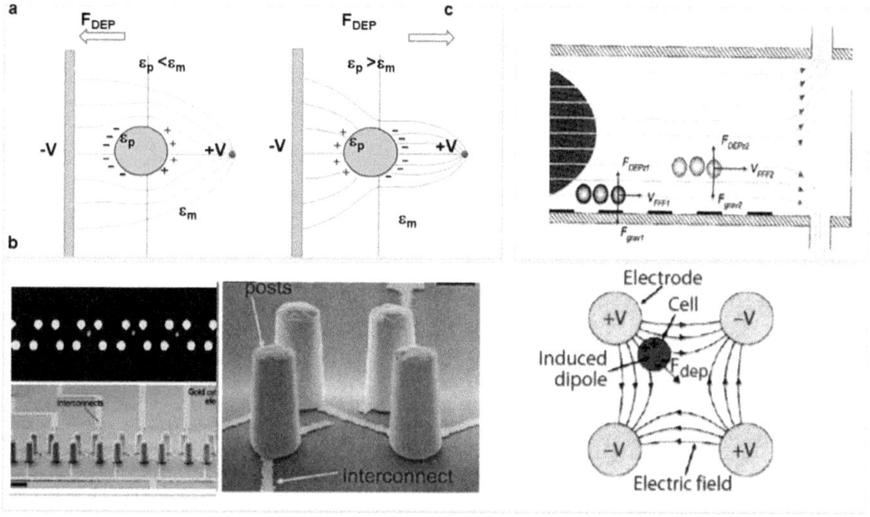

Figure 14.5 (a) Schematic of the working principle of dielectrophoresis. (b) Improvement of separation by dielectric levitation and concentration of target cells on the surface of electrode arrays made by quadrupole posts. (c) Hyperlayer trap combining fluidics and dielectrophoresis. (b) Reproduced from ref. 11 with permission from American Chemical Society, Copyright 2002. (c) Reproduced from ref. 12 with permission from Elsevier, Copyright 2000.

into distinct flow streams based on the difference in dielectric properties and thus separated in time or space. This can be applied for the effective separation of a very heterogeneous sample (*e.g.* blood). Integration of negative DEP can be used for separating cells with different characteristics in order to avoid the crossover frequency, so that several steps of separation need to be included. In the absence of flow, the efficiency of separation is improved by dielectric levitation and concentration of target cells on the surface of electrode arrays (Figure 14.5b).[11] This is possible by introducing positive DEP. Positive DEP was used to trap the cells in energy cages of sizes comparable to the cells, for separation from flowing mixtures. Cells that are trapped are in a stable equilibrium and can be released by simply turning off the electric field. The same quadruple structure in Figure 14.5b has also been used to create both regions of high field strength and regions of low field strength, thus facilitating both *p*-DEP and *n*-DEP, and generate a non-uniform field, which has been used for concentrating and separating bacteria with the aim of detecting pathogenic bacteria in water.

DEP trapping is particularly interesting for separating live and dead bacteria (*e.g. Escherichia coli*) owing to the huge difference in conductivity in the respective cell membranes, where live cells have a membrane conductivity that is 10^4 times lower than that of dead cells.

By integrating hydrodynamics and DEP, it is possible to develop the hyperlayer technique of separation. The DEP forces are applied to the cells flowing in microfluidic channels from an electrode array located at the bottom of the channel. Every cell is subjected to sedimentation forces that are proportional to the weight, dielectrophoretic forces that decrease exponentially with the distance between the cells and the electrodes and drag forces that are proportional to the fluid velocity in the channel (Figure 14.5c).[12] Equilibrium positions at different heights inside the channel for different cells are the result of the balance between changing levitation and constant sedimentation forces. For each cell, different velocities along the channel are the consequence of the parabolic velocity flow profile across the channel. As a result, cells from a heterogeneous mixture can be separated along the channel and captured at different time intervals at the outlet.

The aforementioned examples show how DEP methods can be used for the precise trapping of cells, but another advantage, and arguably the most important, is that the heat generated is negligible. Any increase in the temperature of the cell can cause protein denaturation and permanent damage. Sometimes the contrast between the electrical properties of different cells is less clear, resulting in more difficult separations. Also, application of the dielectric principles can be difficult to implement.

14.2.4 Optical Trap

In optical trapping, a tightly focused laser beam is used to trap the cells with very high precision; see also the discussion of optical tweezers in Section 5.4 in Chapter 5. The laser beam carries a momentum that is transferred to the cell when the beam hits the object. The Gaussian profile of the laser beam will cause the object to be drawn into the centre of the beam and, if a tightly focused laser beam is used, the particle can also be trapped in the direction of the laser beam. This technique was first demonstrated in 1987,[13] when for the first time it was observed that particles in a laser beam were observed to move away from the light due to the radiation pressure. Subsequently, a single-beam gradient trap was created, which is the standard technique used today, and demonstrated the trapping of single tobacco mosaic viruses and *E. coli* bacteria.[14]

The particle range that can be handled oscillates from a few ångstroms up to 10 μm and the maximum achievable trapping forces are in the region of a couple of hundred piconewtons. The resolution in force measurements is now as low as 50 fN in commercial instruments.

These original tests were performed using an argon ion laser, but the laser beam caused damage to the cells. To avoid damaging the cells, infrared (IR) and near-IR lasers have been used, simultaneously lowering the absorption in the cell and the power used to create the trapping force. In fact, most commercial systems for performing optical trapping use near-IR lasers in the range 800–1100 nm to minimize the risk of damaging the sample.

An example of very high-precision manipulation was presented in 1999, when optical tweezers were used to tie knots on different biofilaments such as DNA or actin by attaching a polystyrene bead at the extremity of the filament.[15] The experiment aimed to measure and compare the force required to rupture the filament for different filaments. One of the extremities of the DNA filament was moved and the other was fixed.

Approaches using light to separate cells from complex mixtures are particularly attractive because mechanical contact between cells and surfaces is avoided and the potential for activation is also reduced. However, although cell handling can be very precise, only one cell at a time can be manipulated. The introduction of diode lasers allowed the use of such technology to be extended to the manipulation of a large number of samples. The separation can also take place by sorting by refractive index.[16,17] Optical trapping devices are easily reconfigurable by adjusting the interference pattern. Additionally, the devices do not have narrow channels and clogging does not occur. The increased availability and continuously decreasing cost of solid-state coherent light sources are additional arguments for the use of microscale optical interactions for the separation of cells from blood samples. The greater power input results in a greater increase in the local temperature, which may have an impact on cell viability. Further, the additional equipment needed to control the lasers leads to more complicated multiplexing and higher instrumental costs.

14.2.5 Biochemical Separation

Biochemical differences among cells can be used for creating a selective environment for cells. In fact, biochemical components that can be harmful to some subpopulations of cells are not toxic until certain levels for others. A typical example is an environment that contains ammonium chloride and white and red blood cells. Red blood cells are lysed (disintegrated) within tens of seconds after exposure to ammonium chloride, whereas white blood cells are not. Control of the exposure time permits the selection of which subpopulation of the cells is lysed.

14.3 Manipulation and Analysis

Integrated microfluidic devices permit the identification and separation of the target cell(s). Such devices are complex in terms of fabrication and handling, but are more efficient and accurate because more than one method of separation can be combined in series or parallel.

Flow cytometry represents a serial approach, in which the cells are aligned in a single row, identified by passing one by one in front of a detector module, and immediately separated using a distinct module. The result is a small, portable and cheap device that is discussed in the following section on microfluidic cell sorters.

The arrays represent the parallel approach, in which the cells are simultaneously identified and separated into pure subpopulations. The separations result in an array format that can process a large number of cells, with a huge impact on the throughput.

14.3.1 Microfluidic Cell Sorters

In a conventional flow cytometer, the cells of interest are labelled with a fluorescent dye, so on passing through an array of excitation laser beams and optical detectors, the cells are recognized according to the wavelength of the emitting dye and sorted using switching electric fields. The efficiency of these systems is high, but the equipment is usually larger than 1 m^3 in volume and the operation is time consuming. Also, the most critical step is the sample preparation, involving concentrating the initial sample and labelling. This operation can easily alter the target cells.

Progress in the fields of microfluidics, optical and flow control structures has stimulated the implementation of miniaturized cytometers as an extension of fluorescence-activated cell sorting (FACS). The miniaturized system includes the optical elements (*e.g.* waveguides, lens and optical couplers) and the microfluidic channel for focusing the cells in the centre of the channel for identification; finally, the systems allow the sorting of living and fixed cells.[18] The sample is focused hydrodynamically; however, the selection of particles is achieved not by breaking the stream into droplets, which can alter the cells, but rather through electroosmotic current flow by switching the fluid stream to one or another channel at the downstream junction. Alternatively, electrokinetic focusing, switching and optical tweezers can also be used. During the flow inside the microfluidic channel, cellular parameters such as the deformability and adhesivity can be measured or used for separation.

For example, an optical trap inside the narrow channel can separate normal and less deformable cells. Optical traps can be implemented to hold a cell in a precise position for impedance measurements. An impedance signal, which can be measured at multiple frequencies, allows the quantification of membrane capacitance, cell size, cytoplasm resistance and ionic channels; hence a device that can measure such parameters is able to differentiate a complex mixture of cells. In this way, normal and cancer cells can be characterized according to size and impedance, and analogously red and white blood cells can also be differentiated, with an efficiency and purity above the standard of conventional centrifugal methods. The capacitance measurement of cells has been shown to differentiate cells on the basis of their DNA content.[19] The sorting of the cells can also be achieved by binding the specific cell receptor with magnetic beads, eventually bearing the antibody, which allows the beads to flow into the microfluidic channel and be captured in precise locations or to deviate the tagged cells away from the main stream and into separate channels.[20] Figure 14.6 shows an example of a microfluidic cytometer.[21]

Whereas cytometry requires only incremental improvements to fabrication, the opposite is true for cell arrays. Microfabrication plays a fundamental role because only techniques working at the same scale as the cell size can achieve both the precise positioning of the cells in the array format and gentle extraction of target cells without altering them in the process.

14.3.2 Microarrays for Cell Sorting

Microarrays work in parallel for multiple cell separations, which is very important for applications based on manipulating and analysing a large number of cells or where high-throughput analysis is required. In fact, microarray-based cell sorting has the potential

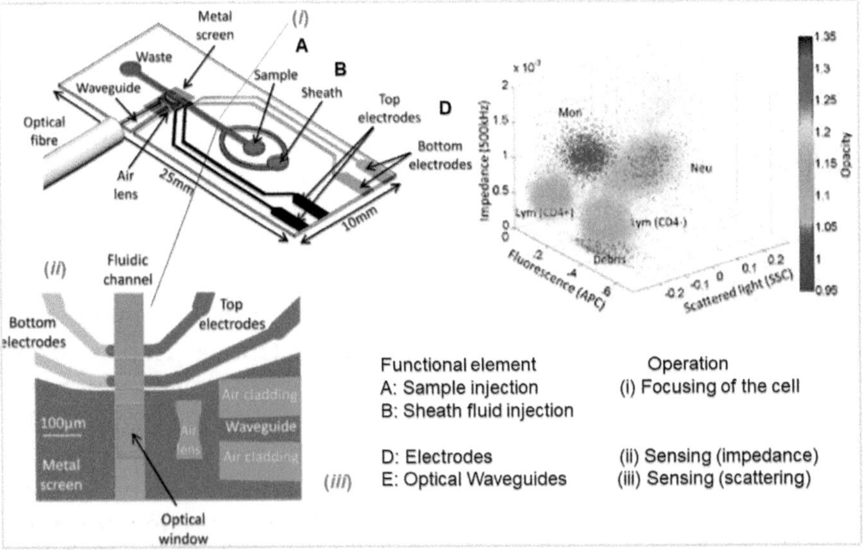

Figure 14.6 Details of a microfluidic cytometer including the necessary operations. Right: a possible set of results from the cytometer. Reproduced from ref. 21 with permission from the Royal Society of Chemistry.

for higher throughput and productivity than DEP and optical trapping. Microarrays allow multiple simultaneous measurements that can be easily repeated, unlike the single-pass, single-measurement principle upon which FACS relies. Microarray devices would allow the separation of cells based on the individual dynamic responses to stimuli or drugs, or permit long-term observation of individual cells, in addition to the analysis of the morphology of cells and organelles.

The first step of the operation includes the positioning of the cells in the array, which is carried out by passive sedimentation in microscale wells or chambers from a cell suspension. The material used for fabricating the wells is important, as it must ensure specific adhesion to the target cells. For example, the specific and selective attachment of cells on patterned antibodies allows immunophenotyping and the simultaneous sorting of specific cell populations into multiple subpopulations.[22] The selection of a patterned surface coated with a specific protein restricts the application of the array to a small group of cells, but it guarantees a high-purity subpopulation for analysis.

The variety of approaches to cell separation, using different principles and a wide range of selection criteria, underlines the growing interest and opportunities in the area of cell separation using microscale techniques. Figure 14.7 depicts how a fully realized concept of a lab-on-a-chip system, with the complete procedure of specific cell capture and analysis in the form of gene and/or protein expression, which could be used for clinical diagnostics and medical research, was integrated in a unique platform.[23,24]

14.4 Cell Sample Analysis

After the selected target cell has been trapped by one of the techniques mentioned, the sample must be analysed. The environment within a microfluidic system can be controlled with high precision, hence nutrients, stimulating biological factors and chemical and physical stimuli can be finely tuned and the cellular response tracked. The real problem lies in the sample preparation; this preliminary step is usually carried out before introducing the cellular sample inside the lab-on-a-chip system, as this operation leads to some more or less important modifications of the sample properties.

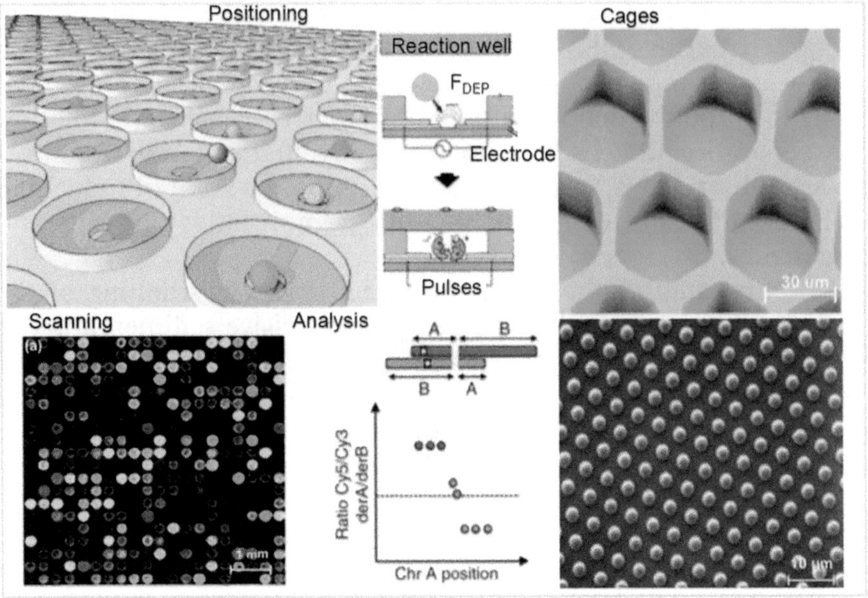

Figure 14.7 Schematic of a cell array based on dielectrophoretic caging in microwells. After the positioning, the array is scanned and analysed. Reproduced from ref. 22, with permission from the Royal Society of Chemistry.

Lab-on-a-chip systems have been used for lysis of cells, quantitative biochemical analysis of soluble components at the single-cell level, analysis of nucleic acids and proteomic analysis. Table 14.1 summarizes the applications of lab-on-a-chip systems for cellomics and cellular biology.[1]

The application of combi-microfluidic devices integrating full cell analysis is difficult and often the interfaces between methods are incompatible. For this reason, many devices still show modular architecture that allows the different modules to operate according to different principles.

14.5 Organ-on-a-chip

Microfluidic devices have also been used for the culture of adhesive cells, organoids and tissue. The purpose behind this is to replace the animal experiments that are currently used in drug discovery, drug metabolism and toxicity studies. Unlike animal experiments, such *in vitro* studies can use human cell lines, which may represent the human *in vivo* situation much better than animals. The standard

Table 14.1 Main applications of lab-on-a-chip systems for cell analysis.

Application	Analysis
Clinical diagnostics	• Detection of pathogens (*e.g. Mycoplasma pneumoniae*) • Islet functionality • Morphological changes and response • Magnetic resonance
Cancer research	• Immunophenotype • Morphology • Electroporation
Drug discovery and screening	• Drug screening, chemodrug assay, drug resistance, multidrug resistance • Cytotoxicity and viability • Response to neurotoxin • Chemoselectivity
Stem cell research	• Culturing and co-culturing • Differentiation
Assay	• Proteomics (ELISA, Western blot) • Genomics
Neuroscience	• Culture and co-culture • Response to stimuli
Migration and chemotaxis	• Gradient generation • Biochemical and biomechanical forces within 2D and 3D scaffolds • Motility • Specific adhesion and rolling
Cell mechanics	• Adhesion • Shear stress response
Tissue models	• Microbioreactors • Culturing • Vascularization • Viability and metabolic activity

method for cell assays is to use Petri dishes employed for 2D cell cultures. Microfluidic devices open up new possibilities to cultivate cells in 3D, and also combine cultures with fluidic channels for perfusion. Such cell cultures can grow from stem cells into organ-like tissue or organoids.

In terms of microfluidics, these devices are mostly very simple. The device contains a location designed to hold cells, tissue or organoids, a fluidic inlet, outlet and optical access for use with microscopy (see Figure 14.8A). The cells can be held mechanically, for example in a larger cavity on the device, or by incorporating the cells into a gel that is mechanically immobilized. More advanced devices provide gel–water phase boundaries (Figure 14.8B).[27] This is necessary for long-duration cell experiments, *e.g.* 1 week to 1 month, as used in inflammation, infection and drug metabolism studies.

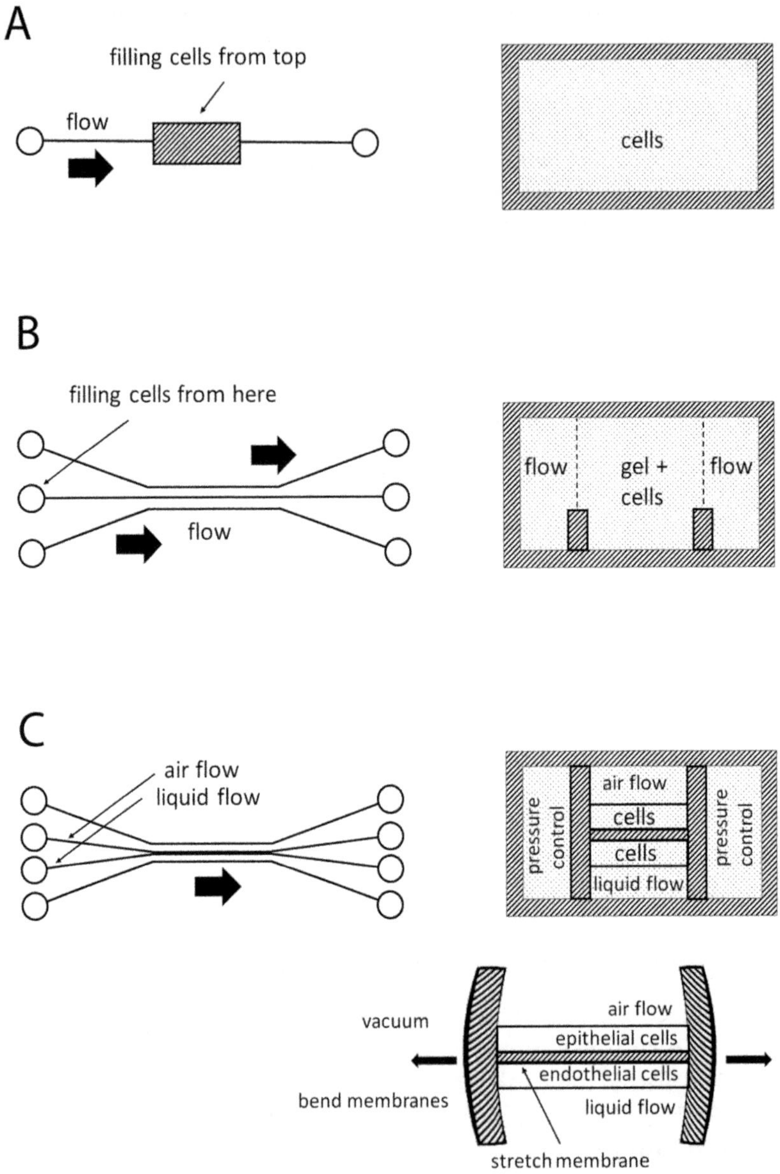

Figure 14.8 Schematic of organ-on-a-chip devices as used for mimicking liver, kidney, gut and lung. The channel layout and channel cross-section are shown. (A) Simple device with a central chamber for the 3D cell culture. (B) Device with parallel flow and cells encapsulated in a hydrogel. (C) 'Breathing' lung device. The inset shows how the membranes move if the pressure in the side channels is lowered.

A particularly interesting design was used for mimicking a breathing lung-on-a-chip (Figure 14.8C). A central flexible membrane was sandwiched between two identical structured polydimethylsiloxane (PDMS) devices that contained a central channel with flexible side walls and two pressure control channels on either side. When the pressure in the side channels was lowered by evacuating the air, the side walls of the central channel would bend and the membrane was stretched. In this way, a 'breathing' behaviour could be obtained. Epithelial and endothelial cells were placed on the outside and inside of the membrane to mimic the behaviour of the lung.[26]

There have been many biological experiments in addition to the lung-on-a-chip that cannot be described here in detail. The list of applications includes the mimicking of the gut, using human intestinal epithelial cells,[27] mimicking of the liver, using stem cell-derived human hepatocytes,[28] and mimicking of the kidney, using human primary renal proximal tubule epithelial cells.[29] The current trend is towards integrating multiple organs on a single device, which is then called body-on-a-chip or human-on-a-chip.

References

1. G. B. Salieb-Beugelaar, G. Simone, A. Arora, A. Philippi and A. Manz, *Anal. Chem.*, 2010, **82**(12), 4848.
2. D. Mark, S. Haeberle, G. Roth, F. Von Stetten and R. Zengerle, *Chem. Soc. Rev.*, 2010, **39**, 1153.
3. J. El-Ali, P. K. Sorger and K. F. Jensen, *Nature*, 2006, **442**, 403.
4. G. M. Whitesides, *Nature*, 2006, **442**, 368.
5. S. Takayama, J. C. McDonald, E. Ostuni, M. N. Liang, P. J. Kenis, R. F. Ismagilov and G. M. Whitesides, *Proc. Natl. Acad. Sci. U. S. A.*, 1999, **96**(10), 5545.
6. D. Di Carlo, J. F. Edd, D. Irimia, R. G. Tompkins and M. Toner, *Anal. Chem.*, 2008, **80**(6), 2204.
7. E. W. Kemna, R. M. Schoeman, F. Wolbers, I. Vermes, D. A. Weitz and A. Van Den Berg, *Lab Chip*, 2012, **12**(16), 2881.
8. S. C. Hur, A. J. Mach and D. Di Carlo, *Biomicrofluidics*, 2011, **5**(2), 022206.
9. D. Di Carlo, N. Aghdam and L. P. Lee, *Anal. Chem.*, 2006, **78**(14), 4925.
10. A. Winkleman, K. L. Gudiksen, D. Ryan, G. M. Whitesides, D. Greenfield and M. Prentiss, *Appl. Phys. Lett.*, 2004, **85**(12), 2411.
11. J. Voldman, M. L. Gray, M. Toner and M. A. Schmidt, *Anal. Chem.*, 2002, **74**(16), 3984.
12. J. Yang, Y. Huang, X. B. Wang, F. F. Becker and P. R. Gascoyne, *Biophys. J.*, 2000, **78**(5), 2680.
13. A. Ashkin, J. M. Dziedzic and T. Yamane, *Nature*, 1987, **330**, 769.
14. A. Ashkin, *Phys. Rev. Lett.*, 1970, **24**(4), 156.
15. R. W. Applegate, J. Squier, T. Vestad, J. Oakey and D. W. Marr, *Opt. Express*, 2004, **12**(19), 4390.

16. Y. Arai, R. Yasuda, K. Akashi and Y. Harada, *Nature*, 1999, **399**(6735), 446.
17. M. P. MacDonald, G. C. Spalding and K. Dholakia, *Nature*, 2003, **426**, 421.
18. Z. Wang, J. El-Ali, M. Engelund, T. Gotsaed, I. R. Perch-Nielsen, K. B. Mogensen, D. Snakenborg, J. P. Kutter and A. Wolff, *Lab Chip*, 2004, **4**(4), 372.
19. L. L. Sohn, O. A. Saleh, G. R. Facer, A. J. Beavis, R. S. Allan and D. A. Notterman, *Proc. Natl. Acad. Sci. U. S. A.*, 2000, **97**(20), 10687.
20. Y. Zhou, Y. Wang and Q. Lin, *J. Microelectromech. Syst.*, 2010, **19**(4), 743.
21. D. Spencer, G. Elliott and H. Morgan, *Lab Chip*, 2014, **14**(16), 3064.
22. S. H. Kim and T. Fujii, *Lab Chip*, 2016, **16**(13), 2440.
23. K. M. Weerakoon-Ratnayake, S. Vaidyanathan, N. Larkey, K. Dathathreya, M. Hu, J. Jose, S. Mog, K. August, A. K. Godwin, M. L. Hupert, M. A. Witek and S. A. Soper, *Cells,* 2020, **9**(2), 519.
24. S. M. Gribble, B. L. Ng, E. Prigmore, T. Fitzgerald and N. P. Carter, *Nat. Protoc.*, 2009, **4**(12), 1722.
25. M. Jang, P. Neuzil, T. Volk, A. Manz and A. Kleber, *Biomicrofluidics*, 2015, **9**(3), 034113.
26. D. D. Huh, *Ann. Am. Thorac. Soc.*, 2015, **12**(Suppl. 1), S42.
27. H. J. Kim, D. Huh, G. Hamilton and D. E. Ingber, *Lab Chip*, 2012, **12**, 2165.
28. A. Shlomai, *et al.*, *Proc. Natl. Acad. Sci. U. S. A.*, 2014, **111**, 12193.
29. E. J. Weber, *et al.*, *Kidney Int.*, 2016, **90**, 627.

15 Development of Lab-on-a-chip Systems for Point-of-care Applications

15.1 Introduction

The polymerase chain reaction (PCR) method was first described in 1986 by Kary Mullis and co-workers.[1] What was interesting was that he, in his own words, only used a sequence of known steps, thus achieving one of the greatest inventions of the last century. Rightfully, Mullis was awarded the Nobel Prize in Chemistry in 1993 for his invention. The Nobel Prize was based on two aspects, the PCR itself and the extraction of an enzyme capable of surviving at high temperatures. Mullis used Taq polymerase enzyme from the bacterium *Thermus aquaticus*,[2] which is able to survive and reproduce in hot water. PCR is a phenomenal method that has revolutionized many fields, such as medicine, genetics, forensic science, paternity recognition, genealogy, *etc.*

Why is PCR so important? PCR can also be considered as a highly specific biological or biochemical replicator of deoxyribonucleic acid (DNA) (Figure 15.1).[3] The PCR solution, also known as a PCR master mix, has to contain the sample of DNA which we are trying to replicate, polymerase enzyme, oligonucleotides (primers) specific to the DNA sequence we are looking for, free nucleotides adenine (A),

Microfluidics and Lab-on-a-chip
By Andreas Manz, Pavel Neužil, Jonathan S. O'Connor and Giuseppina Simone

Published by the Royal Society of Chemistry, www.rsc.org

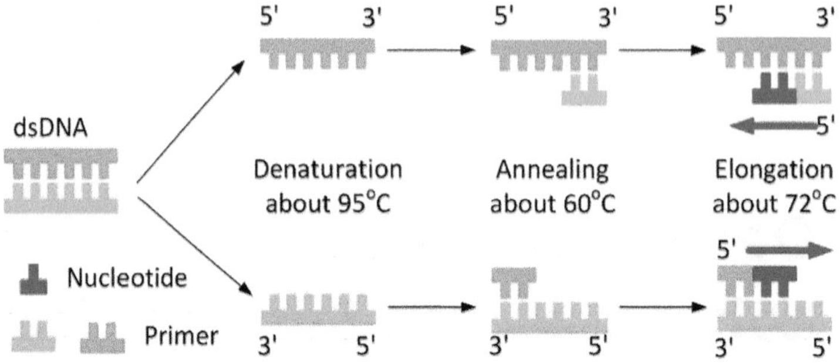

Figure 15.1 Principle of PCR.

thymine (T), cytosine (C) and guanine (G), bivalent cations Mg^{2+} or Mn^{2+}, and buffer solution to stabilize the polymerase and correct the pH. PCR itself is a sequence of temperature cycling steps of a DNA template in the PCR master mix. The first step is to heat the solution to a high temperature, typically ~95 °C, to activate the polymerase enzyme. During that step, each molecule of double-stranded DNA (dsDNA) melts into two molecules of single-stranded DNA (ssDNA). This step is also called denaturation. The master mix is then cooled to an annealing temperature (between ~50 and ~60 °C) at which the primers anneal to the ssDNA, assuming that there is a DNA template with a sequence complementary to the primers in the master mix. In the next step, the solution is heated to ~72 °C and, assuming that the pH of the solution is favourable, in the presence of a bivalent salt, polymerase enzyme and free nucleotides A, C, G and T, the two DNA sequences are combined into two dsDNAs. These three temperature steps are repeated 30–40 times. For example, if the process is repeated 40 times, one would expect that the number of DNA copies in the sample would be replicated 2^{40} times. This is an incredible number, almost 1.1×10^{12}. In reality, the number of replicas is not that high, as during the PCR the master mix becomes depleted and the replication stops. Also, the PCR efficiency is never 100% as the actual number of replicas is smaller. Nevertheless, the number of replicated copies is still phenomenal. The PCR method also has high specificity because if there is no DNA specific to the primers, no replication occurs. Once the PCR is completed, the product (if there is any) has to be detected by its analysis. This is typically done by hybridization using suitable targets or by electrophoresis, either using gel or a capillary. This short description demonstrates the power of the technique.

Consider an application where we need to diagnose the presence of a certain type of virus that has its genetic information coded in ribonucleic acid (RNA) and not in DNA. To do this, first we have to prepare the specific complementary DNA (cDNA). In the presence of an enzyme called reverse transcriptase in a suitable environment and at a temperature of ~60 °C, the RNA becomes reverse transcribed into cDNA. Once the genetic information is in the format of cDNA, these DNA molecules can be replicated and their number can be increased to a detectable level using the PCR process. As mentioned earlier, the presence of the DNA product (called an amplicon) after the PCR has to be confirmed by another end-point detection method.

Regardless of which end-point detection method is used, there are two drawbacks. First, the result obtained is only qualitative as it does not give any information on how many copies were present in the sample at the beginning, *i.e.* no quantitative information can be acquired. The second drawback is the necessity for sample manipulation after the completion of PCR as the sample has to be transferred from the PCR system to another one to detect the presence of the amplicon and its properties.

In 1993, a method of quantitative PCR (qPCR) was published by Higuchi *et al.*[4] They added the fluorescent intercalating dye SYBR Green I to the PCR master mix. This dye does not produce any fluorescence in an aqueous environment. Once it binds to the dsDNA, which is hydrophobic, SYBR Green I starts to emit green light when illuminated by blue light. The optimum excitation wavelength is ~490 nm and maximum emission wavelength is ~515 nm. The amplitude of the fluorescence signal is proportional to the number of dsDNA copies in the solution. During the PCR sequence, the fluorescence amplitude is measured in each cycle at the end of the extension step at ~72 °C. Once the measurement is completed, one can plot a graph with the cycle number on the x-axis and fluorescence amplitude on the y-axis, typically having a sigmoid shape. The fluorescence monitoring is conducted during the PCR in real time, hence this method has also been named real-time PCR.

This method is also quantitative. We can perform a set of measurements with a known number of copies of the DNA template. Typically, the range of DNA copies differs by many orders of magnitude, from 1 (statistically the most probable number) to at least 10 000 or more.[5] Once the PCRs are completed, the PCR amplification curves are plotted and an artificial threshold value is set. The intercept values of the threshold and the PCR amplification curve are then plotted as a function of log(concentration). The result is called the PCR standard curve and it is the basis for further quantification of samples with an unknown number of DNA copies.

The slope of this curve shows the efficiency of PCR amplification. When we then perform PCR with a sample containing an unknown number of copies of the DNA template, we create a PCR amplification curve, determine the threshold value, and from the normalization curve we can conclude what the number of DNA copies in the original sample was.

This quantitative PCR method can also be used to determine the efficiency of reverse transcription (RT) by comparing different samples or different RT conditions by determining DNA copy numbers by the real-time PCR method. SYBR Green I dye produces fluorescence only in the presence of dsDNA. During the PCR process, each molecule of dsDNA melts at the melting temperature (T_M) into two molecules of ssDNA and at that moment the SYBR Green I stops exhibiting fluorescence properties. The value of T_M depends on the length (number of base pairs) of the DNA and its sequence. Once the PCR is completed, a melting curve analysis (MCA) is performed. The sample is slowly warmed from ~72 to ~93 °C while its fluorescence amplitude is monitored. We then plot $-dF/dT$ as a function of T, and there should be a peak at approximately the predicted T_M of the expected amplicon, the precise value of T_M being affected by the master mix composition, bivalent salt concentration, *etc.*, and it can vary by a few degrees. A single peak on the graph of $-dF/dT = f(T)$ shows that there is only a single amplicon in the PCR product. This method is not as precise as capillary electrophoresis, but its great advantage is its convenience. The sample is analysed immediately once the PCR is completed using the same system with no sample manipulation. Typically, the MCA is part of the PCR protocol.

Typically, for any clinical sample at least three tests should be conducted concurrently: a clinical sample, a negative control and a positive control. The sample for the negative control is just water with no template to determine if there is any sample-to-sample cross-contamination. It should exhibit no amplification curve. The positive control sample contains either RNA or DNA, depending on the template of the clinical sample. The purpose of the positive control is to verify that the PCR or RT-PCR process was successfully conducted and everything, such as master mix composition and (RT) PCR protocol, was conducted successfully. Both false-positive and false-negative results can have severe consequences, hence their elimination is an important step.

15.2 Fundamental Considerations

We decided that the new system for point-of-care (POC) applications to be developed must have the following properties (Figure 15.2):

- have the ability to detect the presence of viruses or bacteria using RT-PCR (or qPCR) method;
- be economical and thus affordable in countries with low gross domestic product per capita;
- have disposable parts coming into contact with the sample to avoid sample-to-sample cross-contamination;
- have an efficiency for either DNA or RNA that is comparable to those of commercially available systems;
- have a small size for easy transportation and lower power consumption.

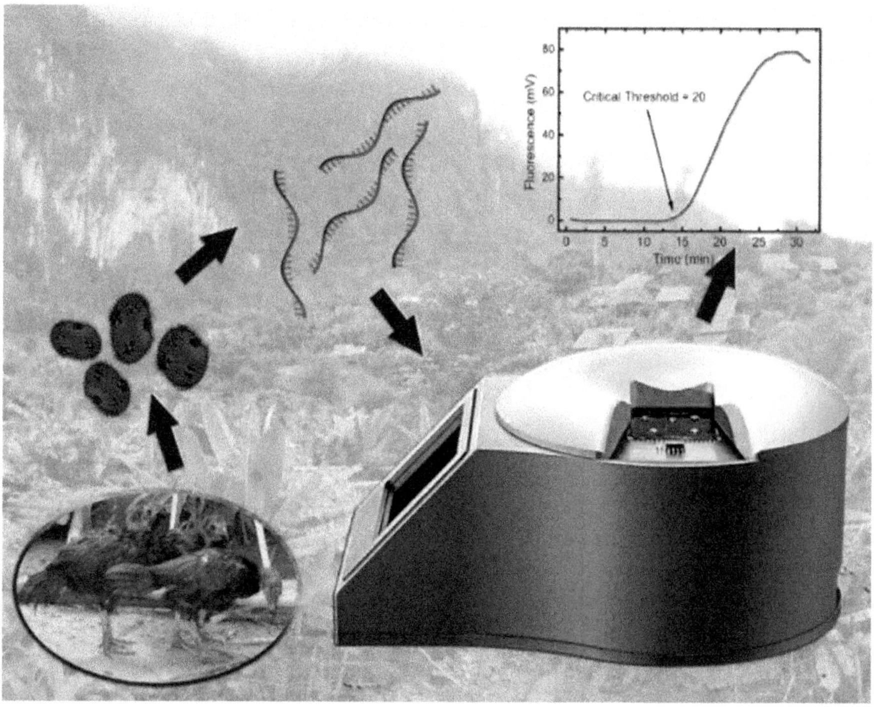

Figure 15.2 Our first idea for the entire system is shown here. We would extract virus from a poultry sample, release its RNA and use RT-PCR to detect its presence. Reproduced from ref. 14 with permission from the Royal Society of Chemistry.

The new system also had to be fast. We expected that these newly developed systems would also be used at airports. Travellers showing signs of fever during the SARS (severe acute respiratory syndrome) outbreak could be tested in 10–15 min to confirm the infection.

We also decided that the system had to be able to detect RNA, which meant performing RT-PCR. This requirement is obvious, as viruses such as SARS and H5N1 and H7N9 avian influenza viruses have their genetic information coded in RNA and not in DNA. From an engineering point of view, that does not matter at all. PCR begins with a hot start step. An RT-PCR has an additional RT step before the hot start, consisting of keeping the sample at ~61 °C for a few minutes. Engineers achieve this by just adding one extra step to the temperature control programme. Molecular biologists know that a sample containing RNA is not simple, as RNA is rather unstable compared to DNA, and the system has to be rigorously tested to establish whether the RNA stability is adequate.

The system had to be portable, which means that its weight had to be significantly less than 1 kg. Finally, there was a requirement to power the system with a 12 V dc power supply. Different countries have different effective ac voltage standards in their distribution networks, from 100 V in parts of Japan to 230 V in Europe, but there is one voltage practically universally available throughout the entire world, namely the 12 V dc supplied by a car battery.

15.2.1 Virtual Reaction Chamber (VRC)

In order to provide a disposable device, we decided to use glass, which is cheap, it can be processed by the same mass production techniques as silicon and the surface chemistry at the SiO_2 surface is well known and fairly simple. Our system can be then considered as one of open-surface microfluidics.

We proposed a system consisting of a micromachined silicon chip integrated with a heater and a temperature sensor. The sample in the form of a VRC would be separated from the silicon chip by a glass microscope cover-slip. This glass, typically with a thickness of 170 μm, would be placed on the silicon chip and thermally connected with the silicon chip by a thin layer of mineral oil. We used a sample with a volume between ~0.1 and ~5 μL, which was found to be a practical range. Handling samples smaller than ~0.1 μL is complicated and it is also the smallest volume that one can dispense using a manual pipette. Volumes greater than ~5 μL have high heat capacities, hence the system would be too slow.

The glass microscope cover-slip was coated with a fluorosilane, namely (heptadecafluoro-1,1,2,2-tetrahydrodecyl)trimethoxysilane, (tridecafluoro-1,1,2,2-tetrahydrooctyl)trichlorosilane (FOTS) or 1*H*,1*H*,2*H*,2*H*-perfluorodecyltriethoxysilane (FAS-17), typically by a chemical vapour deposition process at ~150 °C. The water contact angle at this surface was as high as ~115° and mineral oil exhibited an angle of ~65° (Figure 15.3).[6] These hydrophobic and oleophobic properties ensured that both liquids did not spread out and remained where we dispensed them. The water-based sample was pipetted into an oil droplet with typically a volume three times greater than that of the sample. The surface tensions of the two liquids with a ratio of 1:3 resulted in a self-aligned system with the water sample in the centre of the oil droplet. Figure 15.3 shows an oil droplet on the glass surface, covering the PCR sample shown in green color to form a virtual reaction chamber, and Figure 15.4 shows an entire VRC-based system on a heater.

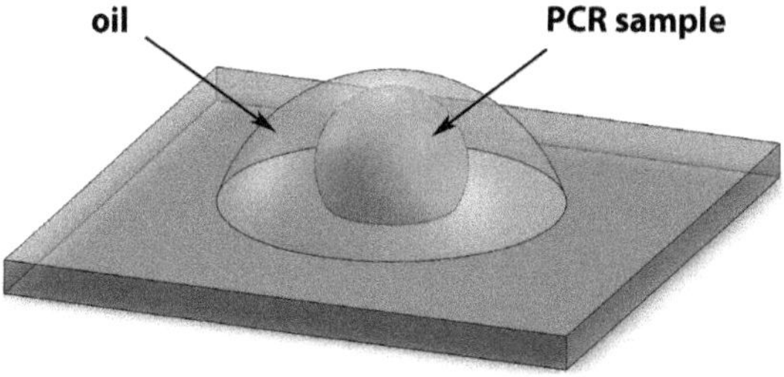

Figure 15.3 Two images overlapping each other. Inside is the water droplet covered with oil on a hydrophobic coated glass surface.

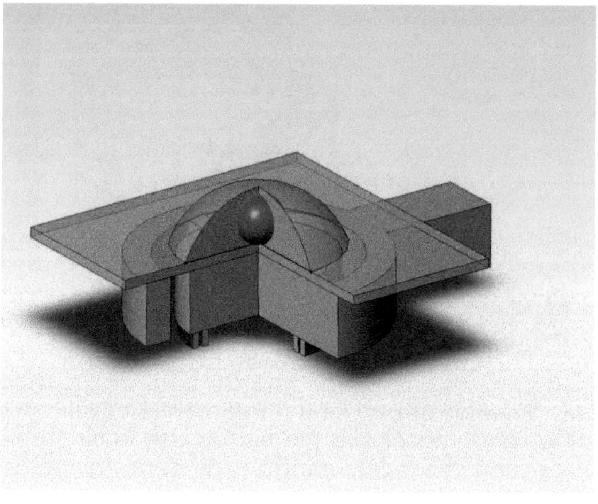

Figure 15.4 VRC system obtained using CAD software. Reproduced from ref. 8, https://doi.org/10.1093/nar/gkl416, under the terms of a CC BY license, https://creativecommons.org/licenses/.

15.2.2 MicroPCR System

As in the previous section, we assumed that glass with a thickness of 170 μm will be placed on a silicon chip made by the micromachining technique for microelectromechanical systems (MEMS), as shown schematically in Figure 15.5. The silicon chip should have the heater and the resistor on the backside made of sputtered or evaporated metal. This metal had to fulfil several functions, hence choosing it was rather critical. We planned to use it as a resistive temperature detector (RTD) type of sensor, so its temperature coefficient of resistance (TCR) was an important parameter. It should also serve as a heater and we planned to use a relatively low-voltage power supply, so we needed material with a low sheet resistance value. Also, processing this material should be simple. Finally, we needed to solder the fabricated chip to the supporting printed circuit board (PCB). Only platinum and gold fulfilled these requirements. Processing of gold was significantly easier than that of platinum so we chose gold as the material for the heater and sensor.

We designed the silicon chip in a double-doughnut shape (Figure 15.5). The thermal properties of the shape were modelled using the finite element analysis (FEA) software ANSYS and, after several optimizations, we found that the temperature along the inner doughnut at the steady state is practically constant (Figure 15.6). We also performed transient analysis and found that the structure should have a very fast heating/cooling rate of about a

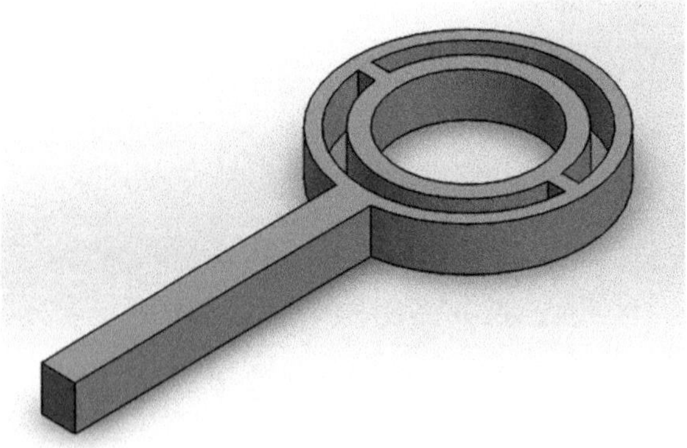

Figure 15.5 CAD drawing of part of a micromachined silicon chip. Both the heater and sensor are placed on the back side of the silicon. This silicon shape including the air gaps ensures a minimal temperature gradient along the inner doughnut shape required for high PCR performance.

few tens of degrees per second. Heating is almost always easier than cooling and it is conducted by dissipated Joule heat (P) in the structure:

$$P = \frac{V^2}{R} \tag{15.1}$$

where V is the voltage at the heater and R is its resistance. Solving the thermal heat balance equation gives us the heating rate:

$$\frac{\Delta T}{\Delta t} = \frac{P}{H} \tag{15.2}$$

where H is the heat capacity of the system. We calculated the value of H and assumed that perhaps we might not need active cooling, hence just passive cooling should be sufficient. This should be possible as silicon is a highly thermally conductive material with a thermal conductance λ_{Si} of 157 W m^{-1} K^{-1}. In such a case, cooling of the system would follow a first-order exponential equation controlled by the thermal time constant (τ) of the system:

$$\tau = \frac{H}{G} \tag{15.3}$$

where G is the thermal conductance of the silicon beam. The thermal conductance can be derived using the equation

$$G = \frac{wt}{L}\lambda_{Si} \tag{15.4}$$

where w, t and L are beam width, thickness and length, respectively. A fast system with a small t has to have a small heat capacity and/or a high value of G. The latter is problematic as systems with high values of G are also power demanding:

$$P = G\Delta T \tag{15.5}$$

Hence the solution of having a high G is not practical for POC systems, and having a small H is a better option. We decided that we would try to keep the heat capacitance and thus the sample size minimal to achieve ultimate heating/cooling rates.

The layout of the heater had a hole in its centre. From the very beginning we decided that that we had to be able to measure the fluorescence amplitude from the sample as a function of a cycle number, *i.e.* performing real-time PCR. We decided that the integrated optical system has to be underneath the sample, which is why we designed the heater with a hole in its centre. We knew that the minimal number of samples is three or four. One has to act as a negative control and one as a positive control and one or two represent clinical samples. We then designed the chip to accommodate four samples a time. The temperature distribution at four heaters was simulated and we found that the micromachined silicon with thermal isolation provided by air gaps was an excellent solution as we did not observe significant cross-talk (Figure 15.6).

The electronic circuit for heating and temperature sensing was fairly simple. The dissipated power was controlled using the pulse-width modulation (PWM) technique with a closed-loop system, using a proportional integrative derivative (PID) control method. The dissipated power was controlled by switching the transistor Q1 on and off (Figure 15.7). We chose the PWM method owing to its power efficiency. When the Q1 transistor was on, full power was used to heat the heater. When the transistor was off, there was no power dissipation in the heater. The on/off switching had to be much faster than the τ

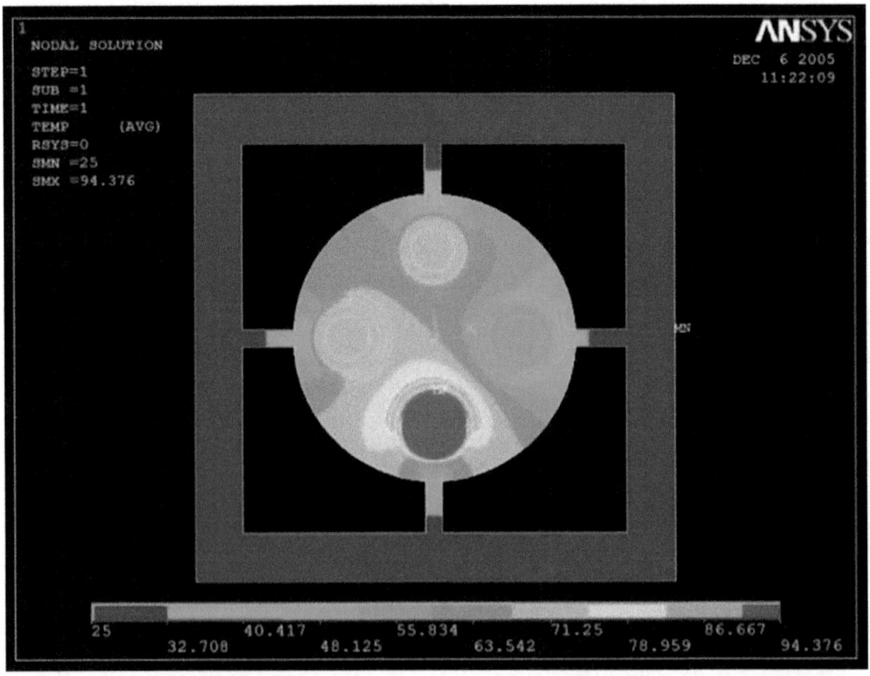

Figure 15.6 Temperature distribution on the silicon chip as modelled by ANSYS. We assumed four heaters with the temperature set to ~56 °C, two heaters at ~72 °C and one heater at ~93 °C.

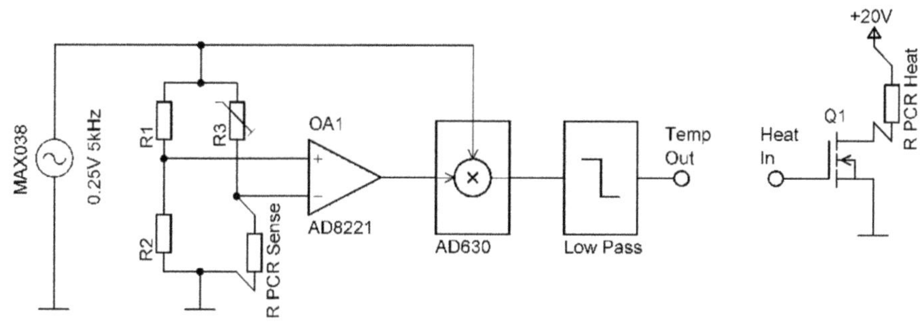

Figure 15.7 Temperature measurement and its control. Reproduced from ref. 15 with permission from the Royal Society of Chemistry.

value of the system. This was easy to achieve as τ was expected to be between ~0.2 and ~2 s and we used a square-wave signal with 10 kHz pilot frequency for the PWM.

First, temperature sensing was performed using a conventional dc-powered Wheatstone bridge. Its output signal was proportional to its dc bias voltage. We used 5 V to power it and found that the dissipated power in the sensor warmed up the silicon heater by at least ~3 °C. The only solution was to lower the bias voltage for the Wheatstone bridge without affecting the sensitivity of the temperature measurement. We therefore decided to power the bridge by an ac voltage with a demodulator and low-pass filter at the bridge output. This system is called a lock-in amplifier and it can also be considered as a very narrow bandpass amplifier, *i.e.* selective to the ac signal frequency used to power the bridge.

Once the original considerations were completed, we fabricated the chips. We used double-sided polished silicon wafers with a crystallographic orientation of <100>, a diameter of ~100 mm and a thickness of ~450 µm. First, we deposited a low-stress layer of SiO_2 with a thickness of ~1 µm using the plasma-enhanced chemical vapour deposition technique followed by a thin layer of chromium of thickness ~20 nm for adhesion of a subsequent ~200 nm thick layer of gold. Both metal layers were deposited by evaporation under high vacuum using an electron gun power source. We used such a thin layer of Cr to ensure that the resulting TCR was practically that of Au with Cr only marginally affecting it. We performed contact lithography and removed the unwanted Au and Cr layer using a back-sputtering or ion milling process. The photoresist was removed and a layer of SiO_2 was deposited a second time, now to mechanically protect the Au/Cr layer. We performed a second lithography and using HF–NH_4F (1 : 6) solution the contact pads for further gold

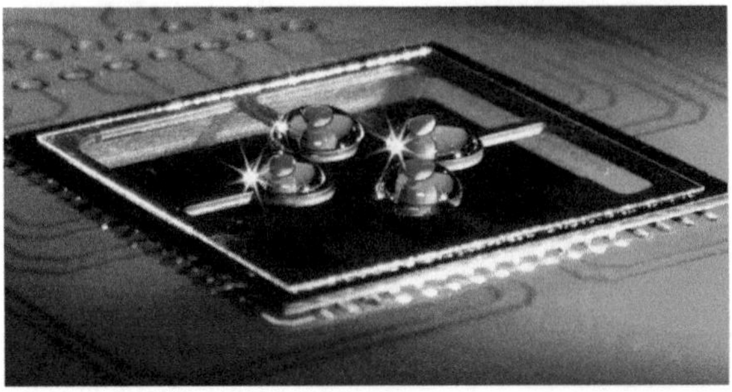

Figure 15.8 Fabricated PCR chip with four samples, here highlighted in blue. The chip is mounted on the PCB and the samples are separated from the chip by a glass cover-slip with a thickness of ~170 μm. Reproduced from ref. 7 with permission from the Royal Society of Chemistry.

soldering were opened. We removed the photoresist and applied a third lithography, this time with photoresist with a thickness of ~10 μm. The next step was again SiO_2 etching to open access to the silicon chip and by deep reactive ion etching we etched through the entire silicon chip. Once the photoresist had been removed and the chips thoroughly cleaned, we soldered them to custom-made PCBs and they were ready for testing.

The fabricated chip (Figure 15.8) had a size of ~24 × 24 mm. We placed a glass microscope cover-slip with a size of ~22 × 22 mm and a thickness of ~170 μm on the silicon chip. The glass was only placed on the silicon chip and we dispensed a few microlitres of mineral oil at their interface. Due to capillary forces, oil was sucked in between them and temporarily glued the glass to the silicon chip. This thin layer of oil also served as a thermal bridge between the two materials.

We used an infrared camera operating at wavelengths in the range 8–14 μm to check the non-uniformity of the temperature at the glass (Figure 15.9). Next, we measured the thermal parameters of the fabricated chip and we were ready for PCR.

We first tested this new PCR concept using a fragment of the human genome called glyceraldehyde-3-phosphate dehydrogenase (GAPDH) with a length of 208 base pairs (bp) using 940 copies of the template in the master mix solution. Specific oligonucleotides were forward 5′-CTCATTTCCTGGTATGACAACGA-3′ and reverse 5′-GTCTACATGGCAACTGTGAGGAG-3′. The PCR master mix was prepared in a volume of ~50 μL based on the manufacturer's protocol.

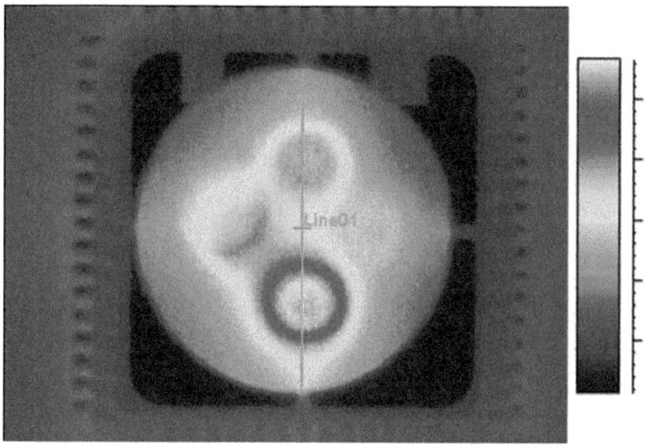

Figure 15.9 IR image of a PCR chip with the temperature of the heaters set to 56, 72 and 95 °C. Reproduced from ref. 7 with permission from the Royal Society of Chemistry.

The PCR master mix was meant for standard non-quantitative PCR. We then added a solution of SYBR Green I dye in a ratio of 1:10 000 and also bovine serum albumin (BSA) with a final concentration of ~1%. The presence of SYBR Green I allowed us to monitor the PCR in real time and BSA prevented the DNA template from incubating at the glass surface to inhibit the PCR.

A VRC with a sample with a volume of a few microlitres covered with oil was illuminated *via* a blue filter with a metal halide light source using an upright microscope with a long working distance objective lens. The emitted fluorescence collected by the same lens after passing it through a green filter interacted with a photomultiplier tube (PMT) module and its output voltage was recorded by an oscilloscope (Figure 15.10).[7] The PCR protocol consisted of three steps: denaturation, annealing and extension at temperatures of ~94, ~55 and ~72 °C, respectively. The duration of each step was ~60 s to ensure that the PCR would indeed be completed. Heating from ~72 to ~94 °C took ~0.5 s and cooling from ~94 to ~55 °C took ~2 s, corresponding to a heating rate of ~40 K s^{-1} and a cooling rate of ~−20 K s^{-1}.

Once the PCR was completed, we immediately performed MCA (Figure 15.11) and also capillary electrophoresis to ensure that the PCR amplification is specific and we did not create unwanted products, such as primer–dimer formation. Here we demonstrated for the first time that PCR can be performed at a silicon chip to take advantage of its excellent thermal properties.

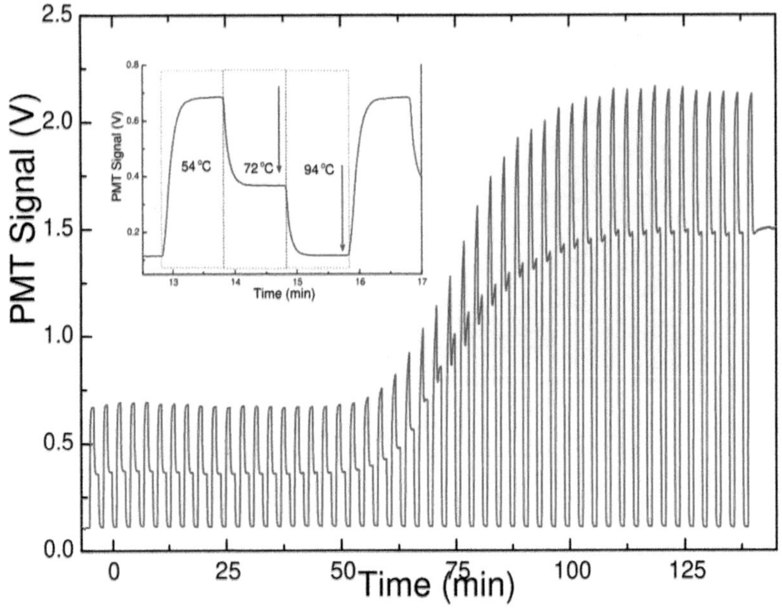

Figure 15.10 PCR amplification curve using continuous fluorescence monitoring and (inset) a single PCR step. Reproduced from ref. 7 with permission from the Royal Society of Chemistry.

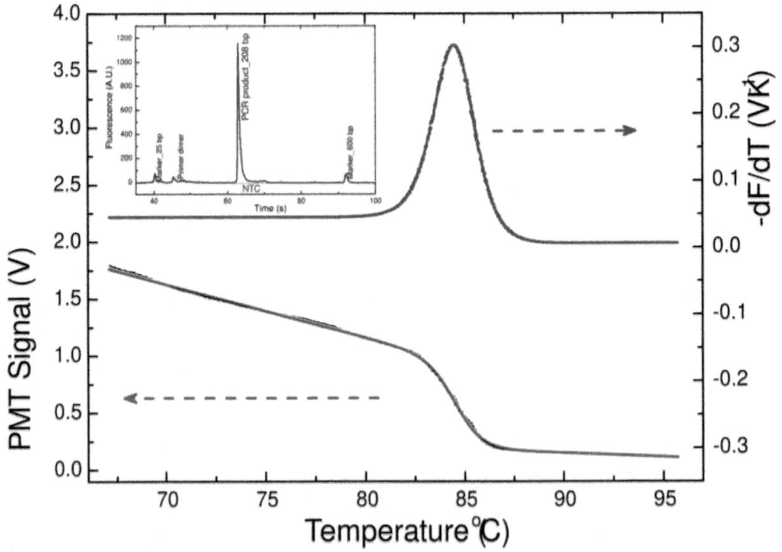

Figure 15.11 Melting curve analysis (MCA) and (inset) capillary electrophoresis analysis of the PCR product. Reproduced from ref. 7 with permission from the Royal Society of Chemistry.

15.2.3 NanoPCR System

The previous system was simplified and we also decreased the volume of the VRC (Figure 15.12). These changes allowed us to shorten a single PCR cycle to ~8.5 s, corresponding to performing 40 PCR cycles in ~340 s.[8] The system in principle remained the same.

The heater was again made with a patterned thin film of gold integrated with micromachined silicon. This time the double doughnut shape was abandoned and we used only a single doughnut configuration. Owing to this simplification and also increasing the power supply voltage we were able to achieve a system with a time constant of ~0.27 s, with a heating rate of ~175 K s^{-1} and a cooling rate of ~−125 K s^{-1}.

As with the large PCR system, we pipetted the sample onto the glass surface and covered it with oil. Here we used a sample volume of ~100 nL and an oil volume of ~600 nL (Figure 15.13). We found that the ultimate speed of the PCR is limited by the heat transfer between the heater and the sample through the glass and sample-surrounding oil, which took ~1.5 s. Once the denaturation was shortened below ~1.5 s, the PCR amplification efficiency was dramatically lowered. Instead of the use of SYBR Green I dye we also used TaqMan chemistry, considered at the time to be one of the fastest methods with a 6-carboxyfluorescein (FAM) probe using a protocol with only two temperature steps.[9]

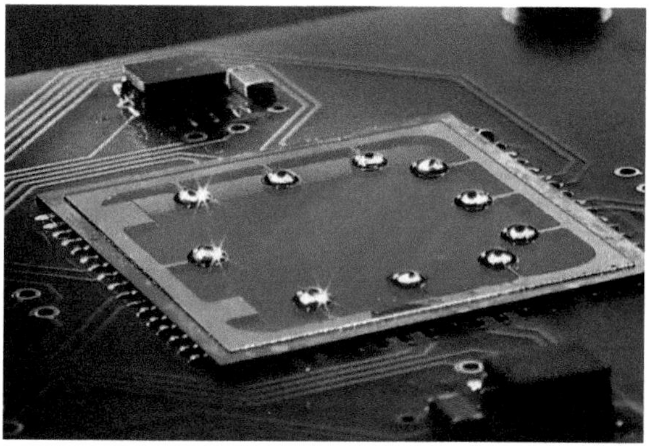

Figure 15.12 Miniaturized version of the micro-PCR system, so-called nanoPCR, with VRCs with a sample volume of ~100 nL. Reproduced from ref. 8, https://doi.org/10.1093/nar/gkl416, under the terms of a CC BY license, https://creativecommons.org/licenses/.)

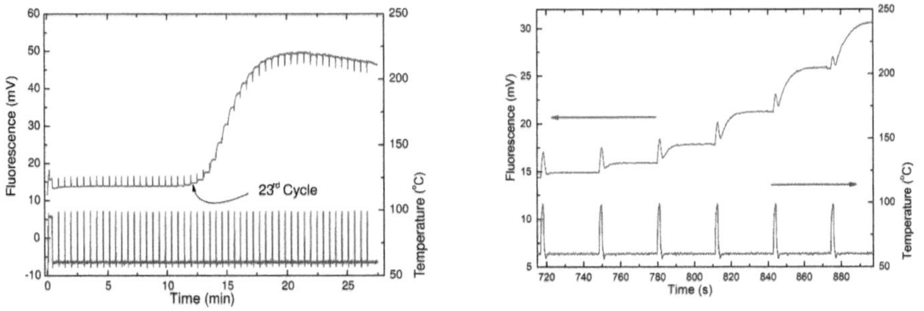

Figure 15.13 Continuous fluorescence monitoring using the nanoPCR system. Reproduced from ref. 8, https://doi.org/10.1093/nar/gkl416, under the terms of a CC BY license, https://creativecommons.org/licenses/.

We mixed a solution for real-time PCR using TaqMan Fast Universal PCR Master Mix 2, ~0.5 μL of TaqMan Assays-by-Design containing oligonucleotides and FAM probe for cDNA of green fluorescent protein (GFP) – forward primer 5′-CACATGAAGCAGCACGACTT-3′, reverse primer 5′-CGTCGTCCTTGAAGAAGATGGT-3′, probe 5′-FAM-CATGCCC GAAGGCTAC-BHQ-1-3′ – and ~2 μL of cDNA prepared by reverse transcription from transgenic gene glial fibrillary acidic protein (GFAP)–GFP. The length of the PCR amplicon was 82 bp.

This system performed very well and it was also exceptionally fast owing to simplification of the protocol to only two temperature steps: denaturation and annealing/extension. We found empirically that the shortest time required for complete denaturation was ~1.5 s. At that point this was the time-limiting factor. Nevertheless, this PCR was considered at that time to be one of the fastest in the world, at least from the heating and cooling rate point of view.[10]

The next natural step was to perform RT-PCR in a VRC format. Therefore, we prepared a single-step RT-PCR master mix supplied by a commercial company for vials with glass surfaces having an optimized concentration of BSA. We chose artificially prepared RNA of H5N1 avian influenza virus as at that time avian influenza was a hot topic. Later we used other types of influenza such as H1N3 swine flu and H7N9 avian flu and also Ebola virus.

We tested the ultimate speed of the RT-PCR (Figure 15.14). Reverse transcription was conducted at ~61 °C for ~2 min followed by a hot start at ~95 °C for ~30 s. The hot start step had two functions: it destroyed reverse transcriptase enzyme and it also activated second enzyme polymerase. Subsequently we ran 40 cycles of the PCR, each step consisting of three temperature sequences, denaturation, annealing and extension, at the usual temperatures of ~93, ~56 and ~72 °C. A

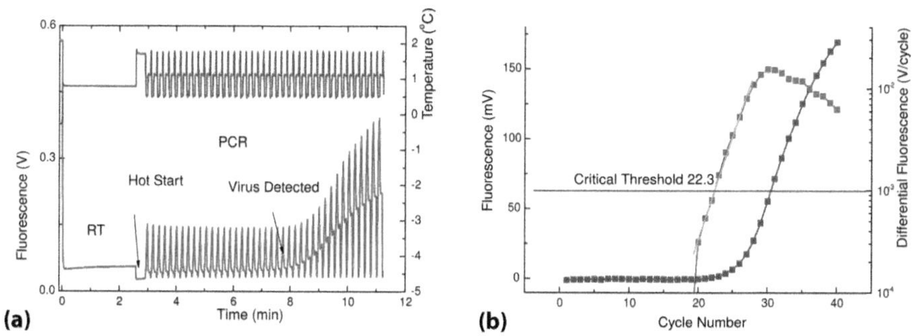

Figure 15.14 (a) Fluorescence and temperature as a function of time for RT-PCR of RNA of H5N1 avian influenza virus and (b) extracted amplification curve of the PCR with fluorescence as a function of cycle number.

single PCR step was completed in ~14.5 s. After completion of the PCR we also conducted MCA. With a volume of ~200 nL of sample covered with ~600 nL of mineral oil we were able to perform a single PCR cycle in ~14.5 s. The entire RT-PCR was completed in less than 12 min.

Electrical and thermal parameters of both microPCR and nanoPCR are given in Table 15.1.

15.2.4 Integrated Optical System

Both previously described PCR systems required an external microscope to monitor fluorescence. This was obviously only a temporary solution as a fluorescence microscope is an expensive, bulky and delicate instrument completely unsuitable for portable applications. Also, it was highly disproportionate, as the PCR system occupied a volume of a few cm^3 with a weight of 30–50 g and the microscope was at least 20 times larger and 100 times heavier (Figure 15.15). It needed optics, a sturdy body, a light source such as a metal halide lamp, a fluorescence system consisting of three optical filters, such as excitation, a dichroic beamsplitter and detector, a PMT with power supply and a recording system. We decided to develop a system with parameters comparable to those of a fluorescence microscope but small enough to be integrated into the PCR system for POC applications.

We collected suitable elements, designed the optical housing according to the size of the optical element and put everything together.[11] The entire system except the housing was based on off-the-shelf elements. The focusing lens had a diameter of ~6.35 mm and a numerical aperture of ~0.68. Other components had similar sizes. We

Table 15.1 Electrical and thermal parameters of microPCR and nanoPCR.

Parameter	MicroPCR	NanoPCR
Sensor resistance (@ 25 °C)	320 Ω	427 Ω
Heater resistance (@ 25 °C)	110 Ω	141 Ω
Temperature response of the sensing system	11 mV K^{-1} (dc)	30 mV K^{-1} (ac)
Heat conductance	4.4 mW K^{-1}	0.42 mW K^{-1}
Heat capacitance	6.6 mJ K^{-1}	1.5 mJ K^{-1}
Thermal time constant	1.74 s	0.28 s
Heating rate from 60 to 95 °C	50 K s^{-1}	175 K s^{-1}
Cooling rate from 95 to 60 °C	−30 K s^{-1}	−125 K s^{-1}

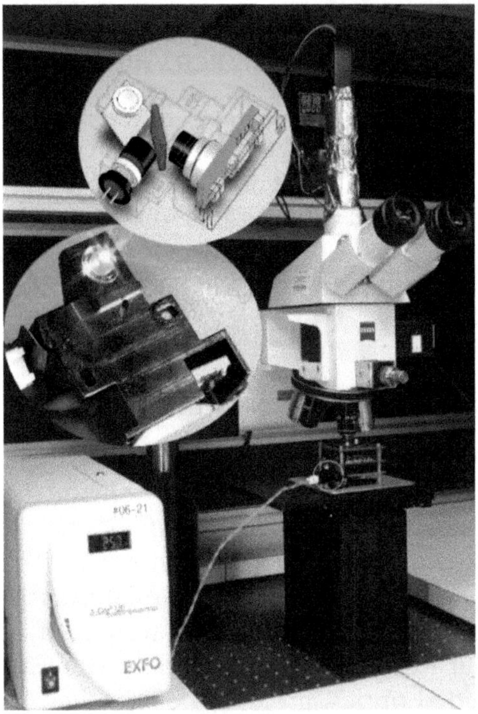

Figure 15.15 Fluorescence microscope used for our work with the PCR system underneath it. Reproduced from ref. 11 with permission from the Royal Society of Chemistry.

employed a turquoise colour LED with a principal emission of ~490 nm. It was difficult to obtain a diode with a high optical power output; there are very few manufacturers of this type of model as 490 nm is an unpleasant looking blue-green colour. Nevertheless, its principal emission wavelength was favourable to induce fluorescence from a fluorescein isothiocyanate (FITC)-type system. The filter set was a multilayer interference type with about 150 layers. It has an exceptionally

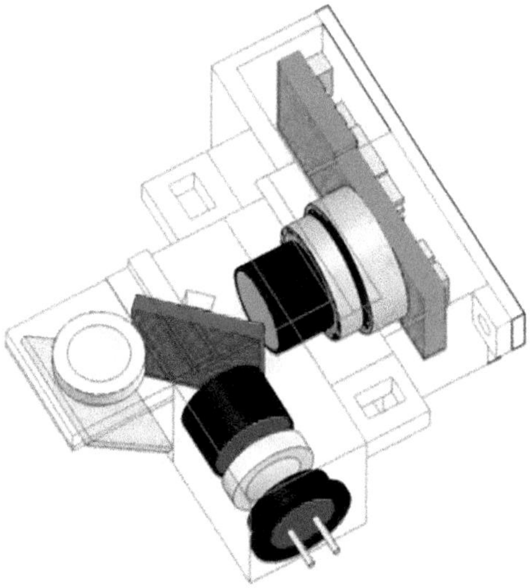

Figure 15.16 CAD drawing of the miniaturized fluorescence detection system. Reproduced from ref. 11 with permission from the Royal Society of Chemistry.

sharp cut-off, a transparency of ~0.98 and light suppression outside the transparent region of better than ~10^6. We used a conventional silicon photodiode with a luminous detection area of ~7.4 mm^2. We also built-in a transconductance amplifier to minimize ambient noise. It was based on an ultra-low-noise diFET operational amplifier. The entire optical housing had a size of ~20 × 20 × 10 mm (Figure 15.16).

We found very quickly that the amplitude of the fluorescence signal from the system was insufficient for any meaningful work even with the high magnification factor that we used in the transconductance amplifier system. The system was also very sensitive to ambient light and we had to operate in total darkness or inside a black box, making it impractical.

We therefore added an electrical modulator for the LED and a demodulator and a low-pass filter to process the photodiode output signal. This system is called a lock-in amplifier and in principle it was very similar to the one that we use for temperature measurement (Figure 15.17). It worked as an extremely narrow bandwidth amplifier/filter. Only a signal with the correct frequency to which the system was locked could pass further. The lock-in amplifier typically improved the signal-to-noise ratio by a factor between 10^3 and 10^7. Light in our laboratory was either daylight, in principle a dc light source, or came from fluorescent tubes, typically with a frequency of 50 Hz. Therefore, we carefully chose an odd lock-in amplifier frequency to make sure that it

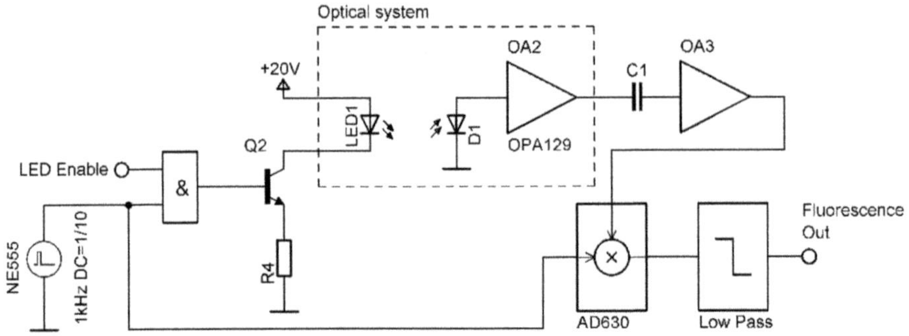

Figure 15.17 Schematic of the electronics of the optical detection system with lock-in amplifier. Reproduced from ref. 11 with permission from the Royal Society of Chemistry.

was not a higher harmonic of 50 Hz. Typically, we operated the lock-in amplifier at a frequency slightly above 1 kHz. The optical system was calibrated and we determined that the minimum concentration of fluorescein the we could detect was 1.96 nM. We also demonstrated that we could perform MCA, showing that the sensitivity of the system should be sufficient to perform real-time PCR.

15.2.5 Complete System with Sample Preparation I: Detection of RNA of H5N1 Avian Influenza Virus

The World Health Organization suggests that the time frame to discover and quarantine a pathogen outbreak is about 20 days and if that is not done the disease outbreak can get out of control and become pandemic. Disease discovery in developed countries and large cities in developing countries is a relatively simple task. However, in rural areas in countries with limited economic resources it is difficult if not practically impossible. The availability of small, economical and easy-to-use infectious disease diagnostic systems is then essential to limit the chance of a pandemic.

In previous sections, various parts of a real-time PCR system that we were developing were considered. Even with all those parts integrated together to form a small real-time PCR system, it is still not sufficient for practical disease diagnoses. Real-time PCR can detect the presence of a purified RNA or DNA template in buffer solution. Real samples such as saliva and blood contain genetic material inside cells or in viruses plus also PCR inhibitors. The virus (cell) has to be opened, RNA (DNA) released and PCR inhibitors removed. The virus (cell) opening is called lysis and it can be enzymatic, chemical or thermal. The sample has to be processed in a reasonable quantity to ensure that even if the virus is present in only a small quantity the system is able to diagnose its presence and avoid a false-negative result. A typical volume of sample to be processed is ~140 µL, which is too large for processing on the glass surface.

The entire process was again performed on a highly hydrophobic glass surface with a set of VRCs having different functions for PCR, such as RNA lysis, binding RNA to the surface of SiO_2, debris removal and RNA purification, and finally real-time RT-PCR was performed. In the first VRC we incubated the RNA at the SiO_2 shell surface of superparamagnetic particles. Using a strong magnet, we moved the particles from VRC to VRC to perform the entire process sequence.[12]

A sample of 40 µL of blood was spiked with a known number of copies of RNA of H5N1 avian influenza virus and the sequence of all steps described above was performed to detect the presence of RNA. The entire sequence lasted only ~28 min. It started with preconcentration of RNA, where we were able to lower sample volume by a factor of 500 without losing RNA from the sample. We demonstrated that our system was as sensitive as the commercial system. We even achieved a better performance as it was 4.4 times faster and the system was estimated to be at least 200 times cheaper. The RNA detection efficiency was ~99% and we were able to demonstrate that the system is capable of detecting a single molecule of RNA (Figure 15.18).

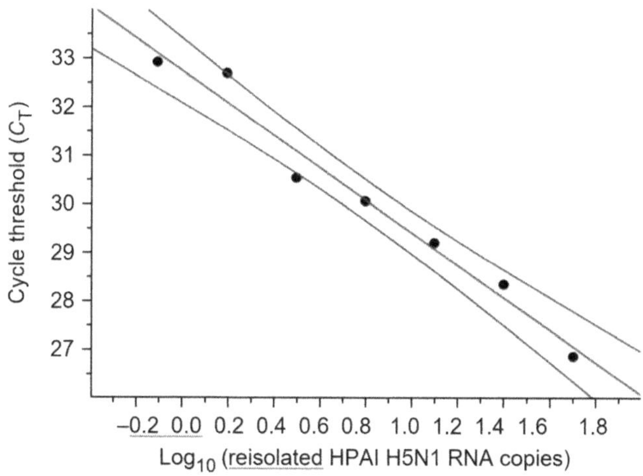

Figure 15.18 Normalization curve of RT-PCR using RNA of H5N1 avian influenza virus. Reproduced from ref. 12 with permission from Springer Nature, Copyright 2007.

15.2.6 Complete System with Sample Preparation I: Detection of RNA of SARS Virus

We also prepared a system to detect SARS virus. In fact we did not really detect the actual SARS virus, but its RNA was transfected into other cells and the entire process was performed using harmless material. It was impossible to work with the real virus for safety and also formal and other reasons. The ability to work with the SARS virus requires a laboratory equipped based on a standard for biosafety hazard level 3 or 4, which we did not have. Therefore, we used cells from cell line THP-1 transfected with GFP. We spiked blood with those cells, isolated them, preconcentrated them 100×, purified them, opened the cell membrane (performed lysis) to release RNA and conducted real-time RT-PCR. The entire process took only ~17 min. Here, we used space domain PCR. We used 3–4 heaters kept at constant temperature and with magnetic force we moved the sample between zones at different temperatures, performing the RT-PCR.[13]

15.3 First-generation Real-time PCR

In previous sections, the development of very fast PCR and also a small optical fluorescence detection system was described. We subsequently combined them together, including a single chip controller, and created the first generation of miniature real-time PCR (Figure 15.19). It had a built-in temperature control system for fluorescence monitoring, graphical LCD control and generation of all required

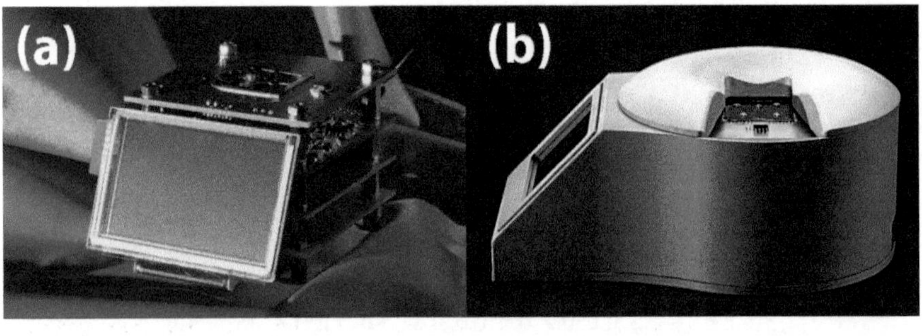

Figure 15.19 (a) Photograph of a first-generation real-time PCR system. With the housing removed one can see the individual PCBs with electronics. (b) The entire system in its protective cover after anodic oxidation treatment. Reproduced from ref. 14 with permission from the Royal Society of Chemistry.

voltages from a single 12 V DC power supply.[14] The optical system was fairly small but even with its small size we were able to integrate only a single optical unit. Because of that limitation, we were able to process only a single sample at a time. Nevertheless, we were able to demonstrate the detection of RNA from a segment of H5N1 avian influenza virus using single-step real-time RT-PCR. We again employed a commercial single-step RT-PCR master mix. Specific oligonucleotides were supplied by our collaborators at the Genome Institute of Singapore (GIS), A*STAR, Singapore. The forward primer was 5′-TGCATACAAAATTGTCAAGAAAGG-3′ and the reverse primer was 5′-GGGTGTATATTGTGGAATGGCAT-3′.

The entire process took ~35 min including reverse transcription and MCA. The expected melting temperature of the amplicon was ~76 °C and we measured ~75.8 °C, which is almost a perfect match. This confirmed that we had achieved specific cDNA replication and that we did not have different unwanted PCR.

15.4 Second-generation Real-time PCR [Universal Lab-on-a-chip (LOC) System]

When we compared our real-time PCR and other LOC systems we found that there were many similarities. Temperature regulation is very common, regardless of whether it is for PCR in the approximate range of 50–95 °C or for cell culturing at ~37 °C. Also, there is some kind of product detection, with either an optical method such as fluorescence, reflection or absorption or an electrical method using electrochemical or conductometric sensors. Often there is sample manipulation, sometimes a high-voltage power supply is required for compound separation and always the results have to be displayed.

We developed an entire family of LOC systems based on a single universal platform,[15] representing a portable application-specific lab-on-a-chip (ASLOC) instrument that was easy to reconfigure for a wide variety of LOC applications. The top three PCBs were application specific whereas all other PCBs were the same. The reconfiguration meant replacing the top PCBs with new ones and uploading specific software for that particular application to process the signals (Figure 15.20). The system was controlled by a single chip controller and the embedded program could be uploaded with a USB interface.

The core of the system was a four-channel lock-in amplifier for detection of LOC products. It contained four programmable pulse

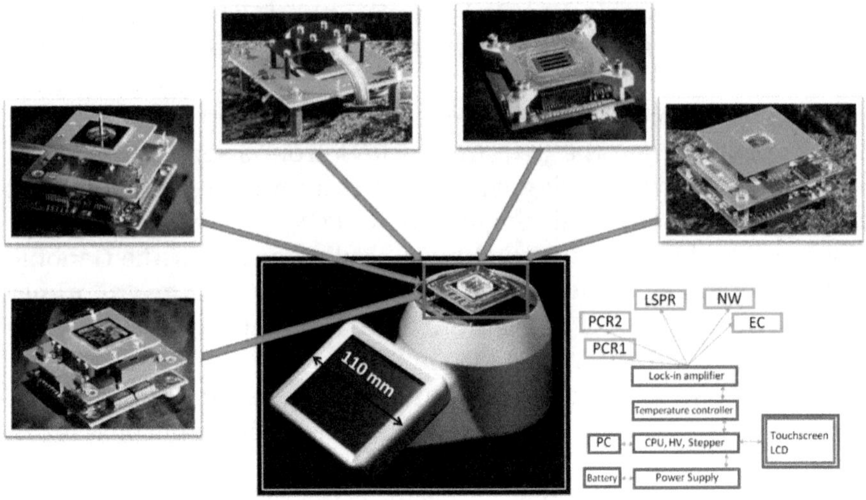

Figure 15.20 Second-generation modular LOC system. Reproduced from ref. 15 with permission from the Royal Society of Chemistry.

Figure 15.21 Photograph of second-generation LOC system of with a chip having 64 nanowires and a microfluidic chip made of polydimethylsiloxane (PDMS). Reproduced from ref. 15 with permission from the Royal Society of Chemistry.

generators, each of them operating at a different frequency. *Via* bipolar junction transistors we generated current pulses that powered directly LEDs or *via* current/voltage converters and attenuators we biased nanowires or electrochemical cells (pseudo reference electrodes) (Figure 15.21). The composite voltage from the transconductance amplifier was further amplified after dc offset removal and separated by four demodulators into individual compounds based on the modulating frequencies.

Each channel had in fact two signal generators (based on a micro-controller), one to interrogate the LEDs (nanowires, electrochemical cells, *etc.*) and the other for demodulation. Both operated with identical frequency and duty cycle but there was a controllable phase shift between them. We found that the signal processed *via* LED, fluorescence or photodiode is delayed by ~28 μs compared with the signal from the generator fed directly into the demodulator. Compensating for the delay using the second signal generator improved the signal efficiency by ~30%. All three parameters, frequency, duty cycle and delay of second signal generator, were remotely controlled by an external PC using the RS-232 protocol.

The processed signals were either loaded on an oscilloscope and displayed there during the development phase or processed by embedded software and displayed on its LCD.

The optical sensing properties of this LOC were demonstrated with two-channel real-time PCR in the time domain and also single-channel operation in the space domain. We also used a modified optical head to detect the effects of localized surface plasmon resonance (LSPR) (Figure 15.22)[16] on a nanostructured gold surface. This

Figure 15.22 LSPR system. Reproduced from ref. 16 with permission from American Chemical Society, Copyright 2008.

system's capability to monitor biosensors based on changes in electrical parameters was also tested using an array of nanowires and electrochemical sensors. Here we added an analogue multiplexer (switch) 1 × 16 into each channel, thus expanding its capability from four to 64 sensors (Figure 15.21).

15.5 Third-generation Real-time PCR

15.5.1 Technical Concept

We soon realized that the entire concept of second-generation LOC was not very successful. Having a universal LOC that is capable of performing practically any LOC method does not make much sense. A laboratory focused on electrochemical methods is unlikely to use PCR, nanowires or *vice versa*. Hence the concept of modular LOC does not offer anything extra and the system is rather complicated owing to its versatility. Our main interest was still real-time PCR and second-generation LOC was capable of performing only two reactions at a time and required six optical filters. The concept was just too expensive and two PCRs at a time were not sufficient for practical applications. We therefore set the requirements for a new third-generation system:

- performing concurrently four real-time PCRs;
- negative and positive control plus two samples;
- palm size;
- capable of detecting RNA, *i.e.* performing RT-PCR.

We took advantage of our knowledge gained in the course of previous work and developed a new real-time PCR system (Figure 15.23). With its dimensions of approximately 100 × 60 × 33 mm and a weight of ~75 g it is thought to be, practically without any competition, the smallest real-time PCR system in the world.[17,18] We again used the concept of VRCs placed on a disposable glass cover-slip above the micromachined silicon chip. In comparison with the previous systems, we simplified both the optics and the electronics. The four VRCs required only five optical filters in total instead of the expected 12. The two systems shared the same excitation filter and dichroic mirror and there was only a single emission filter for all four systems (Figure 15.24). There was only one temperature controller as we used serial–parallel combination of all four heaters and four sensors.

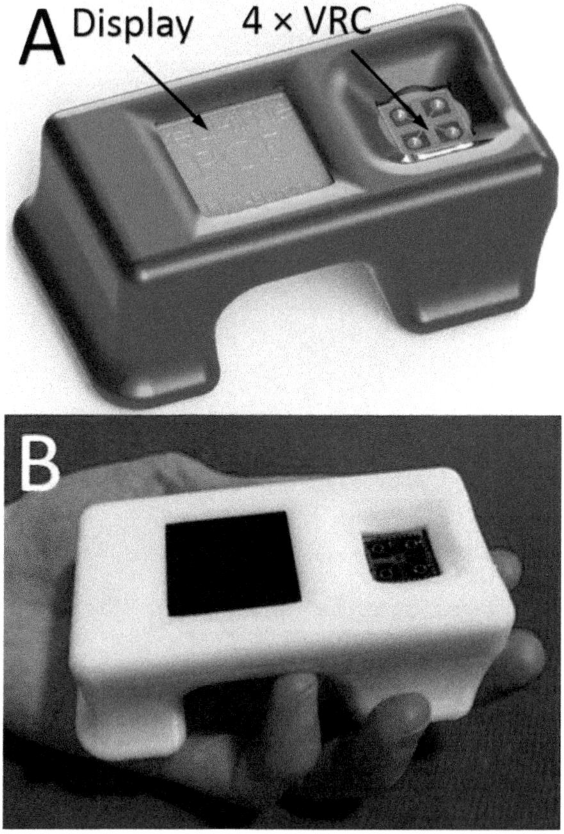

Figure 15.23 Third-generation real-time PCR. (A) CAD drawing and (B) fabricated system. Reproduced from ref. 17 with permission from the Royal Society of Chemistry.

There was also only one electronic system to detect the fluorescence. All four photodiodes were connected to the same lock-in amplifier and only the active one was selected by powering its LED and only one LED was powered at a time. The fluorescence amplitude was measured only once during each cycle at the end of an extension step. During the RT section of the protocol the fluorescence was not measured at all. So, did we really need a dedicated system for fluorescence measurement as the same one was used to measure the PCR temperature? The single lock-in amplifier was then used for both temperature measurement and fluorescence signal processing. The lock-in amplifier was normally used to monitor temperature. Only at the end of the extension step was the closed feedback loop used to control temperature disconnected for the last ~2 s. We used

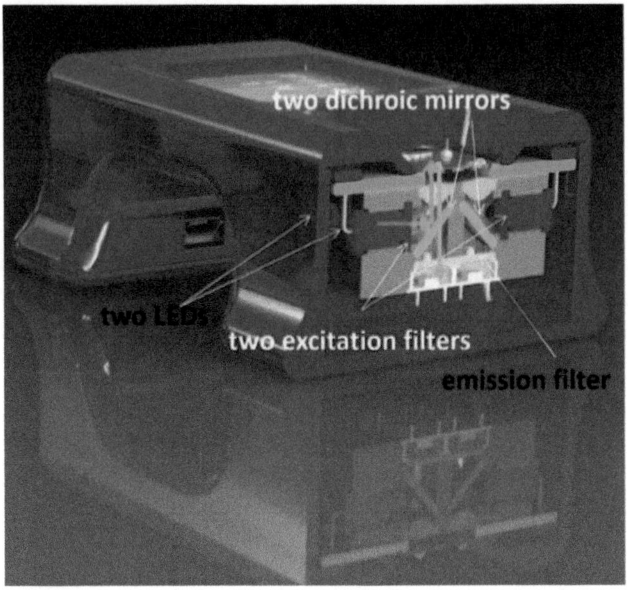

Figure 15.24 Cross-section of the optical part of the system showing the light path. Reproduced from ref. 17 with permission from the Royal Society of Chemistry.

an average duty cycle of the PWM measured during the controlled part of the extension step and the lock-in amplifier was utilized for fluorescence measurement, 0.5 s per each VRC.

We used this system to analyse four samples containing cDNA of H7N9 avian influenza virus each with a volume of ~200 nL. One VRC was used for the negative control, one was used for the positive control and two emulated actual samples (Figure 15.25). We also obtained a standard PCR curve showing that the system was capable of detecting a single cDNA copy with an efficiency of 0.91 ± 0.05 (mean ± standard error).

15.5.2 Applications

We report here the utilization of the recently developed palm-sized RT-PCR device to detect RNA of Ebola virus using single-step RT-PCR.[19] The device performed four PCRs concurrently each with a sample volume of ~100 nL. One was a positive control with both Ebola and GAPDH RNA and one was a negative control. The last two positions were used to measure the GAPDH and the Ebola content of a sample. Comparison of the critical thresholds from the two samples provided relative quantification. The entire process, which consisted of reverse transcription, PCR amplification and MCA, was conducted

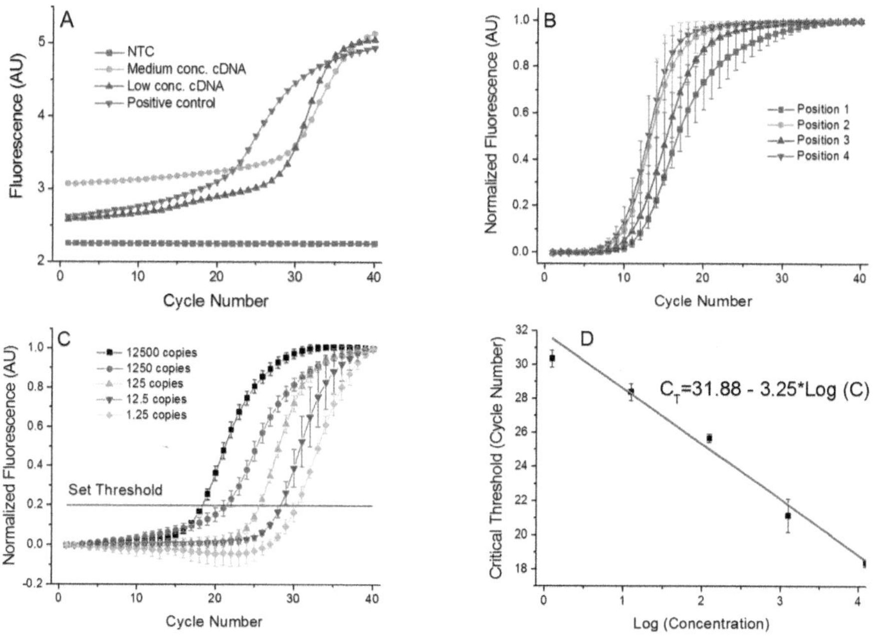

Figure 15.25 Results from the third-generation real-time PCR system showing the detection of cDNA of H7N9 avian influenza virus. (A) Amplification curves for cDNA, (B) demonstrating the uniformity of the units, (C) amplification with different initial concentrations, and (D) a standard curve for quantification. Reproduced from ref. 17 with permission from the Royal Society of Chemistry.

in less than 37 min. The next step will be integration with a sample preparation unit to form an integrated sample-to-answer system for POC infectious disease diagnostics.

15.5.3 Development of PCR Technique

15.5.3.1 Multiplexing I

The sample during the PCR is heated from the annealing temperature (typically ~56 °C) to the extension temperature (typically ~72 °C) and to the denaturation temperature (typically ~93 °C). When the sample was heated to the denaturation temperature, the dsDNA molecule melted into two ssDNA molecules, which was the same process as used during MCA. In the presence of an intercalating dye such as Eva Green, we performed continuous fluorescence and temperature monitoring, obtaining two relationships for fluorescence (F) and temperature (T) as a function of time (t):

$$F = f(t) \tag{15.6a}$$

$$T = f(t) \tag{15.6b}$$

We were able to eliminate t from both equations, resulting in F as a function of T for each cycle of PCR, called a dynamic MCA. We analysed all 40 PCR cycles, thus performing multiple quantitative PCR using a single fluorescent channel. The only requirements were the ability to perform continuous fluorescent monitoring, a melting temperature of each amplicon differing by at least 3 °C and a small sample volume to limit the internal sample temperature gradient during the transition period. NanoPCR under an optical microscope fulfilled all of the requirements. We demonstrated the application of this method to the multiple detection of cDNA of haemagglutinin (HA) and neuraminidase (NA) genes from H7N9 avian influenza virus.[20] We therefore improved the precision of the diagnosis of this dangerous virus and also the determination of the safety hazard, as this depends on the ratio of the two genes.[21]

15.5.3.2 Multiplexing II

It has been mentioned several times that our real-time PCR was conducted typically with an intercalator dye such as SYBR Green I or Eva Green. Once the sample with dsDNA had been warmed above the DNA melting temperature, the fluorescence amplitude diminished sharply. The ultrafast PCR system was based on a FAM probe, where the dsDNA melting had no influence on the fluorescence amplitude. We combined two types of oligonucleotides, one pair specific to DNA I with the FAM probe and the other pair specific to DNA 2 without the FAM probe. We then performed real-time PCR in the presence of an intercalator. Both amplicons contributed to the fluorescence amplitude at ~72 °C owing to the intercalator binding into their dsDNA and also the FAM probe from DNA 1. Once the sample temperature exceeded the melting temperature of both amplicons, only the FAM probe contributed to the fluorescence signal. We then obtained two sets of curves, at the end of the extension step and at the end of the denaturation step. Only the amplicon related to DNA 1 contributed to the second set, which we then subtracted from the composite signal in the extension step.[22]

15.6 Future Work and Conclusion

During outbreaks of infectious diseases, centres for disease control need to monitor particular areas. Considerable effort has been invested in the development of portable, user-friendly and cost-effective systems for POC diagnostics that could also create an 'Internet of things' (IoT) for healthcare *via* a global network. However, at present an IoT based on a functional POC instrument is not available. Here we have developed a fast, user-friendly and affordable IoT system based on a miniaturized PCR device. We demonstrated the system's capability by amplification of either cDNA or RNA of dengue fever virus (DENV). The resulting data were then automatically uploaded *via* a Bluetooth interface to an Android-based smartphone and further sent wirelessly to the global network, instantly making the test results available anywhere in the world. The IoT system presented here could become an essential tool for healthcare centres to tackle outbreaks of infectious diseases identified either by DNA or RNA. Its use can be of wider scope, since it can be modified to be compatible with other sensing methods.

The next natural step would be digital PCR[23] based on a simple platform, either using a CMOS (complementary metal oxide semiconductor) imager similar those for web cameras, cellphone cameras with signal processing by the phone CPU or performing the amplification directly on the CMOS imaging chip with a serial output signal for subsequent processing.[24]

References

1. K. Mullis, *et al.*, *Cold Spring Harbor Symp. Quant. Biol.*, 1986, **51**, 263–273.
2. K. B. Mullis, H. A. Erlich, D. H. Gelfand, G. Horn and R. K. Saiki, Process for amplifying, detecting, and/or cloning nucleic acid sequences using a thermostable enzyme, *US Pat.* US4965188A, 1990.
3. Z. Fohlerová, *et al.*, *Sci. Rep.*, 2020, **10**(1), 1–9.
4. R. Higuchi, C. Fockler, G. Dollinger and R. Watson, *Bio Technol.*, 1993, **11**, 1026–1030.
5. LifeScience, LightCycler® 1536 System Performance Data. https://lifescience.roche.com/en_cz/articles/lightcycler-1536-system-performance-data.html.
6. P. Neužil, W. Sun, T. Karásek and A. Manz, *Appl. Phys. Lett.*, 2015, **106**, 024104.
7. P. Neuzil, J. Pipper and T. M. Hsieh, *Mol. BioSyst.*, 2006, **2**, 292–298.
8. P. Neuzil, C. Zhang, J. Pipper, S. Oh and L. Zhuo, *Nucleic Acids Res.*, 2006, **34**(11), 1–9.
9. P. M. Holland, R. D. Abramson, R. Watson and D. H. Gelfand, *Proc. Natl. Acad. Sci. U. S. A.*, 1991, **88**, 7276–7280.
10. C. S. Zhang and D. Xing, *Nucleic Acids Res.*, 2007, **35**, 4223–4237.
11. L. Novak, P. Neuzil, J. Pipper, Y. Zhang and S. Lee, *Lab Chip*, 2007, **7**, 27–29.
12. J. Pipper, *et al.*, *Nat. Med.*, 2007, **13**, 1259–1263.
13. J. Pipper, Y. Zhang, P. Neuzil and T.-M. Hsieh, *Angew. Chem., Int. Ed.*, 2008, **47**, 3900–3904.
14. P. Neuzil, *et al.*, *Lab Chip*, 2010, **10**, 2632–2634.
15. P. Neuzil, *et al.*, *Lab Chip*, 2014, **14**, 2168–2176.
16. P. Neuzil and J. Reboud, *Anal. Chem.*, 2008, **80**, 6100–6103.
17. C. D. Ahrberg, B. R. Ilic, A. Manz and P. Neuzil, *Lab Chip*, 2016, **16**(3), 589–592.

18. M. Johnson, https://www.genomeweb.com/pcr/study-describes-worlds-smallest-qpcr-platform, *GenomeWeb*, 2015.
19. C. D. Ahrberg, A. Manz and P. Neužil, *Anal. Chem.*, 2016, **88**, 4803–4807.
20. C. D. Ahberg, A. Manz and P. Neuzil, *Sci. Rep.*, 2015, **5**:11479, 1–7.
21. M. Johnson, https://www.genomeweb.com/pcr/european-researchers-enable-single-dye-multiplex-qpcr-melt-curve-analysis-each-thermal-cycle, *GenomeWeb*, 2015.
22. C. D. Ahrberg and P. Neužil, *Sci. Rep.*, 2015, **5**, 12595.
23. B. Vogelstein and K. W. Kinzler, *Proc. Natl. Acad. Sci. U. S. A.*, 1999, **96**, 9236–9241.
24. H. Li, *et al.*, *Sens. Actuators, B*, 2019, **283**, 677–684.

Subject Index